MANUAL ON FRUIT BREEDING

MANUAL ON FRUIT BREEDING

Suneel Sharma

Former Professor
Department of Horticulture
CCS Haryana Agricultural University, Hisar

Om Prakash Pareek

Former Director
ICAR Central Institute for Arid Horticulture
Bikaner

SCIENTIFIC PUBLISHERS

Published by
SCIENTIFIC PUBLISHERS (INDIA) 5 A,
New Pali Road, P.O. Box 91 Jodhpur 342 001
(INDIA)
E-mail: info@scientificpub.com
Website: http://www.scientificpuboline.com

ISBN: 978-93-92590-21-4 (HB)
ISBN: 978-93-92590-22-1 (PB)
ISBN: 978-93-92590-23-8 (EBook)

Printed in India

PREFACE

Fruit breeding activity started from the prehistoric times with the development of the 'practice of fruit growing'. The domestication, introduction and exchange of the identified fruit types continued over generations. Primitive tribes identified and classified plants on the basis of their usefulness for edible or medicinal uses. The unique choice genotypes from the natural seedling population and those arising from spontaneous hybridization between wild races were identified. These were fixed through vegetative propagation by the farmers. Organized breeding activities, however, could be initiated during the 19th century. The book on 'Advances in fruit breeding' by Jules Janick and James N. Moore (1975) provides considerable details on the advances made in fruit breeding and created considerable awareness on this subject.

Efforts on characterization, nomenclature and classification of fruit plants, based on resemblances and differences, was recognized (Waugh, 1912; Hedrick, 1925; and Zielinski, 1955). Bentham and Hooker's 'natural system' (1862-63) had put them into distinct groups of dicotyledons, gymnosperms and monocotyledons. On the basis of the classification by Angiosperm Phylogeny Group (APG III,2009), strictly on monophyletic groups, 364 fruits belonging to 10 phyletic groups and 71 families have been put under the class Magnoliopsida (Angiosperms) and 3 fruits belonging to 3 families under the class Gymnospermae (Pareek and Sharma, 1993). Obviously there are a large number of fruits having distinctly different genetic makeup each associated with peculiar problems such as heterozygosity, long juvenile phase, outbreeding, self-incompatibility, apomixis, parthenocarpy, dichogamy, sterility, etc. Thus each fruit requires a special breeding effort. In most of the fruits, the work usually runs over a long period and thus the process is time consuming. Fruit breeding, which so far remained most prominently in the public domain, is now being practiced more actively by private companies and large marketing organizations as they earn patent and license for their cultivars. This has enormously broadened the current scenario of the stakeholders in fruit breeding endeavors. The breeding objectives have undergone a sea change. From the objective 'to develop dwarf, precocious and regular bearing cultivars' now the aim is 'to develop cultivars having improved productivity, quality and resistance to biotic and abiotic stresses along with developing suitable rootstocks'. The breeding techniques have now been enriched considerably. Molecular information is now available. Molecular linkage maps have been produced for many commercial species and the first quantitative trait loci are being tagged and selected through marker assisted breeding. Regeneration and transformation systems are available for many fruits. Potentially useful genes have been cloned and characterized. As a result, fruit breeders now have a richer tool box than they had earlier. With rapid advances, considerable information on different aspects of breeding in tropical, subtropical and temperate fruits has become available. Therefore, need was felt to put this information in a collated format of a Manual for systematic use.

The Manual is intended to be a brief compilation of information available related to breeding of fruit crops. This has been presented in five sections: 1. Prologue giving historical account of origin of fruits, their culture, domestication and breeding starting from the prehistoric era, achievements made, major objectives and the problems being faced; 2. Germplasm resources, their conservation and management; 3. Breeding methods; the techniques and achievements made in each fruit belonging to: 4. Tropical and subtropical, and 5. Temperate region. Effort has been made to integrate the conventional and biotechnological breeding approaches. The Epilogue at the end points out the thrust areas such as introduction and domestication of new crops, intergeneric and interspecific hybridization, polyploidy and mutation breeding, marker assisted selection, molecular genetics, genomics, biotechnology and CRISPR/Cas9 system, etc. that may contribute in future fruit breeding efforts. List of references has been given along with those suggested for future reading.

The information required for compiling the Manual has been drawn from a number of texts, monographs, books and periodicals. The line diagrams and photographs of the fruits to illustrate the Manual were collected from several national and international sources. The cooperation of several organizations to use line diagrams/drawings of plant species from their published work is gratefully acknowledged. Special thanks are due to Late Dr. J. W. Purseglove, Longman Scientific and Technical, London, Late Dr. H. H. Hume, Macmillan Publishing Co., New York, Dr. S. E. McGregor USDA, Arizona, Dr. Christopher Menzel, Regional Office for Asia and the Pacific, FAO, Bangkok, Dr. Chih Cheng T. Chao, Department of Botany and Plant Sciences, University of California, Riverside, California and Dr. Robert R Krueger, National Clonal Germplasm Repository for Citrus and Dates, Riverside, California, Dr. R. K. Pathak, Former Director, CISH, Lucknow, India, Dr. L. Albrigo, CABI, UK, Drs. C. Wongs-Aree, King Mongkut's University of Technology Thonburi (KMUTT), Thailand and S. Noichinda, King Mongkut's University of Technology North Bangkok (KMUTNB), Thailand, Floyd Zaiger, Zaiger Genetics, Modesto, California, USA and ProMusa. The Alliance of Bioversity International and CIAT (Formely Bioversity International), Rome, Italy.

We hope that the Manual will prove useful not only to students, teachers and researchers engaged in breeding of fruit crops but also for all those interested in improving the productivity and quality of fruits under the available resource conditions. The task would not have been accomplished without the encouragement and support of the family members especially Mrs. Deepali Sharma and Mrs. Kamla Pareek.

Suneel Sharma
O. P. Pareek

CONTENTS

1

Prologue

Archaeological remains of fruit and their pictorial and literary evidences suggest ancient origin of fruit culture. Prehistoric humans were nomadic hunter-gathers who followed the migration of animals and the ripening of foods, such as berries, in order to feed themselves. The fruit tree culture originated/occurred in various locations including the Fertile Crescent, Asia, and the New World in the late Neolithic and Bronze Age. About 8000 years ago, a period known as the Second Neolithic Revolution occurred in the Fertile Crescent that involved a change from villages to urban communities. Zohary and Spiegel-Roy (1975) proposed that fruit culture, in contrast with mere collection, originated 4000 to 3000 BCE. The development of fruit culture in the Fertile Crescent evolved at two loci: the Tigris Euphrates civilization of Mesopotamia and the Nile Valley culture of Egypt. The activities related to fruit breeding also started along with the beginning of fruit culture. The domestication process is traced in pre-history and antiquity, where useful fruit species were chosen and cultivated.

Centres of origin of crops and diversity regions provide clues to the geographic distribution of centres of plant domestication. These centres would obviously be located in areas with maximum diversity and from where agriculture could spread to more regions of crop plant diversification. De Candolle (1886) was perhaps the first to indicate the regions where the initial plant domestication might have taken place. Nikolai Ivanovich Vavilov (1951) gave more elaborate account of the cradles of agriculture. He believed that time is the only factor that influenced the dispersal and variation of a species. Based on this hypothesis, Vavilov (1951), named eight Centres of Origin of crop plants. Zhukovsky (1965) proposed 12 mega gene centres of crop plant diversity. The new areas added to Vavilov's eight centres were Australia, Africa and Siberia followed by revision of the boundaries. Zeven and Zhukovsky (1975) elaborately dealt on this in 'Dictionary of Cultivated plants and their centres of diversity' listing species for different mega gene centres and the range and extent of distribution of genetic/varietal/specific diversity. Zeven and de Wet (1982) preferred the term 'Region' to 'Centre'. These twelve regions have by and large wider coverage and more acceptability and list the prominent crop plants for which rich diversity occurs.

1. **Chinese-Japanese Centre:** Vavilov called it 'East Asian Centre of Origin'. Sometimes this region is associated with the primary region of diversity of fruit crops in Amur-Ussuri area. China is one of the richest regions contributing many crops particularly fruit trees. Orange/citrus, peach, apricot, litchi, kiwifruit, Chinese jujube, Japanese plum, loquat, Asiatic or sand pear, water chestnut, wampee and Chinese chestnut are important fruit crops that originated in this region. Japan is a secondary centre of diversity for several crops.

2. **Indo-Chinese-Indonesian Centre:** Vavilov called it Tropical Asian Centre of origin. This region is important for crops such as mango, banana, rambutan, durian, breadfruit, *Citrus*/lime, pummelo, grapefruit, jamun and mangosteen.
3. **Australian Centre:** Australian Centre was not described by Vavilov, but Zhukovsky (1970) marked it for the domestication of several plant species/crops. The region is important for origin of crops such as *Musa* spp., macadamia nut and Fei banana.
4. **Hindustani Centre:** The important fruit crops that originated in this region are mango, banana, jackfruit, ber, bael, aonla, phalsa and *Citrus* species.
5. **Central Asian Centre:** Vavilov called it Southwestern Asian Centre. Important crops of this centre are apple, walnut, almond, pistachio, grapes, plum and species of pear and apricot.
6. **Near Eastern Centre:** Vavilov called it Near Eastern Centre. Important fruit crops found in this region are European hazelnut, almond, sweet chestnut, *Malus* sp., *Prunus* sp., sweet cherry, *Pyrus* sp., fig, pistachio nut, grapes, plum, pear, apricot and pomegranate.
7. **Mediterranean Centre:** Vavilov called it Mediterranean Centre. Darlington (1956) suggested the name Mediterranean Region. The crops that have been domesticated in this region are olives, grapes and carob.
8. **African Centre:** Darlington (1956) described two areas in Africa, Ethiopia and Central Africa as regions of origin. The most important West African region could be subdivided into A (the Senegambian subcradle), B (the Central Niger subcradle), C (the Benin subcradle) and D (the Adamawa subcradle). The crops of African origin are tamarind, *Ensete ventricosum,* and *Phoenix* sp.
9. **European-Siberian Centre:** Darlington (1956) was the first to refer Europe as a region of origin of crop plants. Zhukovsky (1968, 1970) recognized it as a mega centre of diversity but of a relatively small importance. Important fruit crops that have developed in this region are peach, pear, plum, apricot, apple, almond, walnut, pistachio, cherry, *Fragaria* sp., *Ribes* sp., *Rubus* sp. and gooseberry.
10. **South American Centre:** It was first restricted to the Andes and was described by Vavilov (1949/50) as the Andean Centre of Origin. Vavilov divided it in two areas, a. Peru, Ecuador and Bolivia, and b. on the island of Chiloe and Chile. Zhukovsky (1968) created the mega centre of South America. The fruit trees originated from this region are papaya, cashew, passion fruit, cherimoya, Brazil nut, pineapple, *Feijoa* and *Solanum* sp.
11. **Central American and Mexican Centre:** Important fruit crops that have been domesticated in this region are papaya, white sapote, soursop, mamey sapote, avocado, pineapple and guava.
12. **North American Centre:** This centre is home of important crops like pecan nut, American chestnut, *Fragaria virginiana, Prunus* sp., *Rubus* sp., blueberry, *Vaccinium macrcarpon* (cranberry) and *Vitis* spp.

Fruit tree domestication and cultivation came after establishment of grain agriculture, probably during the fourth millenium BC. Although the pattern has been similar in many fruit trees, domestication of some fruits preceded others. Vegetative propagation (grape, olive, date, fig and banana) or polyembryony (citrus, mango) facilitated the process. Major changes that occurred during domestication are self fertility (peach, sour cherry), hermaphroditism (grape), elimination of bitterness (almond), lower resin (mango) and thornlessness (many fruits). Polyembryony, persistance of syconia (fig), parthenocarpy and seedlessness (banana, citrus) were other notable changes.

A number of popular fruits and nuts including various brambles, *Vacciniums*, pecan, and kiwifruit were domesticated during 18th to 20th century. Many others (lingonberry, various cacti and durian) although were extensively collected and marketed but are still in the process of domestication. Selection of elite natural variants was done with further improvement through identification of promising types arising from recombinants after natural intercrosses. Simultaneously, adoption of practices such as grafting, rootstocks, pruning and training, pollination, irrigation, storage and processing, often unique to each crop, were developed to increase productivity, improve quality and extend availability of the produce.

Domestication of fruits involves a combination of events including species selection, recurrent selection of elite clones and vegetative propagation coupled with techniques such as irrigation, pruning and training, controlled pollination, etc. Fruit crops have a number of characteristic features considered important during selection. Noteworthy is the obvious appeal of natural taste which may be desirable or delicious depending on the combination of sweetness, acidity and aromatic constituents. The taste and colour of fruits are traits associated with seeds dispersed in nature very often mediated through mammals. Subsequent improvement of fruit crops resulted from continual selection in seedling populations, especially of natural inter-crosses, elite wild or introduced clones. Most fruits are highly cross-pollinated and generally have long juvenile phase and longevity. Some have ability to propagate through vegetative organs such as off-shoots, cuttings and nucellar seed. Many grower-selected clones are still grown and are very popular.

Genetic changes associated with domestication of fruits include breakdown of dioecy, loss of self-incompatibility, induction of parthenocarpy and seedlessness, polyploidy and allopolyploidy, loss of toxic substances, ease of vegetative propagation, and loss of spines, thorns or pubescence. The changes that occur due to selection process include increase in fruit size, sugar content, and in shelf life. The contributing factors to these changes are interspecific recombination, polyploidization and generations of sexual recombinants. Many fruits differ from their wild progenitors in a few characters that have appeared as a result of mutations. Normally mutations are not advantageous to the plant in its natural setting as they reduce fitness, but would have been spontaneously selected by humans owing to some desirable characters. For example, the change from bitter to sweet seed in almond, and seeded to seedless fruit along with parthenocarpy in banana and plantain, citrus, fig, grape, persimmon and pineapple may have negative fitness in nature but is of very positive selection value. Parthenocarpy is valuable as it eliminates the need for pollination. Seedlessness in grape, banana, and citrus is considered a valuable edible quality. In dioecious fruits such as strawberry (*Fragaria* spp.), wild grape and papaya, induced mutations resulted in hermaphroditism.

Other major changes that occurred under domestication process are self fertility (e.g. peach, sour cherry), loss of spines (e.g. brambles, pineapple, pome fruits and citrus), loss of pubescence on fruit (e.g. peach), and in growth habit (e.g. pome and stone fruits). Colour sports have become increasingly important, especially in apple, pear and grapefruit. Some of these mutations are not heritable because they do not occur in the appropriate meristematic layer.

Interaction of genetic changes with development of cultural practices often evolved a unique growing technology for each fruit species. The history of two recent domesticates, cranberry and kiwifruit, may illustrate this. Both cranberry and kiwifruit were widely appreciated and had entry in commerce from wild stands long before domestication. Cranberry had been collected from the wild in North America since colonial times, but became cultivated only in 19th century when a series of practices to grow under aquatic conditions were developed. The kiwifruit, a dioecious vine native to China has been appreciated since the 8th century in China and probably much earlier but was never cultivated there. From seed introduced into New Zealand from China, the preferred pistillate and staminate clones (Hayward and Bruno, respectively) were selected. A cultivation system involving training and pruning on trellis with provision for pollination was worked out by nurserymen and growers. When introduced to England and North America in the beginning of the 20th century, the technology was transferred and kiwifruit became a world fruit crop in less than 25 years. In both cranberry and kiwifruit, the early elite selections of wild plants were of high quality and could be propagated by cuttings (cranberry) and grafting (kiwifruit). Thus selection, combined with vegetative propagation became a key domestication technique in these crops. However, even after 100 years, the old selections still dominate.

The related species of kiwifruit -- the yellow-fleshed *Actinidia chinensis* and the small-fruited *A. arguta* (also known as Tara fig) are potential new crops. The latter is widely planted in home gardens. The related blueberry (especially lowbush types) in Maine and lingonberry in Sweden, although harvested from the wild, are popular. Blueberry had greater species diversity than cranberry or lingonberry and had more promise because the fruit could be consumed fresh as well as processed. While domesticates of cranberry and kiwifruit have changed little from their wild forms, the blueberry has undergone remarkable transformation due to interspecific hybridization and ploidy manipulation. The culture of blueberry was dependent on the understanding that the *Vacciniums* are acid loving species and required ammonium form of nitrogen. Intensive selection and breeding with various species of different ploidy levels transformed this crop for wide adaptation. On the other hand, lingonberry was never domesticated, probably because there was no shortage of fruit collectable from the wild. In other fruits, such domestication efforts probably occurred much earlier owing to unique set of problems and prospects.

Further upsurge of interest in fruit tree evolution involving discovery, knowledge and use of wild relatives stems from the growing concern and consciousness of their genetic vulnerability. This imposes a search of wild relatives for breeding, especially for disease tolerance and use as rootstocks.

The fruits that were domesticated first appear to be date palm, olive, grape, almond, fig and pomegranate. Huxley (1978) believes that the first plant to be manipulated was the grapevine and that wine was probably first made in 6000 to 4000 BCE. Asian temperate pome fruits (apple, pear, quince, and medlar) and stone fruits (apricot, cherry, peach and plum) were fully domesticated by antiquity. Citrus fruits were domesticated early in China but reached the West in waves starting with citron. Tropical Asian fruits (mango and banana) and those from the Americas (avocado, papapa and pineapple) were developed in prehistory. Last five centuries witnessed much give and take in the world among the economically useful flora of different regions. In the process of colonisation during the sixteenth century and later, the Portuguese, British, French and Dutch introduced many plants in different regions. Such exchanges were also made by pilgrims and explorers or through introduction by invaders. For example, Mohammedan rulers introduced a variety of plants in India, such as cherries and grapes from Afghanistan and Iraq. By the seventeenth century, the Portuguese introduced New World crops, such as guava, custard apple, pineapple and cashew. In the last quarter of the eighteenth century, British East India Company introduced tea, litchi and loquat from China and mangosteen from Malaysia. Thus, fruit breeding activities aimed at improvement of fruit crops involves the use of methods such as introduction, selection, hybridization, polyploidy, mutation and biotechnology techniques.

As an organized activity, fruit breeding is a 19th century innovation. Throughout the millennia, genetic improvement of these fruit species has been achieved through selection by growers, first from natural seedling populations and then by plantations of unique genotypes fixed by vegetative propagation. Spontaneous hybridization between wild races and cultivated clones was critical to the early domestication of fruits. Breeding and genetic studies in tropical fruits are underdeveloped compared with those in temperate crops. Most improvements have come from selection of chance seedlings; open pollinated seedlings, or mutations, although cross breeding by controlled pollination has been done in pineapple, papaya and avocado.

The processes that were followed during domestication have now been intensified with emphasis on increased adaptability through hardiness, lower chill requirement, photoperiod insensitivity and resistance to biotic stress, plant architectural modifications and selection of colour mutations. Molecular techniques hold promise in increasing the efficiency of genetic improvement by selection using molecular markers and insertion of genes through transgene technology.

Fruit breeding phases broadly occurred in four stages, viz. (i) Pre-Mendelian era before 1900, (ii) Mendelian era - 1900 to 1920, (iii) Post- Mendelian era – from 1921 to 1950 and (iv) modern era – from 1950 to the present. The fruit breeding milestones achieved during these stages are given in Table 1.

Table 1. Fruit breeding milestones.

Year	Major findings/results	Institute/ Author
9000 BCE	First evidence of plant domestication in hills above Tigris River.	
4000 BCE	Date, fig, olive and grape are earliest fruit crop domesticated.	
700 BCE	Hand pollination of date palm was used by Assyrians and Babylonians.	
1338	Bhokri, Fakhri and Sahebi cultivars of grape introduced in Aurangabad (Daulatabad).	Muhammed Bin Tughlaq
1492	Introduction of avocado, papaya, pineapple, tomatillo in Europe.	Christopher Columbus
1493	Pineapple discovered in Guadeloupe.	Columbus
1600	Introduced pineapple, custard apple and cashew from Brazil during the 16th century.	The Portuguese
1700	Guava was introduced in the early 17th century.	The Portuguese
1714	*Fragaria chiloensis* was brought from Chile in France.	Amédée-François Frézier
1750	The garden strawberry was bred in Brittany, France by crossing *Fragaria virginiana* with *Fragaria chiloensis.*	
1769	Discovered breadfruit in Tahiti.	Captain James Cook
1759-1838	First to improve fruit crop by selection from genetic recombinants derived by inter-pollinations of clones.	Thomas Andrew Knight
1765-1842	Systematic breeding of European pear was first carried out.	Jean Baptiste Van Mons, a Belgian physician
1798 -1877	Developed Early Rivers, Early Transparent Gage, Monarch, Czar and President plum varieties.	Thomas Rivers
1806	Spent nearly 50 years in planting apple seeds throughout the wilderness of Pennsylvania, Ohio, Indiana, and Illinois thereby creating variability in apple.	John Chapman
1810-1895	Explained the role of juvenility in fruit growing.	John J. Thomas (1810–1895)
1823	Beurré d Anjou cv. of pear was identified.	Van Mons

1830-1893	Developed apple varieties- Barnack Orange, Laxtons Advance, Laxtons Early Crimson, Laxtons Epicure, Laxtons Fortune, Laxtons Favourite, Laxtons Herald, Laxtons Leader, Laxtons Pearmain, Laxtons Peerless, Laxtons Reward, Laxtons Royalty, Laxtons Superb, Laxtons Triumph, Laxtons Victory, and Lord Lambourne and plum varieties such as Early Laxton, Laxtons Gage, and Laxtons Supreme; strawberry--King of the Earlies, Noble, and Royal Sovereign.	Thomas Laxton
1849 -1926	Introduced 113 varieties of plums from Japan and other parts of world.	Luther Burbank
1850	Europeans introduced apple in Ootacamund, Tamil Nadu.	
1859	Apple was introduced in Harsil, Uttarkashi, Uttarakhand.	F.IE. Wilson
1890	Cross made between *J. hindsii* or *J. californica* and *J. regia* and named the resulting hybrid Paradox.	Luther Burbank
1893	Citrus crosses were made at USDA Subtropical Laboratory in Eustis, Florida.	W.T. Swingle
1916	Introduced Delicious group of apples in India.	Samuel Evan Stokes
1905	Fruit research work in India was started in the Departments of Botany at six Agricultural Colleges at Pune, Coimbatore, Lyallpur, Nagpur, Sabour and Kanpur. Almost at the same time, the Imperial Agricultural Research Institute was set up at Pusa (Bihar) and the Provincial and Central Departments of Agriculture were organized which were to look after the work on horticultural crops.	
1905	698 apple cultivars described in book on "Apples of New York".	S. A. Beachs
1905	Varietal improvement programme in pomegranate was started at Ganeshkhind Botanical garden, Pune.	
1907	Guava improvement work was initiated at Ganeshkhind Fruit Experiment Station, Pune.	
1908	Report on fig breeding in USA.	Walter Swingle
1909	Citrangequat- a trigeneric citrus hybrid of a citrange and a kumquat developed at Eustis, Florida.	Walter T. Swingle

1908-1918	Chinese Jujube cultivars were imported into the United States by U.S. Department of Agriculture (USDA) from 1908 to 1918.	Frank N. Meyer
1910	High yielding monoembryonic seedling of the variety Mulgoba was selected in Florida and named as Haden.	Capt. Haden
1911	Initiated planned hybridization in mango at Pune	Burns and Prayag
1912	Systematic evaluation work on citrus started in India at College of Agriculture, Pune.	
1913	The possibilities for genetic improvement of annonas realized.	P. J. Wester
1914	Pineapple breeding was started at Honolulu, USA.	
1915	Kinnow mandarin, the first generation hybrid between King and Willow leaf mandarin developed at California Citrus Experiment Station.	H. B. Frost
1917	Breeding programme for sweet cherry improvement was started at the Fruit Growing Institute in Plovdiv, Bulgaria.	
1920-1963	At South Haven experiment station in Michigan State University, Michigan-peach cvs. Halehaven, Kalehaven, Redhaven, Fairhaven, Sunhaven, Richhaven, Glohaven, and Cresthaven were introduced.	Stanely Johnston
1920	Pomological Station was established at Coonoor near Ooty to study the adaptability of temperate fruit varieties.	
1922	Banana breeding programmes through hybridization started in Trinidad, West Indies.	
1927	Sardar guava developed through a selection from Allahabad Safeda.	Cheema and Deshmukh
1928	116 grape varieties introduced from USA, Australia, Afghanistan and Russia at Lyallpur.	Sardar Lal Singh
1929	The initiative by the Imperial Council of Agricultural Research to provide financial assistance to the Provincial Governments gave boost to research activities through schemes sanctioned to the State Governments for work on citrus dieback, fruit preservation, nutritional value of fruits and control of San Jose scale in temperate fruits.	
1930	Identified Anab-e-Shahi from the collections of Nawab Baquer Ali Khan.	Sh. R. S. Pillay

1933	Cabezona cultivar of pineapple was identified as a natural triploid.	Collins
1936	A soft seeded selection made from open pollinated seedlings of hard seeded Alandi and was renamed as Ganesh in 1970.	Dr. G. S. Cheema
1938	Described about 30 heritable mutant forms in Smooth Cayenne pineapple.	Collins and Kerns
1940	Punjab Agriculture College and Research Institute, Lyallpur introduced Kinnow mandarin from California.	
1940	Redhaven peach cultivar was developed at Michigan State University, Michigan.	Stanley Johnson
1946	Systematic introduction and exchange of plant genetic resources of horticultural plants started by Division of Botany, IARI, New Delhi.	
1948	Significant variation among trees of the same clone in a mango orchard with respect to fruit shape, size, colour and quality was ascribed to bud mutations.	Naik
1949	Breeding work on banana started at the Central Banana Research Station, Aduthurai.	
1949	Apricot genebank was established in Romania.	
1951	First mango hybrid Mahmood Bahar and Prabha Shanker developed at Sabour.	
1952	The Fruit Experimental Station at Kallar began breeding work on jackfruit to develop short, compact. Many branched and productive trees having precocious bearing.	
1954	Kinnow cultivar introduced at Punjab Agricultural University, Regional Fruit Research Station, Abohar.	J. C. Bakhshi
1954-58	Hybridization work on citrus was carried out at Fruit Research Station, Kodur, India.	
1954	Hybridization in sapota was attempted at Agricultural College and Research Institute, Coimbatore.	Sambamurthi and Ramalingam
1954	Reported no inbreeding depression in papaya during selfing of gynodioecious population.	Hamilton
1955	Reported scoring technique in banana to indicate the relative contribution of two wild species in constitution of a cultivar.	Simmonds and Shepherd
1956	Monograph on Citrus classification was published .	Bhattacharaya and Datta

1958	FHIA was established by the United Fruit Company in La Lima, Honduras, which initially focused on dessert banana.	
1958	Mango hybrids namely Swarna Jehangir, Neeluddin, Neeleshan and Neelgoa were released from the Fruit Research Station, Kodur.	Bhujanga Rao and Rangacharlu
1958	Hybridization work in grapes was started at IARI, New Delhi to develop seedless varieties with early maturity, high yield and better quality with resistance to biotic stress.	
1958	Storey classified papaya flowers into eight categories.	Storey
1959	Gynodioecious variety of papaya Coorg Honey Dew was selected from the local type Honey Dew.	Aiyappa and Nanjappa
1960	Mango varieties Dashehari, Langra, Samar Bahisht Chausa and Ratul are result of chance selection.	Singh
1960	Developed a botanical key to the genera and species and the major characteristics of five cayenne clones for selection of appropriate parents for pineapple breeding.	Collins
1961	Gave exhaustive classification of the forms of Citrus.	Tanaka
1962	Reported the presence of self incompatibility in Dashehari mango.	Singh *et al.*
1962	Imported Anab-e-Shahi and Perlette cultivars from California.	Sardar Partap Singh Kairon, Chief Minister of Punjab
1963	Reported successful interspecific hybridization in the Genus *Mangifera.*	Mukherjee
1965	Pollen sterility can be induced by applying growth regulators in grapes to replace emasculation for production of male sterile lines.	Iyer and Randhawa
1966-67	Systematic work for breeding of uniform papaya varieties was started at IARI Research Station, Pusa, Bihar.	
1968	IIHR, Bengaluru was established and started grape breeding programme to develop superior varieties for table, raisins, wine and juice.	
1969	Solo and Bush of Hawaii and Hortus Gold of South Africa are bonafide varieties of papaya.	Storey

1970	Pusa Seedless grape variety was released from IARI.	
1970	Muttam Varikka was reported by selection in jackfruit Varikka type from Kerala.	Srinivasan
1970	Cultivars Dashehari, Langra and Chausa of mango were found self incompatible.	Sharma and Singh
1971	All India Coordinated Fruit Improvement Project (AICFIP) was launched by ICAR to conduct research on banana, citrus, pineapple and papaya.	
1971	Mallika, world's first regular bearing mango hybrid released for commercial cultivation, from IARI, New Delhi.	
1972	TNAU released papaya variety CO-1 evolved by sibmating Ranchi.	
1972	Citrus rootstock hybridization was started at Central Horticultural Experiment Station, Chethalli.	
1972	Systematic breeding programme was started at CHES, Chethalli for developing an ideal rootstock variety for citrus.	
1972	Central Mango Research Station, Lucknow was established to conduct research on different aspects of mango.	
1974	CO-2, variety selection from Baramasi of sapota was released from TNAU.	
1975	Spontaneous hybridization between wild races and cultivated clones was critical to the early domestication of fruit crops	Zohary and Spiegel Roy
1976	Breeding of acid lime for resistance to canker was started at Tirupati and Parbhani.	
1978	All India Coordinated Research project on arid zone fruits was started.	
1979	TNAU released papaya variety CO-2 as pure line selection from a local type.	
1979	Central Horticulture Experiment Stations were established at Gujarat and Ranchi to initiate work on the fruit crops of western and eastern India	
1979	Mango hybrid Amrapali was released by IARI	
1980	An efficient technique of mango hybridization was developed at IARI, New Delhi.	Singh *et al.*

1980	Four grape hybrids Arkavati, Arka Kanchan, Arka Shyam and Arka Hans were released from IIHR.	
1980	Released Oroblanco an early maturing seedless grapefruit type.	Soost and Cameron
1980	Sapota variety PKM-1, a selection from Guthi released from Periyakulam, TNAU.	
1980	*Fusarium* wilt Tropical Race 4 (TR4) was identified in Taiwan.	
1981-82	IBPGR mounted several Commissions to collect ten fruit tree species in several countries to build up genetic resources.	
1981	Four superior clones of Dashehari mango were located.	Singh and Chadha
1981	Reported the use of isozymes to note the genetic variations in mango cultivars.	Gan *et al.*
1981	Developed a dwarf mutant line by treating papaya seed with 15K gamma rays.	Ram and Majumdar
1981	Attempted inter-specific hybridization in genus *Carica.*	Subramanyam and Iyer
1981	Mango hybrid PKM-1 and sapota selection PKM-1 were released from TNAU.	
1982	A separate division of Fruits and Horticultural Technology was established at IARI, New Delhi	
1982	Reported induction of embryogenic mango cultures from nucellar tissues of some polyembryonic cultivars after 1-2 months from ovule cultures excised from 40-60 days old fruitlets.	Litz *et al.*
1983	Regular bearing mango hybrid variety Ratna was released from Dapoli.	
1983	Hybrid papaya varieties CO-3 and CO-4 were released from TNAU.	
1984	Banana hybrid CO-1 developed through multiple crossing was released from TNAU.	
1984	Clonal selection G-137 from Ganesh was made in pomegranate.	RFRS, Ganeshkhind, Pune
1984	Application of *in ovulo* embryo culture in grapes was reported.	Emershad and Ramming
1985	Isolated high yielding clones of Langra mango variety.	Singh *et al.*
1985	Melogold, an early maturing seedless hybrid grapefruit type was developed.	Soost and Cameron

1985	Papaya CO-5 was released as a selection of Washington variety from TNAU.	
1985	NRC on citrus was established at Nagpur.	
1985	International network for improvement of banana and plantains (INIBAP) was established in France.	
1986	Mango hybrids Neelphonso, Neeleshwari and Neeleshan were released.	
1986	Papaya CO-6, selection from a Pusa Giant variety, was released from TNAU.	
1986	Papaya varieties Pusa Dwarf, Pusa Delicious and Pusa Majesty were released from IARI.	
1986	Cariflora a dwarf papaya variety tolerant to PRSV became popular in Florida.	Conover *et al.*
1986	Numerical taxonomy for citrus classification was attempted based on leaf phenolic compounds.	Ulubelde and Tan
1986-87	A selection P-23 from the orchards of Muskat variety of pomegranate was made around Kolhar/ Shirdi region.	MPKV, Rahuri
1987	The Hong kong wild Kumquat, *Fortunella hindsii* was reported as tetraploid.	Soost
1987	Described two breeding systems for papaya - the Hawaiian system and the Queensland system.	Aquilizan
1987	Eksotika was released through backcross breeding in papaya.	MARDI, Malaysia
1987	Reverted sexually ambivalent male (SAM) used for development of inbred lines in breeding of dioecious papaya varieties.	Aquilizan
1987	Apple variety KKL-l, selection from Parlins Beauty was released from TNAU.	
1987	Taiwan Banana Research Institute started a selection program for resistance to Fusarium wilt in Cavendish bananas using somaclonal variation.	Hwang and Ko
1988	Regularity in bearing and dwarfness are controlled by recessive genes in mango.	Sharma and Majumdar
1988	The bunch bearing observed to be dominant over single fruit bearing in mango.	Sharma and Majumdar
1988	Identified two promising clones of acid lime, Vikram and Pramalini, in Maharashtra.	
1989	International Musa Testing Programme (IMTP) Phase I started to evaluate germplasm from the FHIA breeding programmes in Honduras.	

1989	USDA/ARS released Ambersweet [(Clementine × Orlando) × Sweet Orange.	USDA/ARS
1990	Kagzi Kalan, lemon variety released from IARI	
1990	PAU Selection- a selection from two strains of Baramasi lemon released from Ludhiana.	
1990	Mango variety PKM-2 and acid lime variety PKM-1 were released from TNAU.	
1990	Jai Devi, an acid lime variety released from Priyakulam,Tamil Nadu.	
1990	Somatic hybridization via protoplast fusion proved established technique for citrus improvement.	Grosser and Gmitter
1990	Aprium and Pluots were developed through intricate hybridization.	Chris Floyd Zaiger
1991	Higher amount of phenolics in the apical bud is associated with reduction in vigour and dwarfing in mango.	Iyer
1992	Regular bearing seedless mango variety Sindhu with papery stone was released from Dapoli.	
1992	Androgenic banana plants were obtained from the wild clone Long Tavoy of *Musa acuminata.*	Bakry and Horry
1992	Mango selection Paiyur-1, jackfruit selection Palur-1 (PLR-1) and sapota hybrid PKM-2 were released by TNAU.	
1992	Three mango hybrids, Arka Aruna, Arka Puneet and Arka Anmol were released by IIHR, Bengaluru.	
1992	Arka Neelamani, a black seedless grape variety was released from IIHR, Bengaluru.	
1993	Genus *Mangifera* divided into two subgenera -*Mangifera* and *Limus.*	Kostermans and Bompard
1993	IPGRI initiated the project "Promoting conservation and use of tropical fruit species in Asia".	
1993	Use of RAPD genetic markers to identify mango cultivars and to determine genetic relationships among cultivars.	Schnell and Knight
1993	In papaya, hybridization using female flowers obviated emasculation and resulted in four times higher seed production compared to that by use of hermaphrodites.	Chan and Mak
1993	Dessert bananas FHIA-O1 and FHIA-02 and cooking variety FHIA-03 were released.	
1994	Sapota hybrid PKM-3 was released from TNAU.	

1994	Grape hybrid Arka Majestic, developed by crossing Angur Kalan and Black Champa, was released from IIHR, Bengaluru.	
1994	Grape hybrid Arka Soma, developed by crossing Anab-e-Shahi and Queen of Vineyards, was released from IIHR.	
1994	Grape hybrid Arka Trishna, developed by crossing Bangalore Blue and Convent Large Black, was released from IIHR.	
1994	Grape hybrid Shweta Seedless, developed by crossing Anab-e-Shahi and Thompson Seedless, was released from IIHR.	
1994	Kagzi Lime selection Sai Sharbati released from Rahuri.	
1994	The first reliable protocol reported by co cultivation of etiolated epicotyl segments with *Agrobacterium tumefaciens* for transformation of citrus relative *Poncirus trifoliata.*	Kaneyoshi *et al.*
1995	Mango hybrid Arka Neelkiran was released from IIHR.	
1995	Hisar Safeda and Hisar Surkha guava hybrids were developed at CCS HAU, Hisar.	
1995	ACC synthase and alternative oxidase genes were cloned to control mango ripening.	Cruz-Hernandez and Gomez-Lim
1995	Aonla selection BSR-1 was released.	TNAU, Coimbatore
1995	Transgenic banana plants were produced through particle bombardment of embryogenic suspension cultures.	Sagi *et al.*
1995	PCR-RFLP analysis of cpDNA successfully used for phylogenetic analysis of fruits like *Prunus* spp.	Badenes and Parfitt
1996	A banana mutant Novaria was derived from gamma radiation of the clone Grand Naine.	Mak *et al.*
1996	Characterization of citrus genome was done through analysis of restriction fragment length polymorphisms (RFLPs).	Liou *et al.*
1996	IMTP- International *Musa* Testing Programme-Phase II was started.	
1996	Jackfruit selection PPI-1 was released.	TNAU
1996	Swarna Roopa variety of litchi was developed through clonal selection.	HARP, Ranchi
1996	Grape varieties Pusa Urvashi and Pusa Navrang (tenturier) were released.	IARI

1997	Avocado variety TKD-l and papaya hybrid CO-7 were released.	TNAU
1998	Mutagenic variety Osa Gold Pear was developed and popularized in Japan.	
1998	Guava hybrid Arka Amulya was released by IIHR.	Subramanyam and Iyer
1998	A gynodioecious advanced generation papaya hybrid Arka Surya was released from IIHR.	
1998	Banana hybrid BRS-1 developed by crossing Agniswar and Pisang Lilin.	BRS, Kannara, KAU
1998	In mango, polyembryony genetically inherited and thought to be influenced by a single gene.	Aron *et al.*
1998	Banana hybrid BRS-2 developed by crossing Vannan and Pisang Lilin.	BRS, Kannara, KAU
1998	Lalit, a selection from local guava types was released	CISH, Lucknow
1998	Pineapple hybrid, H-7 developed by crossing Kew with Ripley Queen.	BRS, Kannara, KAU
1998	Pineapple hybrid, H-8 was developed by crossing Kew with Ripley Queen.	BRS, Kannara, KAU
1998	Sun Up, first genetically engineered variety of papaya developed in 1991. Rainbow- a transgenic hybrid between SunUp and Kapoho Solo was commercialized in Hawaii.	Gonsalves
1998	First attempt for the identification of molecular markers that coded sex in papaya.	Somsri *et al.*
1999	Mango hybrid Konkan Ruchi was released from Dapoli for pickling purposes.	
1999	Selection Lalit developed from half sib population of Apple guava.	
2000	Mango hybrid Ambika, developed from a cross between Amrapali and Janardhan Pasand, was released from CISH, Lucknow.	
2000	Mango hybrid Sonpari (with Alphonso as female parent) was released.	AES, Paria, Gujarat
2000	Hybrid sapota CO-3 was released.	TNAU
2000	Grapefruit mutant cultivar Rio Star was developed and popularized in USA.	
2000	Pummelo is the female parent of sour orange, sweet orange and grapefruit. Sour orange is the maternal parent of *Citrus limon*. Citron is the male parent in the origin of many Citrus hybrid species including limes and lemons.	Nicolosi *et al.*

2000	FAO/IAEA developed database having 48 mutant cultivars of fruit crops including 10 of apple and 8 of sweet cherry.	
2001	Shahi variety of Litchi was identified through germplasm evaluation.	AICRP-STF
2001	Hamdan and Sulaiman were promising selections in walnut.	SKUAST , Srinagar
2001	Mango hybrid Pusa Peetambar, developed by crossing Amrapali x Lal Sundari, was released.	IARI, New Delhi
2001	Report of first successful PRSV resistant Tainung-2 papaya through replicase mediated resistance.	Chen *et al.*
2002	GM hybrid papaya was developed from cross between Kamiya and Rainbow F_2.	
2001	The Global Musa Genomics Consortium (MGRC) was created to support banana improvement.	National Science Foundation, Arlington, USA
2002	Poamoho Gold hybrid of papaya was developed by back crossing F_1 hybrid of Kapoho and Rainbow F_2.	Fitch *et al.*
2002	Mango varieties Pusa Arunima and Pusa Surya were released.	IARI
2002	Cryopreservation shown to be a powerful tool for the sanitization of banana plants infected by viruses.	Helliot *et al.*
2002	Transgenic plants of table grape cultivars Silcora and Thompson Seedless produced via organogenesis.	Mezzetti *et al.*
2003	Custard apple variety APK(Ca)-1 and sapota selection PKM (Sa)-4 were released.	TNAU
2004	Jackfruit variety Konkan Prolific, Karonda variety Konkan Bold and Jamun variety Konkan Bahadoli were released.	Dr. BSKKV, Dapoli
2004	Guava variety Pusa Srijan, useful as an extremely dwarf rootstock was released.	IARI, New Delhi
2004	Swarna Manohar cultivar identified through selection from land races of jackfruit collected from Jharkhand, Bihar, West Bengal, UP and Odisha.	IVRC, Varanasi
2004	Swarna Poorti cultivar selected from the land races of jackfruit collected from Jharkhand, Bihar, West Bengal, UP and Odisha.	IVRC, Varanasi

2004	A high density genetic map of papaya using 54 F_2 plants derived from Kapoho and Sun Up cultivars was constructed with large number of markers.	Ma *et al.*
2005	Udhayam banana evolved by selection from ABB Pisang Awak sub group.	NRC for Banana, Tiruchirappalli
2005	Selection Shweta was made from half sib population of apple guava.	CISH, Lucknow
2005	Off-season guava cultivar TRY(G)-I released.	TNAU
2006	An acid lime cultivar Balaji, developed by pure line selection released.	Periyakulam, Tamil Nadu
2007	Intergeneric hybridization between *Carica papaya* var. Arka Surya and *V. cauliflora* was successfully done.	Dinesh *et al.*
2007	Jackfruit selection PLR(J) 2 and sapota selection PKM(Sa) 5 were released from TNAU.	
2007	Walnut promising selection--- Sarwari, K-12, KN-5, SH-23 and SR-11.	Dr. YSPUHF, Solan
2007	Konkan Lemon variety was released from Dapoli.	
2007	Modi apple developed by crossing Liberty x Gala apple cultivar.	CIV (Consorzio Italiano Vivasti), Italy
2007	First genome sequences produced for a fruit crop-two high quality draft genome sequences reported for grapevine.	Jaillon *et al.*
2007	Aonla selection Goma Aishwariya made.	Godhra station of CIAH, Bikaner
2007	Ber hybrid Thar Sevika developed by a cross Seb x Katha.	CIAH, Bikaner
2007	Ber cultivar Thar Bhubharaj developed by selection from Bhusavar area of Bharatpur, Rajasthan.	CIAH, Bikaner
2008	Mango cultivar Arunika, developed from a cross between Amrapali and Vanraj, released from CISH.	CISH, Lucknow
2008	Manila tamarind variety PKM (MT)-l released from TNAU.	
2008	Independence (Self fertile) almond cultivar was developed by Zaiger's Inc. Genetics.	Gary Neil Zaiger
2008	Citrus somatic hybrid having resistance to tristeza virus evolved.	Mourao Filho *et al.*
2008-09	Gujarat Junagadh Custard Apple 1 (GJCA 1) developed by Clonal selection.	Junagadh Agricultural University, Junagadh
2009	Suvarna mango hybrid released from Dapoli.	Dr. BSKKV Dapoli, Maharashtra

2009	Banana cultivar Konkan Safed Veichi released from Dapoli.	
2009	Guava hybrid Arka Kiran released from IIHR.	IIHR, Bengaluru
2009	Clonal selection CITH Apricot-1, CITH Apricot-2 and CITH Apricot-3 were made at CITH, Srinagar.	Sofi *et al.*
2010	A selection Goma Yashi made in bael.	Godhra station of ICAR CIAH, Bikaner
2010	Mango cv. Konkan Raja released from Dapoli.	
2010	A jamun selection Goma Priyanka made.	Godhra station of ICAR CIAH, Bikaner
2010	A pomegranate selection Goma Khatta made.	Godhra station of ICAR CIAH, Bikaner
2010	A tamrind selection Goma Prateek made.	Godhra station of ICAR CIAH, Bikaner
2010	Swarna, a jackfruit selection made.	UAS, Bengaluru
2011	Mango hybrids Pusa Lalima and Pusa Shreshth released.	IARI, New Delhi
2011	Breaking of the intergeneric hybridization barrier in *Carica papaya* and *Vasconcellea cauliflora.*	Jayavalli *et al.*
2012	Golden Delicious apple proved to be a *prepotent* parent to develop new cultivars.	Janick
2012	Three somatic hybrids, an allotetraploid hybrid and two cybrids (2n and 4n), were obtained by protoplast fusion of Valencia sweet orange (*Citrus sinensis* (L.) Osbeck) and Femminello lemon (*C. limon* (L.) Burm.).	Loredana *et al.*
2012	Walnut cultivars CITH-W-6, 7,8,9 and 10 developed by clonal selection at CITH, Srinagar.	Sofi *et al.*
2012	Cultivar CITH Cherry-01 developed through clonal selection from Double (Bigarreau Napoleon) at CITH, Srinagar.	Sofi *et al.*
2012	Cultivar CITH Cherry-02 developed through clonal selection from variety Mishri (Bigarreau Noir Grossa) at CITH, Srinagar.	
2012	CITH Lodh Apple 1 released.	
2013	Non browning Arctic apple produced by Okanagan Speciality Fruits.	
2014	Pusa Aditi and Pusa Trissar grape cultivars were developed and released.	IARI, New Delhi

2014	A clonal selection, mango cv. Arka Neelachal Kesari released.	ICAR IIHR Central Horticultural Experiment Station (CHES), Bhubaneswar
2014	Konkan Samrat released by crossing Alphonso and Tommy Atkins.	Dr. BSKKV Dapoli, Maharashtra, India
2014	Custard apple cultivar Arka Neelachal Vikram of was selected by clonal selection.	ICAR IIHR Central Horticultural Experiment Station (CHES), Bhubaneswar
2015	Pusa Abhinav, Pusa Sharad and Pusa Round cultivars of sweet orange identified and released.	IARI, New Delhi
2015	Pusa Udit and Pusa Abhinav of acid lime identified and released.	IARI, New Delhi
2015	Jackfruit variety Sindoor developed by selection.	KAU, Vellanikkara
2016	Transgenic variety of plum Honey Sweet resistant to plum pox potyvirus (PPV) was developed through genetic engineering.	Scorza *et al.*
2016	Arka Neelachal Kesari of mango and Arka Neelachal Vikram of custard apple were released.	ICAR IIHR Central Horticultural Experiment Station (CHES), Bhubaneswar
2016-2017	Gujarat Junagadh Papaya 1 (GJP 1) developed by mass selection.	Junagadh Agricultural University, Junagadh
2017	Red fleshed apple cv. Mandared developed in Italy.	
2018	Walnut cultivar Liaonimg 1 developed by crossing *Juglans regia* 10103 × 11001.	The Liaoning Institute of Economic Forestry (LIEF), China
2018	Everest Seedless grape cultivar developed.	Bruce Reisch, Cornell Agri Tech, Geneva, New York
2019	Cosmic Crisp apple cultivar developed at the Washington State University (WSU) Tree Fruit Research and Extension Center in Wenatchee, Washington.	Bruce Barritt and Kate Evans
2019	A double-cross hybrid Arka Suprabath developed at IIHR, Bengaluru using parents Amrapali (cross of Dashehari and Neelam) and Arka Anmol (cross of Alphonso and Janardhan Pasand).	
2019	Two jackfruit selections, Siddu and Shankara, made by farmers released.	Karunakaran and Singh

2020	Mandarin cultivar Banzzai developed in Japan as result of multiple crossing between Satsuma, orange, mandarin and grapefruit. It has unique flavour of these citrus species.	
2020	Agricultural Research Service (ARS) released Joy peaches—Rich Joy, Liberty Joy and Crimson Joy.	USDA, Southern Fruit and Tree Nut Research Laboratory, Byron, Georgia.
2020	Robust CRISPR/Cas9-based genome editing of banana has been established and can be applied for developing disease resistant varieties.	Tripathi *et al.*

FRUIT BREEDING OBJECTIVES

The objectives of breeding vary depending on the requirements of each scion or rootstock species.

Scion Cultivars

(a) Dwarf growth stature.
(b) Precocious, regular and prolific bearing.
(c) Attractive fruit colour with pleasant aroma.
(d) High productivity and good quality.
(e) Cross compatible and self fertile lines.
(f) Suitability for processing and export.
(g) Good keeping quality and transportability.
(h) Biotic and abiotic stress resistance.

Rootstock Types

(a) Wide geographic adaptability.
(b) Ease in asexual propagation.
(c) Compatibility with most scion cultivars.
(d) Imparting good fruit productivity and quality.
(e) Free from brittle roots and suckering habit.
(f) Resistance to biotic and abiotic stress.

Major Fruit Breeding Problems:

1. Difficulty to maintain continuity in long term breeding programmes.
2. Long juvenile period delays to see or assess results. Some techniques given below to shorten juvenile phase in seedlings can be adopted:

(i) Continuously growing seedlings at optimum conditions of fertility and under long photoperiods.

(ii) Manual defolitation to induce flowering.

(iii) Cincturing or bark ringing to hasten flowering.

(iv) Grafting on dwarfing rootstock (e.g. apple M 9, M 27 and B 9 in apple) ensuring that stocks are not virus infected.

(v) Exogenous application of plant growth substances like paclobutrazol, alar or ethephon.

(vi) Embryo rescue or seed treatment to break dormancy.

(vii) Use of biotechnological techniques using transgenic approach such as inducing early flowering by constitutive expression of genes such as LFY, API and FT that promote flowering or suppression of genes such as TFLI that delay flowering. Both approaches have been utilized in apple to induce early flowering in the transgenic plants. A stable transgenic line constitutively expressing FT has been used as a rootstock to drive flowering in a grafted scion, to be used as a pollen parent. Another novel approach involves use of one transgenic parent and recurrent selection for early flowering progeny thereby accelerating breeding cycle. By using early flowering trait, breeders have been able to cycle through a generation each year. In apple, it has been possible to induce resistance to primary pathogens like scab and fire blight by crossing early flowering transgenic offspring with elite cultivars or wild relatives. The technology using genetically modified precocious flowering also holds promise in other fruit plants such as citrus and pear.

3. Most fruit crops being highly heterozygous create complexity in breeding because of wide segregation observed in the progenies.
4. Because of narrow genetic base suitable plant material (germplasm) having desirable traits is scarce.
5. Parthenocarpy and seedlessness in fruits such as banana, pineapple etc. hinder cross breeding.
6. In general, genetic analysis in most of the fruit trees has not been attempted.
7. As vegetative methods of propagation take long time to generate ample disease free stocks of established or new cultivars, availability of enough parent plant material becomes a hindrance in breeding programme.
8. Polyembryony or nucellar embryony as observed in mango, citrus and java plum is a great impediment in crop improvement. Since development of embryo may not involve parent embryo sac one does not know whether the seed obtained is as a result of cross and thus no reliable genetic ratio is obtained.
10. Very low seed set/fertility owing to polyploidy in fruits such as pears, *Rubus*, apples, bananas, strawberries and interspecific hybridity as in *Rubus*, strawberries, bananas. Formation of unreduced gametes causes uncertainties in the outcome of controlled crosses.

11. The occurance of polyploidy in ber, banana, mulberry, etc. hinders genetic analysis.
12. Excessive fruit drop as in mango, citrus, grape, etc. is a problem.
13. Presence of single seed in several fruits requires a large number of crosses, e.g. mango, litchi, mahua, etc.
14. Owing to sexual incompatibilities in both self and cross combinations, combining desirable traits in crosses is not possible.
15. Difficulty in selection due to predominance of asexual propagation and new combinations are not obtained owing to the absence of sexual propagation.
16. Knowledge of genetic mechanisms and inheritance pattern of important traits is inadequate in several fruits.
17. Lack of suitable genetic markers makes it difficult to identify successful cross combinations in seedlings/plants.
18. Inadequate information about genetic constitution of most fruit crops and on evolutionary changes over the years.

2 Germplasm Resources

Germplasm resources are raw materials for use in improvement of crop plants. It is a sum total of hereditary material, i.e., the alleles of various genes present in a crop species and its wild relatives. Plant genetic resource consists of i) the entire genetic variability or diversity available in a crop species; ii) land races, modern cultivars, advanced (homozygous) breeding materials, mutants, obsolete cultivars, breeding stocks; iii) both cultivated and wild species, forms and relatives of cultivated crop plants; iv) materials collected from the centres of diversity, gene banks, gene sanctuaries, farmer's fields, markets and seed companies; v) basic material for launching a crop improvement programme; and vi) indigenous materials (collected from within the country) or exotic materials (collected from foreign countries).

Nothing can be achieved only by availability of genetic engineering techniques unless the desirable genes are at our disposal (Mengesha, 1984). But utilization of these genetic materials in breeding programme may require *germplasm exchange* which has implications in intellectual property issues. The breeder is protected from unauthorized use of the cultivars developed from his breeding programme, but in the first place must ensure that he has the right to use the germplasm in his breeding programme. In 1983, the International Treaty on Plant Genetic Resources for Food and Agriculture (PGRFA) was signed by a number of countries (Moore and Tymowski, 2005). All international germplasm exchanges (except the return of germplasm to the country of its origin) are governed by a Material Transfer Agreement (MTA) appropriate to the material (Moore and Tymowski, 2005). International Agricultural Research Centres (IARC) are strategically located in regions of rich crop diversity to facilitate the collection, consearvation, and exchange of germplasm for use by breeders (Mengesha, 1984). These centres have protocols that govern their own germplasm exchange programmes.

GERMPLASM RESOURCE MANAGEMENT

In India, plant genetic resources (PGR) of horticultural crops exhibit enormous variability which can be attributed to the existence of 15 different agro-climatic zones/regions in the country coupled with diverse socio-economic, cultural and ethnic needs. During the 1976 to 2018, with a view to conserve this diversity for future use, a total of 2,697 multicrop or crop-specific explorations were undertaken and a total of 2,71,664 accessions (including 35,562 wild germplasm) were collected from different agro-ecological zones/habitats. In addition, a total of 40,000 trait-specific accessions were introduced during the last decade. Several horticultural crops are recalcitrant and are conserved in field genebank or in *vitro*/cryo genebank. Specific quarantine measures and protocols against exotic and indigenous insect pests and weeds are now available. However, at national level more efforts are required for the development of field genebanks of horticultural crops.

Trait-specific evaluation of the existing variability conserved in genebanks for utilization in the crop improvement programmes as per changing needs is the ultimate goal of managing PGR of horticultural crops. With rising population, changing climate scenario, competition in international market and paradigm shift in the focus from food security to nutritional security, characterization and evaluation of PGR using location specific or trait-specific core reference sets is a necessity for their effective utilization.

The research centres of ICAR-AICRP on fruits explored the local areas for enriching the germplasm collection thereby strengthening the activities of National Active Germplasm Sites (NAGS). In addition, these centres also acted as duplicate safety sites for the germplasm of mandate crops. Since most of these crops are vegetatively propagated, duplicate safety field gene bank is considered to be a major support for the NARS to overcome losses due to national calamities. List of such collections made during last decade and those conserved at different centres is available. In view of shortage of land, elimination of duplicates was started during 12th plan so as to maintain the core collections in the hotspots of mandate crops. The information on germplasm of individual fruit crop maintained at SAU and ICAR institutes is available.

Bioversity International (IPGRI) maintains a number of worldwide databases with summary information on *ex situ* germplasm collections of fruits and nuts. The databases include address and information of organizations holding germplasm with summary information on the type of germplasm that is maintained, such as species names, number of accessions per species, type of accessions, etc. IPGRI attempts to keep these databases up to date by continuously incorporating new information. Updating and collation of *ex situ* germplasm holding data is done in close collaboration with Food and Agriculture Organization of the United Nations (FAO) which provides similar type of data as part of its World Information and Early Warning System (WIEWS). The US National Plant Germplasm Systems repositories house collections of plant material that not only preserves our horticultural heritage, but can be used by fruit breeders to develop new and better varieties. In 1980, the NPGS added clonal genebanks to its network of national repositories. There are now eight genebanks devoted to fruit and nut crops and their wild relatives, which maintain 30,000 accessions representing 1,600 species. Each of these facilities has internationally recognized and globally diverse collections of genetic resources for their assigned crops. Temperate crops, including tree fruits and grapes, are preserved at facilities in Corvallis (Oregon), Davis (California) and Geneva (New York). The facilities are developed in Hilo, Hawaii; Mayaguez, PR., Miami, Fla., and Riverside, Calif. to preserve collections of tropical and subtropical fruit and nut crops.

GENETIC EROSION

Genetic erosion refers to loss of genetic diversity between and within populations of the species over a period of time or gradual reduction in genetic diversity in the populations of a species, due to elimination of various genotypes. It leads to reduction of genetic base of a species due to human intervension and environmental changes. The following five are main reasons of genetic erosion.

1. **Replacement of land races with improved cultivars:** The main features of modern cultivars are high yield, uniformity, narrow genetic base and narrow adaptability. On the other hand, land races and primitive cultivars have low yield potential, more genetic diversity, broad genetic base and wider adaptability. Land races have continually being replaced by modern cultivars and are thus disappearing. This has resulted in reduction in genetic diversity.
2. **Modernization of agriculture:** Clean cultivation and improved crop management practices in modern agriculture has resulted in the elimination of wild and weedy forms of many crops. The weedy forms enhance the genetic diversity through introgression of genes between them and crops.
3. **Extension of farming into wild habitats:** It has resulted in destruction of wild relatives of various crops resulting in reduction of their genetic diversity.
4. **Grazing in wild habitats:** Grazing of animals in the wild habitats destroys the wild and weedy forms of crop plants and thus reduces genetic diversity.
5. **Wasteland areas are cleaned:** Land areas cleaned for civic activities such as hydroelectric projects, air ports, industrial areas and growth and expansion of towns, cities and roads leading to genetic erosion of crop plants.

Four species of mango viz. *M. Pajang, M. Zeylanica, M. lalijiwa* and *M. odorata* heve been listed as endangered.

GERMPLASM CONSERVATION

Germplasm conservation refers to protection of genetic diversity of crop plants from erosion/losses. To achieve safe conservation of the fruit species facing threat of extinction, the best option has to be adopted. The possible conservation strategy involves the use of *in situ* and/or *ex situ* approach.

The Indian National Genebank (NGB) was established at ICAR-NBPGR to conserve the PGR for posterity in the form of seeds, vegetative propagules, in vitro cultures, budwoods, embryos/embryonic axes, genomic resources and pollen. The NGB has four kinds of facilities, namely, Seed Genebank (-18°C), Cryogenebank (-170°C to -196°C), *In Vitro* Genebank (25°C), and Field Genebank, to cater to long term as well as medium-term conservation.

In situ conservation: Conservation of genetic diversity under natural habitat is referred to as *in situ* conservation. The area having the diversity is protected from human interference. This protected area under natural habitat is called a gene sanctuary, also known as Natural Park or biosphere reserve. A gene sanctuary is best located/ established in the centre of diversity or microcentre i.e. within the centre of origin of crop species concerned. National Bureau of Plant Genetic Resources, New Delhi (NBPGR)) is making attempts to establish gene sanctuaries for *Musa, Citrus* and *Mangifera* in the Northeastern region. First gene sanctuary for wild relatives of *Citrus* has been set up in the Garo Hills of Assam. Gene sanctuary for conservation of wild relatives of coffee was set up in Ethiopia in 1984.

This method of preservation (i) protects the loss of genetic diversity caused by human intervention; (ii) not only conserves the existing genetic diversity in the population but also allows appearance of new alleles and new gene combinations during the continued selection and evolutionary period; and (iii) eliminates the risks associated with *ex situ* conservation. However, this method has some limitations, viz., (i) each protected area can cover very small portion of the total diversity and requires several such areas to be able to conserve the entire variability of a single species; (ii) the method involves high cost; and (iii) the establishment and maintenance of such areas poses several management problems.

Few model projects have been successfully implemented in India using on-farm conservation of plant diversity with the participatory approach of famers. This approach provides dual benefit of achieving dynamic conservation and substantial livelihood support to the stakeholders. It is highly beneficial for the small and marginal farmers holding substantial genetic diversity of underutilized fruits in remote and fragile landscapes. Farmers having large land holdings are maintaining rich diversity of local selections/cultivars with attraction of either substantial financial gains or to satisfy socio-cultural or religious needs. An illustration of such conservation through farmer's community participation is that of protecting Kachai lemon in the farm or backyard garden in Okhrul district of Manipur.

Ex situ conservation: *Ex situ* conservation is preservation of germplasm (plants, seeds and explants) outside the natural habitat of species. It requires collection and maintenance of plants and systematic long storage of seeds and explants into gene banks. This method allows preservation of entire genetic diversity of a crop species at one place. Besides, the handling of germplasm is easy. Thus this method of conservation of diversity becomes cheaper. *Ex situ* conservation can be done on-farm, in field gene bank, botanical garden, seed gene bank, pollen bank, meristem gene bank, shoot tip gene bank, cell and organ gene bank and DNA gene bank and by cryopreservation.

Field gene banks: Field gene banks or plant gene banks are land areas where germplasm collections of plants are grown and maintained. These are permanent collections of living plants. Often working collections of living plants are established and maintained for experimental purpose. The plant species that have recalcitrant seeds or do not readily produce seeds are conserved in field gene banks. Plant species conserved in field gene banks can be continuously evaluated for economic characters and can be directly utilized in breeding programme. But field gene banks can not cover the entire genetic diversity of a species. It can cover only a fraction of the full range of diversity. The conserved germplasm is exposed to damage by pathogens and insects and sometimes by natural disasters such as bushfires, cyclones, floods, etc. Thus maintenance of germplasm in the field is expensive.

On-farm conservation: It provides sustainable maintenance of crop's wild relatives, landraces and obsolete cultivars (lines, populations) by growing in conditions under which these materials had originated and evolved under similar cultural practices. This is a dynamic conservation method as it enables continued development and evolution under the influence of regional environment and culture.

Botanical gardens: Botanical garden is the most conventional method of ex-situ conservation. Botanical gardens have traditionally focussed on maintaining species diversity, particularly on survival of wild species that are endangered in their natural habitat. Generally, collections consisting living plant material of only one or a few individuals per species are preserved. These also specialize in medicinal and aromatic plants, ornamentals and tropical species.

Seed gene bank: Seed gene bank is the most widely used and convenient method of *ex situ* conservation. The seed can be conserved under long term, medium term and short term storage conditions. It is relatively safe, requires minimum space and is easy to maintain. Glass, tin or plastic containers are used for storage of seeds.

The sequence of activities in seed gene bank involves collection, seed preparation, drying, packaging, storage, periodic germination tests, seed regeneration, re-storage and documentation at each stage of activity.

Roberts (1973) classified seeds, on the basis of their storability, into two groups, *viz.,* orthodox and recalcitrant seeds.

(a) **Orthodox seeds:** Most crop seeds belong to this category. Such seeds can be easily stored for long periods. The seeds are dried to low moisture content of 5% and stored at low temperature without loss of viability. Their longevity increases in conditions of low humidity and low temperature.

(b) **Recalcitrant seeds:** The viability of recalcitrant seeds (as citrus) drops drastically if their moisture content is reduced to 12-30%. Seeds of tropical crops like cocoa, coffee, rubber, oil palm, mango, mahua, durian, jackfruit, etc. belong to this group. Such seeds are difficult to store. IPGRI recommended that the preferred standards for long-term storage of seed of orthodox species is to dry them to a moisture content below 7% and seal the dried seeds in moisture-proof container such as laminated foil bags, aluminium cans or glass jars and store at -18°C. This is applied only for seed of true orthodox species and their relatives. In case of less known wild species, a temperature of -4°C and moisture content of 7-8% is advisable to begin with.

The Svalbard Global Seed Vault was established in remote island of Svalbard archipelago, Norway on 26th, February 2008. The purpose of Seed Vault is to provide insurance against both incremental and catastrophic loss of crop diversity held in traditional seed banks around the world. It is long term seed storage facility and represents the world's largest collection of crop plant.

Pollen banks: Germplasm of fruit species is conserved in pollen banks for several uses. Pollen preserves the nuclear genes of germplasm and is subjected to less stringent quarantine restrictions, because it rarely transfers pests and diseases (except some viral diseases), it can be used for distributing or exchanging germplasm among locations. Besides, stored pollen is used to improve fruit setting in orchards and to enable success in crosses made in breeding programmes, particularly when there is gap between flowering times of male and female parents. Pollen is also used for basic physiology, biochemistry and fertility studies and in biotechnology studies involving gene expression, transformation and *in vitro* fertilization.

Pollen is stored under classic storage conditions (partial desiccation followed by storage at subzero temperatures). But the pollen remains viable for only a relatively short period and therefore the method is used only to a limited extent for germplasm conservation. However, in some crops, the storage of pollen is possible in appropriate conditions and might be an interesting alternative for long-term conservation of problematic species. When stored at the appropriate relative humidity (RH) and temperature (10 - 30 % RH and 0 - 10°C), depending on species, the pollen of *Citrus* spp., *Fragaria* spp., *Olea europea, Pistacia atlantica, Vitis vinifera, Pyrus* and *Malus* maintained viability for more than one year.

Pollen used for germplasm conservation should, however, remain viable for many years, and thus cryopreservation seems to be the most efficient method. For example, maize pollen could be dried to 50 per cent of its original water content in an air current for one hour and then stored at –196°C in liquid nitrogen. Deep-frozen maize pollen can be used for fertilization after ten years storage. Cryopreservation techniques have been developed for a large number of species, mainly those having desiccation-tolerant pollen, also including for several species having desiccation-sensitive pollen.

Cryobanks of pollen have been established for fruit tree species in several countries. At the Indian Institute of Horticultural Research, Bengaluru (IIHR) a pollen cryobank conserves about 300 accessions of fruit species including citrus, mango, papaya, pomegranate, jackfruit and grape. *Pyrus* pollen is cryopreserved at the National Clonal Germplasm Repository in the United States.

Transmission of organelle genomes via pollen is lacking, and sex-linked genes in dioecious species are lost. Since pollen is an exhaustible reserve; provision must be made for its periodic replenishment in the pollen banks. Thus it is obvious that pollen preservation is only supplemental to the seed or clone conserved to yield the pollen. Over multiple generations there is risk of genetic problems such as loss of alleles through random drift or splitting of adaptive complexes. Therefore, only paternal material is conserved and regenerated and, in order to utilize the germplasm, a recipient female plant is always needed for fertilization. To sum up, storing pollen is useful in the framework of establishment of complementary conservation strategies and may provide a means to overcome the difficulties involved in preserving recalcitrant-seeded species. This is particularly relevant because there is no correlation between seed storage behaviour for a given species and the desiccation sensitivity of its pollen.

In vitro germplasm conservation methods

In vitro conservation offers several distinct advantages. For example, the material can be maintained in a pathogen-tested state, thereby facilitating safer distribution. Further, the cultures are not subjected to environmental disturbances.

In vitro conservation can be done in two ways:

(i) under classic storage conditions (partial desiccation followed by storage at subzero temperatures), viz., *in vitro* slow growth procedure, where germplasm accessions are kept as sterile plant tissues or plantlets on nutrient gels to provide short and medium term storage options, and

(ii) by cryopreservation where tissues are stored at a very low temperature (-196^{0}C) in liquid nitrogen virtually stopping all biological processes to enable long term storage of the plant material.

Cryopreservation: It is done by storage of seeds, pollen, tissue or embryos in liquid nitrogen. This method is used for storage of the material without deterioration over a much greater time period relative to all other methods of *ex situ* conservation.

It offers long term storage for germplasm that cannot be conserved as dry seed at subzero temperatures. Tissue cultures are conserved at very low temperatures, usually at –196°C in liquid nitrogen to facilitate the arrest of mitotic and metabolic activities and thus to guarantee long term preservation of germplasm in a genetically unaltered state.

Cryopreservation protocols, like other *in vitro* culture techniques, need to be developed for each species. The basic prerequisites for cryopreservation protocols are that the plant material (1) survive the freezing procedure before storage and the thawing procedure after storage and (2) be regenerated into whole plants for use, conserving the genetic stability. Explants produced from apical and axillary meristems and somatic embryos have a high potential for genetic stability (Hawkes *et al.,* 2000). Most of the explants employed in cryopreservation (cell suspensions, calluses, shoot tips, embryos) contain high amounts of cellular water and are thus extremely sensitive to freezing injury. Cells have thus to be dehydrated artificially to protect them from damage caused by the crystallization of intracellular water into ice. The first two methods developed achieved this through freeze-induced dehydration, i.e., either ultra-rapid freezing or slow, stepwise freezing of cultures. Often the material is treated with a cryoprotectant such as dimethylsulfoxide (DMSO), glycerol, ethylene glycol, proline, or sucrose, either singly or in combination. Cryoprotectants reduce the size and growth rate of ice crystals, lower the freezing point of intracellular contents, and enable cells to be subjected to very low temperatures without disruption of the cell membrane.

Procedures of Cryopreservation (Engelmann, 2000)

1. **Encapsulation–dehydration:** The encapsulation-dehydration procedure is based on the technology developed for the production of artificial seeds. Explants are encapsulated in alginate beads, pregrown in liquid medium enriched with sucrose for 1–7 d, partially desiccated in the air current of a laminar airflow cabinet or with silica gel to a water content around 20% (fresh weight basis), then frozen rapidly. Survival is high and growth recovery of cryopreserved samples is generally rapid and direct, without callus formation. A modified protocol has been proposed by Sakai *et al*. (2000) in which encapsulation and pregrowth in medium with sucrose and glycerol are performed simultaneously. This technique has been applied to apices of numerous species from temperate and tropical origin as well as to cell suspensions and somatic embryos of several species.

2. **Vitrification:** This technique involves treatment of samples first with cryoprotective substances and then the tissue is desiccated with a cocktail of vitrification solutions, followed by rapid freezing. The vitrified material is thawed in 40°C water and the cryoprotectants and then removed. This procedure has been developed for apices, cell suspensions and somatic embryos of numerous species. It has been applied successfully to embryogenic cultures of mango (*Mangifera indica*) and to dormant shoot tips of persimmon.
3. **Encapsulation–vitrification:** The method is successful for meristems and shoot tips of crops as these are sensitive to direct desiccation. It is a combination of encapsulation–dehydration and vitrification procedures, where samples are encapsulated in alginate beads, then subjected to freezing by vitrification.
4. **Dehydration:** It is the simplest procedure since it consists of dehydrating explants, then freezing them rapidly by direct immersion in liquid nitrogen. This technique is mainly used with zygotic embryos or embryonic axes extracted from seeds. It has been applied to embryos of a large number of recalcitrant and intermediate species. Desiccation is usually performed in the air current of a laminar airflow cabinet, but more precise and reproducible dehydration conditions are achieved by using a flow of sterile compressed air or silica gel. Ultra-rapid drying in a stream of compressed dry air (a process called 'flash drying' developed by Prof. Berjak's group in South Africa) allows freezing of samples with relatively high water content, thus reducing desiccation injury. Optimal survival is generally obtained when samples are frozen with water content between 10 and 20% (fresh weight basis).
5. **Pregrowth:** The pregrowth technique consists of cultivating samples in the presence of cryoprotectants, then freezing them rapidly by direct immersion in liquid nitrogen. The pregrowth technique has been developed for *Musa* meristematic cultures.
6. **Pregrowth–dehydration:** In this procedure, explants are pregrown in the presence of cryoprotectants, dehydrated under the laminar airflow cabinet or with silica gel, and then frozen rapidly. This method has been applied notably to asparagus stem segments, oil palm polyembryonic cultures and coconut zygotic embryos.
7. **Droplet freezing:** The droplet freezing technique has presently been applied to potato, asparagus and apple apices. Apices are pretreated with liquid cryoprotective medium and then placed on aluminum foil in minute droplets of cryoprotectant and frozen slowly (apple) or rapidly in liquid nitrogen.

Cryopreservation of germplasm in gene banks is being carried out in apple, pear, banana and mulberry. Important prtocols have been developed for cryopreservation of *in vitro* derived explants in citrus, grapes, pineapple and papaya.

Meristem gene banks: Germplasm of asexually propagated species can be conserved in the form of meristems. This method is widely used for conservation and propagation of horticultural species.

Shoot tip gene banks: Germplasm can be conserved as slow growth cultures of shoot tips and nodal segments. Their regeneration involves sub-culturing the cultures after every 6 to 36 months. The chief merits of this gene bank are:-

1. Genotypes of the accessions of vegetatively propagated crops and tree species can be conserved indefinitely and free from diseases and pests.
2. The method can be used for crops which either do into produce seeds or the seeds are recalcitrant.
3. Subculture becomes necessary only after relatively long periods (every 6-36 months).
4. For regeneration, subculturing, takes comparatively very short time.

Cell and organ gene banks: A germplasm collection based on embryogenic cell cultures, shoot tips and/or somatic/zygotic embryos is called cell and organ bank. The cell or organ is conserved at -196^{0}C in liquid nitrogen. The techniques for cryopreservation of plant cells and tissues are being rapidly refined. Some such banks have been established, e.g., for potato in Germany.

DNA gene banks: DNA storage is one of the emerging germplasm conservation techniques. DNA or RNA sequences are considered as genetic resources and are now routinely extracted and conserved in DNA banks. The bank preserves and distributes the DNA molecules of particular plant genetic resource and provides associated information. Genetic resource collections in the form of frozen tissues, purified DNA samples, frozen viable cell cultures and derivatives such as RNA, cDNA and genomic libraries represent valuable components of comprehensive storage strategy.

In these banks, DNA segments from the genomes of germplasm accessions are maintained as cosmid clones, phage lysates or pure DNA. The last one is maintained for relatively short periods. These DNA segments can be evaluated and the desired ones may be used to produce transgenic plants. This approach is applicable to the conservation of genetic materials of already extinct species since DNA extracted from well preserved herbarium specimens can often be cloned. However, it is very expensive and highly sophisticated. A worldwide network of DNA banks for threatened/ endangered species has been established.

3

Methods of Breeding

Breeding methods in fruit crops begin with introduction of germplasm which is then catalogued, evaluated, characterized and put to selection process to identify genetic material for various uses. Similarly, clonal material is also identified, collected and evaluated to select useful types. Simultaneously, methods such as hybridization and backcrossing may be done wherever and whenever required. The other breeding tools to bring about improvements are polyploidy, inducing mutations and use of several biotechnological approaches. These are briefly discussed below.

INTRODUCTION

Plant introduction refers to the process of taking a genotype or a group of genotypes into a new environment, where they were not being grown before. Plant introduction is done i) for economic uses: directly as a vaiety, after selection as a new variety or as parent material for improvement, ii) to study evolutionary process, iii) to conserve diversity, iv) for genetic improvement in fruit types, and v) to serve aesthetic interests. Thus plant introduction may provide an entirely new fruit type, e.g. kiwi fruit and blueberry, a superior variety directly or after selection and/or hybridization and protect variability from genetic erosion. By introducing in new disease free area the genotypes are protected from damage. It is a quick and economical method of crop improvement.

Plant introduction may be *primary* or *secondary*. Primary introduction constitutes the process of release of the introduced variety without any alteration for commercial cultivation, when well adapted to the new environment. Whereas, secondary introduction includes the processes where the introduced variety is subjected to selection to isolate a superior type (variety) or is hybridized with local varieties to transfer one or few characters from this variety to the local ones.

The steps involved in the procedure of breeding by introduction are:-

1. Procurement of germplasm is done in the form of gift, exchange, purchase and collection through exploration. The sources of introduction are i) Centers of diversity, ii) Gene banks, iii) Gene sanctuaries, iv) Nurseries, and v) Farmer's fields.
2. Quarantine: Plant introductions are associated with the introduction of foreign weeds, diseases and pests. Therefore, before introduction, the genotypes have to undergo a rigorous quarantine.
3. Cataloguing**:** Each germplasm accession is given an accession number. This number is prefixed in India with IC (indigenous collection), EC (exotic collection) or IW (indigenous wild). Information on the species and variety names, place of

origin, adaptation and on its various features or decision is also recorded. The amount of data recorded during evaluation is huge. Its compilation, storage and retrieval are now being done using special computer programmes.

4. Evaluation is done at research stations located in different eco-climatic zones.
5. Acclimatization.
6. Multiplication and distribution.

ICAR-NBPGR, New Delhi has enabled the introduction of 9684 accessions/ varieties of fruit crops from over 40 countries. Some notable introductions made are presented in Table 2.

Table 2. Achievements made in introduction of fruits

Fruit	Cultivars/Species
Banana	cv. Lady Finger (EC 160160) from Australia. Grand Nain MS (EC 27237) from France and Valery from West Indies.
Citrus	Tarocco Blood from USA, Sun Ramon from Peru, Vaniglia Sanguigno from USA; Thai from Thailand.
Date palm	Tayar, Hatemi. Mejnaj, Khalas, Ruiziz, Khesah from Saudi Arabia, Hayani, Samy, Agolani, Saklote and Amhat from Egypt.
Fig	Genoa White, 0030 Genoa, Flanders, Conadria, Mission and *Ficus* standard from USA.
Grape	Perlette, Thompson Seedless, Beauty Seedless from USA, Kishmish Beli, Kishmish Charni from former USSR; Totlocha from Brazil and Dogridge from USA.
Guava	Beaumont G-35, Indonesia Seedless from Australia, Verdie and M. 25988 from USA.
Mango	Tommy Atkins, Zilets, Haden, Sensation, Julie from USA, Dashehari-35.
Papaya	P141436, Soniyamma, Malinchly from Nigeria, Sunrise Solo (from USA), Wilder, Sunset T88/21 1, *Vasconcellea pubescens* syn. *C. cauliflora* from USA.
Pomegranate	Wonderful from USA, Rannij Gl-8-23 & Shirin Anar, Rannyij G-1-3-34 from former USSR.

SELECTION

This process favours survival and further propagation of plants having desirable characters. The process operating in the nature without human interference is called Natural selection. This process is also carried out by artificial selection through choosing certain individual plants having different productivity or quality characters from a mixed population for the purpose of better crop. In fruit crops, open pollinated seedling selection is practiced from mixed population. However, selection depends upon i) genetic variability and that ii) variation must be heritable. Then different methods of selection, as discussed below, can be followed.

(a) ***Mass selection:*** It can be defined as selection of a number of phenotypically superior plants, pods, or heads or seeds from the field population, harvesting and bulking together their produce for sowing the next year's crop and repeating this process till desired charcters are achieved. During the process off-types or undesirable plants are rogued out of the population; alternatively, desirable plants may be selected and bulked from the existing population to form an improved cultivar.

Mass selection is applicable to both self- and cross-pollinated species, but with different genetic consequences (Allard, 1960). Mass selection may be used to maintain the purity of an existing cultivar after it has been contaminated (by natural out crossing, spontaneous mutation and mechanical mixing). It is used to induce horizontal (durable) resistance into a cultivar, and for adapting a new cultivar to a production region. Mass selection is based on plant phenotype and is hence most effective if the trait of interest is highly heritable (Brown and Caligari, 2008).

(b) ***Progeny selection:*** Selection is made based on performance of progeny instead that of plants. It is usually more effective parituclarly dealing with quantitative characters of low viability.

(c) ***Pure line selection:*** It may be defined as a strain obtained from the progeny of a single self fertilized homogygous individual. It can form part of other breeding methods (e.g. bulk breeding). Pure line cultivars lack yield stability over a wide range of environments and are susceptible to devastation by pathogenic outbreaks (Allard, 1960).

(d) ***Pedigree selection:*** Unlike in mass or pure line selection, this method is based on pedigree i.e. descended from known parents. Consequently, hybridization is used to establish the base population for the breeding programme. To maintain ancestry of the cultivar, the breeder must keep meticulous records such that the parent-progeny can be traced back to an individual F_2 plant from any subsequent generation (Acquaah, 2012). The method is suited to species that allow individual plants to be observed, selected and harvested separately (e.g. peanut, tomato, tobacco).

(e) ***Clone and clonal selection:*** Clone is a progeny of genetically uniform plants produced from a single plant by asexual reproduction. Individuals of a clone may be heterozygous or homogeneous but are stable. Mutation is the only means of creating genetic variability in the clones. Clones often have vigorous growth and water adaptability. The loss in vigour and productivity of the clone with passage of time may be due to mutation and infection of virus and bacteria. This is termed as clonal degeneration

Clonal selection is done by selecting desirable clones from mixed population of vegetatively propogated progeny. It is the only method of breeding in vegetatively propagated crops and is easy and less time consuming. The selected varieties are highly uniform and stable like pure line. To create variability is difficult. Permanent hybrid-heterotic clones on selection may be used as hybrid. But the varieties are highly prone to infection by new races of diseases.

Clonal selection procedure (steps)

1st year- Mixture of clones

Few to several hundred superior clones selected

↓

2nd year- Desirable clones are selected from the selected clones

↓

3rd year- Primary yield trial with standard check-Best clones are selected

↓

4th - 8th year - Multilocation trials with standard check

↓

Last year - Best clones for release are selected and multiplied

Table 3. Important clonal selections (cultivars).

Fruit	Cultivars
Mango	Dasheri-51, C24, C18, C35 and C45 (Dashehari), Paiyur – 1 (Neelam), MA- I (Alphonso), Tommy Atkins (Haden), Pusa Surya (Eldon), and Subhash (Zardalu).
Grape	Tas-A-Ganesh (Thompson Seedless), HS37-6 (Perlette), Rao Sahebi (Cheema Sahebi), Dilkhus (Anab-e-Shahi), Manik Chaman, Pusa Seedless (Thompson Seedless).
Ber	Goma Kirti (Umran).
Pomegranate	G-137 (Ganesh), Bhagwa and Yercaud 2.

(f) ***Recurrent selection:*** Selection of generation after generation with inbreeding of the selects to provide for genetic recombination is called recurrent selection. It is a widely used method for population improvement.

Table 4. Important selections made in fruit crops.

Fruit	Cultivars
Mango	Subash (selection from Zardalu), Pusa Surya (selection from Eldon), Paiyur (selection from Neelum), Tommy Atkins (from Haden), Keitt (from Mulgoba), Golden Nuggets Selection and Brooks (from Sandersha)].
Grapes	Cheema Sahebi, Dilkhus, ARI-302, Pusa Seedless, Tas-A-Ganesh, Manik Chaman, Sharad Seedless, Manjari Kishmish and Manjari Naveen.
Guava	L-49, Arka Mridula, Allahabad Safeda, Pant Prabhat and Lalit.
Aonla	Kanchan (NA-4). Krishna (NA-5). NA-6, NA-7 and Anand I & 3.
Litchi	Swarna Roopa
Papaya	Inbred sel. — CO-1, CO-2, CO-5, CO-6, Pusa Delicious, Pusa Giant, Pusa Dwarf and Coorg Honey Dew.
Kagzi Lime	Vikram, Pramalini, Sai Sarbati, Jaidevi and Sel. 124.

Lemon	Pant Lemon-1.
Bael	NB-5, NB- 9 CISHB-l & 2, Pant Aparna, Pant Shivani, Pant Urvashi and Pant Sujata.
Sapota	C0-2 & PKM-1
Pomegranate	Ganesh, G-137, P-23, P-26, Muscat (Maharashtra), Bassein Seedless, Jyoti and Madugiri (Karnataka), Dholka (Gujarat), Kabul Red, Yercaud-1 & CO-1 (Tamil Nadu).
Cashewnut	Vengurla- 1 and BPP- 1 & 2
Karonda	Kokan Bold and RHR-3, 12, 13 & 16
Kokum	Kokan Amritha
Jamun	Paras (Gujarat) and Kokan Bahado
Ber	Gola, Barnarsi Karka, Banarasi, Sanouri 1,2,3 & 4 and Umran.
Tamarind	Prathisthan, PKM-1, DTS-1 & 2, Yogeshwari, Urigram (Tamil Nadu) and Jagdish (Maharashtra).

HYBRIDIZATION

Hybridization is mating of two plants with dissimilar genotype. Based on the taxonomic relationships of the two parents, hybridization may be classified into two broad groups; intervarietal and distant.

1. **Intervarietal hybridization:** The parents involved in hybridization belong to the same species; may be two strains, varieties or races of the same species, e.g. two varieties of *Vitis vinifera*. Therefore, it is also known as intraspecific hybridization. Intervarietal hybridization is the most commonly used in crop improvement programmes.

   Intervarietal crosses may be simple or complex depending upon the number of parents involved. In a simple cross, two parents are crossed to produce F_1. The F_1 is selfed to produce F_2 or is used in a backcross programme, e.g., A x B → F_1 (A x B). In a complex cross, more than two parents are crossed to produce the hybrid, which is then used to produce F_2 or is used in a backcross. Such a cross is also known as convergent cross because the crossing programme aims at converging genes from several parents into a single hybrid.

2. **Distant hybridization:** Desirable traits like resistance to biotic and abiotic stress in varying environments, composition of metabolites or morphological traits and their responsible genes are often found only in wild species, related species and genera of the cultivars, being treasure of such valuable genes. These are mostly lacking in cultivated species. Wide crosses by distant hybridization help in gathering all desirable traits in one species or nucleus. But there are several problems associated with distant hybridization. Both pre- and post-zygotic barriers are encountered in interspecific or intergeneric hybridization. Attempts are needed to overcome these by one or more of several techniques such as use of compatible parents, ploidy manipulation, bridge crosses, pollen

mixtures, pistil manipulation, somatic hybridization, embryo rescue, growth regulators, etc.

Intergeneric hybridization: Intergeneric hybridization gives an opportunity to combine genomes from distinctly different plants and to introgress traits not found in the main genus of interest, e.g. *Citrus wakonai* x *Citropsis gabunensis* and crosses of *Citrus* spp. with *Poncirus* Raf., *Microcitrus* Swing., *Eremocitrus* Swing. and *Fortunella* Swing. The name *Microcitrus* and *Eremocitrus* have changed to *Citrus*. *Microcitrus* and *Eremocitrus* are considered synonym (Mabberley, 1998). The steps that are to be taken for successful hybridization relate to selection of seed parents, selection of branches and flowers on the selected seed parents, bagging, emasculation and pollination of flowers followed by harvesting and storing the hybrid seeds and then raising the F_1 generation.

Interspecific hybridization: The purpose of interspecific hybridization is to introduce good characteristics from wild species into the cultivated species, e.g. imparting insect pest resistance and produce new cultivar such as Safed Velchi (*Musa acuminata* x *Musa balbisiana*), SH-3386 (*Musa balbisiana* x *Musa acuminata*), Wompa (*Musa acuminata* x *M. schizocarpa; Musa schizocarpa* x *Musa acuminata* ssp. *banksii*, TN13- 125 (*Fragaria nilgerrensis* var. Yunnan x *F.* x *ananassa* cv. Toyonoka).

Procedure of Hybridization

(i) **Selection of seed parent:** The breeder should have clear objectives in making a cross, and the parents should be selected to fulfil these objectives. Parental plants must be selected from the local areas since these are supposed to be the best suited to the existing conditions. So far as possible, varieties which produce fairly numerous seeds and high proportion of gametic progeny should be selected as seed parents. Embryo counts from dissection of seeds, or germination tests, give valuable indications on the suitability of varieties as seed parents. It is sometimes advantageous to use parental combinations that are similar with respect to nucellar embryony, provided the nucellar seedlings of the variety are less similar in leaf characters to the hybrid seedlings.

(ii) **Selecting trees, branches and flowers on seed parent:** Trees and branches which are likely to set fruit should be chosen. Those which have borne a heavy crop should be avoided, especially in varieties with tendency to alternate bearing. In acid lime, which is reputed to bloom throughout the year, main season blooms are more fertile than those blooming in off season. Even within the season earlier blooming flowers are preferred. Larger and terminal flowers should be preferred. Only a limited number of flowers on a tagged branch should be utilized for crossing and all the others including old or new buds should be clipped off. To ensure success in fruit setting, it is advisable to girdle the branch in advance well before the time of the blooming by a single circular cut through the bark. In case of highly sterile varieties, some cultural conditions may boost the fertility.

(iii) Bagging the flowers: The unopened flowers should be bagged using paper bags at least one day prior to the actual crossing operation. For selfing simple bagging is enough.

(iv) Emasculation: Emasculation of flowers should be accomplished before the pollen is shed, at least a day earlier to anthesis. Shedding of pollen takes place soon after the petals separate. In some varieties, it may begin a little earlier. The petals should be gently separated and the anthers alone removed, taking precaution not to injure the other parts of the flower, particularly the pistil. Immediately after emasculation the flowers should be rebagged to avoid entry of foreign pollen.

(v) Pollination: The emasculated flowers are usually pollinated by hand the next morning. The best time for pollination is usually from 1-6 days after emasculation, according to the temperature and the prevalent stage of development of flower at emasculation. Flowers which are nearly mature at the time of emascuclation may, however, be pollinated at once. Pollen should be collected well in advance from the desirable male parent and the anthers are rubbed over the surface of the stigma of the female flowers. The flowers after crossing are rebagged. Every precaution must be taken to prevent access of pollen to the stigma either prior to crossing or after crossing. The crossed flowers should be suitably numbered and tagged.

(iv) Harvesting and storing the hybrid seed: Petal fall, drying stigma and green ovary indicate the success of fruit set. When the fruit reaches pea grain size, the paper bags should be replaced preferably by cloth bags. After the set, fruit is allowed to develop normally and is harvested on ripening.

(vii) Raising F_1 generation: The seeds extracted from the ripe fruits are used to raise F_1 generation.

Table 5. Important hybrids.

Fruit	Cultivars
Mango	Mallika, Amrapali, Pusa Arunima, Pusa Lalima, Pusa Pitamber, Arka Anmol, Arka Puneet, Arka Aruna, Arka Nilkiran, Ratna, Ambika, Sindhu, PKM- 1 & 2 and Sai Sugandh.
Guava	Arka Amulya, Safed Jam and Kohir Safed.
Litchi	Sabour Madhu and Sabour Priya.
Papaya	CO-3 and CO-4.
Sapota	CO-1, DHS-1 & 2, Hybrid 214 & 711, PKM-2 and PKM-3.
Ber	CIAH Hy-I (Seb x Katha), Thar Sevika (Seb x Katha), BCF_1 [Seb x Tikadi) F_1 x Seb] (for fruit fly resistance).
Aonla	Chakaiya x Krishna, Kanchan x NA-6 and NA-6 x NA-9 (for fruit size and weight).
Pomegranate	Mridula (Ganesh x Gulsha Red), Ruby, Amalidana and Phule Arakta.
Apple	Lal Ambri, Sunheri, Firdous, Shireen, Akbar, Ambred, Ambrich and Chaubattia Anupam.

BACKCROSS BREEDING

Backcrossing is the only breeding method used for interspecific gene transfer. This breeding method is undertaken to transfer one or a few specific genes of interest from a source (donor parent) to an adapted cultivar (breeding line), preserving all other qualities (Acquaah, 2012). It makes it possible to transfer i) a dominant gene, ii) a recessive gene or iii) a qualitative gene and to introgress genes from wild gene pools into the domesticated ones.

The method is useful to transfer disease resistance and quality characters and male sterility and fertility restorer genes. Undesirable gene is replaced but all desirable characters of a popular adapted variety are retained. However, sometimes undesirable characters tightly linked with the desirable one are transferred.

For initiating the transfer of genes via a cross, the desirable adapted cultivar, called the *recurrent parent,* is repeatedly crossed (backcrossed) to the F_1 to retrieve all the desirable genes of the adapted cultivar. Backcrossing may be done for 2–5 cycles (BC1 – BC5), depending on how much of the recurrent parent the breeder wants to recover, and how easy it is to observe the trait of interest. For a variety developed by this method, the multiplication testing is not required. Backcrossing, however, involves high quantum of crossing work. The developed new variety differs only in one or two gene from the old one.

POLYPLOIDY

Polyploidy is a condition when individual's genomic cells possess multiples of the basic set of chromosomes in the somatic cell in excess of the diploid number for the species. Such individual is called a polyploid. Polyploidy is an important tool used for improvement in fruit crops. Depending upon genome constitution, polyploidy types may either be autopolyploidy or allopolyploidy.

(a) ***Autopolyploidy*****:** In this condition, the multiplication is of the same genome. These may be:-

- **(i) Autotriploidy:** There are 3 sets of chromosome of the same genome (3x).
- **(ii) Autotetraploidy:** There are 4 sets of chromosome of the same genome (4x).
- **(iii) Autopentaploidy:** There are 5 sets of chromosome of the same genome (5x).
- **(iv) Autohexaploidy:** There are 6 sets of chromosome of the same genome (6x).

Triploid and tetraploid autotriploids have been used for improvement in grape, banana and apple.

(b) ***Allopolyploidy*****:** It is a condition in which complete genome of two or more species combine together. Important natural allopolyploids include wheat, oats, tobacco, cotton, sugarcane and strawberry.

- **(i) Allotetraploidy:** Two copies each of two distinct genomes (2x1 + 2x2)

(ii) Allohexaploidy: Two copies each of three distinct genomes (2x1 + 2x2 + 2x3).

(iii) Allooctoploidy: Two copies each of four distinct genomes (2x1 + 2x2 + 2x3 + 2x4).

Alloploids can be used as a bridging species for plant improvement. These are also useful in i) tracing the origin of natural allopolyploids, ii) creating new species, and iii) interspecific gene transfer.

Induction of polyploidy: Polyploidy can be induced in the plant by treatment with colchicine (*Colchicum autumnale*) at 0.01 to 0.5%. Colchicine is a *spindle poison*. When present in the cell, it induces C-metaphase and C-anaphase causing mitotic chromosomes to assemble at metaphase and divide. At the same time there is no cytokinesis and double the number of daughter chromosomes is included in the nucleus. During S phase, daughter chromosomes duplicate and thus the restitution nucleus has double chromosome content. Consequently the cell becomes tetraploid. The process may continue giving rise to still higher levels of ploidy in the presence of colchicine.

For induction of polyploidy, we may treat seeds, seedlings, cuttings or buds. Application of colchicine may be done on growing tips, meristematic cells or axillary buds. The duration of application may be 24 to 96hrs depending upon the species / plant.

The resulting polyploid plant may usually have i) thicker and shorter stems, ii) fleshy, larger, thicker and deeper green leaves, iii) longer and stronger roots, iv) larger flowers, pollens and seeds, v) longer maturity duration and slower growth rate of fruits and vi) higher water contents than in the diploids.

Table 6. Acheivements in fruit crops by polyploidy.

Fruit	Polyploidy typ	Results
Banana	Triploid	cv. Poovan (AAB), Rasthali (AAB), Peyan (AAB), Thote (ABB), Peykunnan ABB) were used as female parents for developing commercial triploids of interspecific origin (Coimbatore).
	Tetraploid	Gold Finger (FHIA-01) , FHIA-03, BITA- 2, BITA -3, PITA 14
Papaya	Tetraploid	BITA- 1, BITA- 2, BITA 8
Pineapple	Tetraploid by colchicine	Tetraploids produced in Cayenne
Seedless lime (*C. latifolia*)	Triploidy	Dwarf plant & large fruit size
Ber cv. Illaichi	Triploidy	Tall plant & small fruit
Grape	By colchicine treatment	Loose Perlette from Perlette, Marvel Seedless from Delight, Early Giant from Campbell Early
Aonla	By colchicine treatment	Small seeded fruit can also be obtained

MUTATION BREEDING

Sudden heritable change in the characteristic(s) of an organism is called mutation. Mutation may be i) natural without any treatment, i.e. spontaneous or ii) artificial as a result of physical or chemical treatment, i.e. induced. Mutation is generally recessive but dominant mutation may also occur. Mutations are random and recurrent and are usually harmful to the organism. Induced mutations commonly show pleiotrophy, often due to mutation in closely linked gene.

Mutation may provide a valuble source of variation in plant material from which the breeder can make selection. As a breeding method, mutation is applicable to both self- and cross-pollinated species (Ahloowalia *et al.*, 2004). It is commonly used to correct a deficiency in an adapted cultivar.

Induction of mutation: Following steps have to be taken in a procedure to induce mutation:-

(a) Define the objective for the attempt to induce mutation.

(b) Select the variety (locally best accepted) for mutagen treatment.

() Decide which part of the plant will be treated (seeds, pollen grains, vegetative buds, cutting).

(d) Select a mutagen, its dose time and duration of treatment. The available mutagens are listed below:-

(i) Physical mutagens:

(a) Ionizing radiations: β-rays, alpha-rays, fast neutron, thermal neutrons, x-rays, gamma-rays (non particulate).

Gamma rays are sparsely ionising electromagnetic radiations. These are produced by the decay of Cobalt or Cesium, one of which is generally used in gamma chamber for treating the biological material. These are measured in Roentgens (R) or kilo Roentgens (kR = 10^3R). The x-rays are electromagnetic radiations consisting of photons. The plant materials can be treated with x-rays produced by x-ray apparatus which consists the simplest form of an x-ray tube with attached rectifier, transformer, control and dosimeter.

(b) Non-ionizing radiations: Ultraviolet rays (UV rays)

(ii) Chemical mutagens:

(a) Alkytating agents: Sulphur mustard, Nitrogen mustard, EMS (Ethyl methane sulphonate)

(b) Acridine dyes: acriflavin, proflavin, aeridine orange, aeridine yellow, ethidium bromide.

(c) Base analogues: 5-bromo uracil, etc.

(d) Others: Nitrous acid, hydroxyl amine, Sodium azide, etc.

Dosage and period of exposure depends on the material, the stage of development and purpose of the treatment. For an entirely new material the dosage and time should be fixed by a series of trials with different dosages combined with different periods of time.

Induction of mutation using a physical mutagen: For this, a facility with remote control, is needed where the source is kept in a lead shield to serve as a barrier to the rays from coming out at the time of exposure.

(a) ***Treatment of seeds***: The following steps are taken:-

1. Select healthy dry seeds of desired variety, count and put them in seed packets. To obtain a particular seed water content, dessicate the seeds in a particular glycerine ± water solution for 4 days at 22°C.
2. Set exposure time.
3. Expose to gamma rays at different doses. Also keep unexposed seeds to serve as control.
4. Germinate the seeds in separate petridishes on moist filter paper.
5. Record germination percentage for each dose, plot on a graph and determine LD (the dose producing 50% lethality).
6. Again expose seeds to gamma rays at a dose around LD.
7. Sow the seeds immediately in soil.
8. Harvest the seeds separately plant-wise and collect mutant (M_1) seeds.
9. From the M_1 seed of the previous year raise M_1 generation and select mutant plant.
10. In the same manner, raise M_2, M_3 generation and select mutants.

(b) ***Treatment of seedlings and bulbs***: The following steps are taken:-

1. Place young seedlings on a petridish. For bulbs, take healthy ones with a tuft of about 2 cm long healthy roots.
2. Expose to alpha/x-rays for desired dose/period.
3. Transfer them into saw dust or in nutrient medium.

(c) ***Treatment of inflorescences***: The following steps are taken:-

1. Grow plants in flower pots.
2. Expose the young inflorescences to α/x-rays for desired dose/period depending on genotype.
3. Select a flower bud of suitable size and observe pollen mother cells.
4. Allow the inflorescence to develop and observe meiotic stages at intervals of 24 hrs.
5. Endosperm and pollen grain mitosis can be studied in similar manner.

Table 7. Achievements by mutation breeding.

Fruit	Cultivars
Mango	Rosica from Rosica-de–lca, Davis Haden (Haden)
Banana	Highgate from Gros Michel (Semi dwarf), Motta Poovan from Poovan, Novaria from Grande Naine
Orange	Washington Navel Late, Navelin, Navelete

Mandarin	Clausellina, Navelete
Grapefruit	Marsh, Thompson, Star Ruby, Red Blush (Induced)
Papaya	Pusa Nanha
Pear	Starkrinson from Clapp's Favourite
Cherry	Compact Stella from Stella (x-ray)
Almond	Supernova (gamma-ray)

FRUIT BREEDING BY BIOTECHNOLOGY APPROACHES

The novel molecular techniques using strategies such as *in vitro* propagation, embryo rescue, protoplast fusion, marker-assisted selection and recombinant DNA (transgene) technology for genetic improvement in plants emerged as a result of the biotechnological revolution during the last half of the 20th century.

A. *In vitro Propagation*

Micropropagation techniques help the breeder in several ways by i) conservation of plant diversity, ii) easy exchange of germplasm and propagated plant material with minimum requirement of quarantine checks, iii) fast multiplication of true-to-type planting material, iv) disease-free plant production v) making year-round availability of plants irrespective of seasonal constraints, and vi) multiplication of hybrids. It is highly beneficial in dioecious fruit species like date palm and papaya where large-scale production of female plants becomes possible. The regeneration of whole plant is possible from small excised plant parts by using the following techniques.

1. ***Regeneration from existing meristems***: This is also known as axillary shoot proliferation. The existing meristems such as shoot tip or nodal bud are cultured on the different types of media fortified with various plant growth regulators, alone or in combination. The shoot proliferation depends on the type of hormones/regulators used. For axillary shoot proliferation, the commonly used growth regulators are benzylaminopurine (BAP), kinetin, and 2-isopentenyl adenine (2-ip). The regenerants are considered to be genetically stable. The process of shoot tip culture is successful in banana.
2. ***Regeneration from adventitious meristems***: Shoot multiplication either directly or by callus formation can be obtained by inducing adventitious shoot production on mature plant organs such as leaves, internodes/stems, and roots. For initiation of adventitious meristems, a proper combination of growth regulators is needed in culture medium. In general, shoots are formed when a high ratio of cytokinin to auxin is present, and reverse is true for root formation. The plants regenerated via this method are not always genetically stable due to formation of mixoploids. The repeated subculture of callus also reduces its potential and regenerative capacity. Fruits like banana, strawberry, pineapple and grapevine can be micropropagated easily.

B. *Somatic Embryogenesis*

Somatic embryos are bipolar structures, possess both shoot and root meristem, and originate from somatic or vegetative cells. It requires a high level of auxin in culture medium for induction, followed by low auxin and cytokinin concentration in the medium. Somatic embryos may arise directly on explants or via callus formation or liquid suspension cultures. The somatic embryos may act as synthetic seeds after encapsulation, which is an attractive alternative for propagation of plants. Among two methods, viz., hydrated or desiccated, for artificial seed production, the production of hydrated seeds is more popular. In this method, the individual somatic embryo is encapsulated in a water-based gel (hydrogel such as calcium alginate). Embryos developed through tissue culture technique are mixed with sodium alginate and dropped with pipette into a calcium salt (calcium chloride) solution to form calcium alginate capsules. The capsules are washed in water and then placed on culture medium for germination. Artificial seeds have been produced in banana, citrus, mango, apple, olive and kiwi fruit.

C. *Micrografting*

It is difficult to regenerate complete plants from meristem in woody fruit and forest plants. Therefore, as an alternative, micrografting is done to produce virus-free plants. The various steps in micrografting include scion preparation, rootstock preparation, *in vitro* grafting and acclimatization/hardening of the plants. The *in vitro* raised nucellar seedlings are used as rootstock. The scion (meristem 0.1–0.4 mm) is obtained from either young growth of field grown trees, defoliated glasshouse-grown plants, or *in vitro* proliferated nodal segments obtained from mature trees. The grafting is done with the help of stereomicroscope under aseptic conditions. Viruses such as citrus tristeza, peach latent mosaic viroid and pear yellow vein virus have been eliminated via micrografting.

D. *Embryo Culture*

Embryo culture techniques have many significant applications in plant breeding. Immature embryo rescue and culture is a particularly attractive technique for recovering plants from sexual crosses where the majority of embryos cannot survive *in vivo* or become dormant for long periods of time. Overcoming embryo inviability is the most common reason for the application of embryo rescue technique. Now there is a great interest in exploiting interploid hybridization to combine desirable genetic traits of complimentary parents at the triploid level for the purpose of developing improved seedless fruits. However, the success of this approach has been reported in limited number of species due to various crossing barriers and embryo abortion at very early stages. Thus, immature embryo rescue provides an alternative means to recover triploid hybrids, which usually fails to completely develop *in vivo.* The utilization of interploid crosses between a monoembryonic diploid female with an allotetraploid male in a citrus cultivar improvement programme, features a clear and comprehensive illustration of successful protocol for immature embryo rescue and culture. Such protocols covering a complete process from embryo excision to the recovered plant in greenhouse can also be adapted in other fruits like apple, banana, mango, litchi, grape, papaya, peach and persimmon.

E. *Marker Technology*

DNA markers are a very common tool for several purposes including identification and improvement in plant breeding. Isozymes are used to identify the zygotic seedling from the nucellar ones as the nucellar seedling should have same isozyme alleles as that of the maternal parent. These have developed into many systems based on different polymorphism-detecting techniques such as RFLP, AFLP, RAPD, SRAP, SSR, ISSR SNP, SSCP, etc. Molecular markers are used in fruit breeding for characterization of germpalsm (genotyping), variety/cultivar identification, genetic diversity assessment, validation of genetic relationships, marker assisted selection, development of genetic linkage maps, DUS testing and varietal registration, breeding line identification and hybrid testing.

F. *Genetic Engineering Techniques*

Genetic engineering (GM technology) is a powerful tool and has the potential to allow integration of desirable characteristics into the existing genomes. Path has been developed to transfer important desirable characteristics (genes) into plant genome for i) enhancing resistance against drought, salinity, fungal and viral pathogens and pests, ii) to silence the undesirable genes and iii) to improve nutrient acquisition.

GM technology takes less time to clean up the genetic background, i.e., reduce the number of backcross generations; the ideal scenario being the targeted gene editing of the genetic background with no alteration (no off-target effects). Thus GM breeding has the potential to be a very fast technique. But the target gene must be known and sequenced. Also, GM breeding is expensive and requires specialized laboratories, although attempts are being made to develop other cheaper and easier options. The two methods for genetic transformation by GM technology are: i) direct and ii) indirect.

1. ***Direct gene transformation methods:*** In these methods, foreign DNA is directly introduced into the plant genome. Direct DNA transfer methods rely on the delivery of naked DNA into the plant cells. Majority of these methods are known to be simple and more effective. Several transgenic plants have been developed by these approaches. However, the frequency of transgene rearrangements is high, transgene copy number is higher and high frequencies of gene silencing could be observed.

   Particle bombardment (biolistic technology): DNA is coated onto gold or tungsten micro-particles and bombarded at high velocity in a stream of helium into the intact cells or tissues. The biolistic process undergoes two stages: i) coating metal particles (microprojectiles) with nucleic acid, and ii) accelerating the coated microprojectiles to velocities appropriate for penetration of target cells or tissues without excessive disruption of biological integrity.

Biolistic transformation is simple and safe since it does not require the manipulation of genetically modified organisms such as *Agrobacterium.* Moreover, it allows integration of larger transgenes and co-transformation of multiple constructs. In some crops, biolistic bombardment assays have been carried out. Embryogenic suspensions of Chancellor (*Vitis* L. complex interspecific hybrid) were bombarded with tungsten particle coated with plasmid pBI426 encoding GUS and NVFII.

But, in biolistics, cotransfer of large fragments of the vector backbone DNA occurs, which can negatively affect transgene expression.

Electroporation: Electroporation consists of the treatment of plant cells with short high voltage electric pulses. The electric pulse shock causes brief permeability of the plasmalemma for high molecular particles such as DNA. The DNA movement takes place via the pores formed after electric pulses in the cytoplasmic membrane.

2. ***Indirect gene transformation methods:*** Members of the genus *Agrobacterium* are predominantly soil inhabiting and plant associated bacteria. They are primarily recognized as pathogens occurring on numerous plant species, including many agricultural crops. The phytopathogenic strains may cause crown gall disease which is mainly determined by the presence of tumor-inducing (Ti) plasmid in their genome, while strains harboring root inducing (Ri) plasmid are the causal agents of hairy root disease.

   The T-DNA region on the Ti (tumor-inducing) plasmid, bordered by left and right repeats, facilitates the transfer of DNA enclosed by these border regions. Some *Agrobacterium* species carry more than one T-DNA on their Ti plasmids, leading to more than two T-DNA borders from which T-DNA can be processed. In the binary vector system, the virulence (vir) gene functions necessary for transformation are usually provided on a second plasmid to secure that they are not transferred to the plant cell. Transformation efficiency is determined by gene type, tissue type, co-cultivation conditions, *Agrobacterium* strain and selection regime. In another study a highly important grape vine rootstock, 110 Richter (*V. rupestris* x *V. berlandieri*), was a target cultivar for several transformation experiments aiming to introduce agronomically important genes (mainly virus resistance). Transgenic plants of 110 Richter were regenerated from embryogenic cultures co-cultivated with a disarmed LBA4404 strain of *A. tumefaciens*, harboring a binary vector contained chimeric genes for hygromycin resistance (hpt), Kanamycin resistance (npt11), the uidA gene and the CP-GCMV.

   In most transformation systems, selectable marker genes are used to select regenerated transgenic shoots. However, the presence of these marker genes in the transgenic plants becomes undesirable after transformation. Theoretically, the use of a selectable marker gene can be avoided and transgenic plants can be directly selected by molecular analyses. *A. rhizogenes* mediated transformation was applied to *V. rupestris, V. riparia, V. vinifera* and several *Vitis* hybrids infected by different phloem-limited viruses such as grapevine fleck virus, grapevine virus A and grape vine virus B.

G. Cisgenic Technology

The cisgenic technology involves genetic modification of a recipient organism with a gene from a crossable or sexually compatible organism (same species or closely related species). Cisgenic plants can be considered similar to conventionally bred plants but different from transgenic plants. The fruit crops like apple, strawberry and vegetatively propagated crops like potato have been subjected to cisgenic modifications for assessment of different crop improvement objectives such as fruit quality, nutrient content, shelf life and resistance to biotic and abiotic stresses. A Cisgenic apple line of cultivar Gala that confers scab resistance has been produced.

H. Tilling

Tilling is a reverse genetic approach which is most widely used application to identify mutations through nuclease cleavage of DNA mismatches. TILLING has come to refer to the general process of detecting induced polymorphism, whether by nuclease cleavage, dHLPC, high resolution melting or sequencing. In contrast to the forward genetic approaches, where mutants are first identified based on phenotypes, in reverse genetic approaches the candidate genes responsible for the expected phenotype are first discovered. TILLING identifies single base pair changes, or small deletions in specific targeted genes. Owing to the advantages such as high sensitivity and high efficiency besides being a simple procedure, TILLING provides a powerful approach for gene discovery, DNA polymorphism assessment and plant improvement. The modified form known as EcoTILLING detects naturally occurring SNPs, especially in landraces and wild accessions. Similarly, DeTILLING refers to the detection of only deleted segment in the targeted amplified fragment. Additionally, EcoTILLING can be applied for functional genomic studies and mutation breeding and natural allele mining can serve as primary selection method for material to be used in breeding programmes. Hence reverse genetics based technologies could be efficient tool to detect mutation and to the possible use of the detected mutated genes for further genetic improvement. It has been successfully utilized for viral particle purification.

I. CRISPR/Cas9

Until 2013, the dominant genome editing tools were zinc finger nucleases (ZFNs) and transcription activator-like effector nucleases (TALENs). Both are artificial fusion proteins comprising an engineered DNA-binding domain fused to the nonspecific nuclease domain of the restriction enzyme FokI, and they have been used successfully in many organisms including plants. The latest ground breaking technology for genome editing is based on RNA-guided engineered nucleases, which already hold clustered regularly interspaced short palindromic repeat (CRISPR)/CRISPR-associated protein 9 (Cas9) systems. These have shown the greatest potential for rapid and efficient editing of genomes in plant species. CRISPR/Cas9 can target a specific genomic sequence, guided by an engineered 20-nt RNA sequence that binds to its DNA target by basepairing rules pyogenes and holds great potential for analysis of loss-of-function, gain-of-function and gene expression. CRISPR/Cas9 has been applied to the functional characterization of individual genes through loss-of-function, gain-of-function and perturbation of gene expression. In comparison with ZFNs and TALENs, the CRISPR/Cas9 approach looks more attractive because of higher simplicity, accessibility and cost, and versatility, possibility of multiplexing and other aspects, including an easier open access to resources. Although contrasting results have been reported, the off-target effects remain the main concern for a wider application of genome editing to functional genomics and plant breeding. Unlike ZFN and TALEN, in CRISPR/Cas9 technology, recognition of the DNA sequence to be modified is determined not by proteins, but by a chimeric sequence of RNA (single guide RNA – sgRNA), which results from the fusion of the two sequences (crRNA and tracrRNA) present in the natural system (the CRISPR/Cas9 system is used by bacteria to defend themselves against phages), while the cutting of DNA sequence is performed by the associated monomeric enzyme Cas9. The preceding technologies, zinc finger and TALEN nucleases were based on specific polypeptide-to-DNA binding which is tedious to optimize. CRISPR-Cas9 is based on DNA-RNA hybridization which is well mastered. CRISPR-Cas9 now appears as the most efficient system to achieve site-specific genome editing being easy, more reliable and the cheapest. The first report of targeted genome modification in citrus using the Cas9/sgRNA system—a system that holds significant promise for the study of citrus gene function and for targeted genetic modification. In that study successfully employed CsPDS gene into sweet orange, DNA sequencing confirmed that the CsPDS gene was mutated.

BREEDING FOR BIOTIC STRESS RESISTANCE

Resistance of fruit plants to insect pests and diseases depends upon their genotypic make up and the effect of environmental conditions at different stages of growth and development. The genetic make up determines the operative mechanism of resistance and the resulting extent of susceptibility.

Insect resistance: Dealing with the phenomenon of insect resistance, Painter (1951) grouped mechanisms of insect resistance into three categories, viz., *nonpreference*, *antibiosis* and *tolerance*. Non-preferred hosts (fruit plants) lack the characteristics for insect feeding, oviposition or shelter. Antibiotic factors in the hosts have adverse effect on the biology (survival, development or reproduction) of the insect. The tolerant fruit plant can produce good yield even while it supports an insect population that would severely damage a non-tolerant plant and decrease its fruit yield. These mechanisms of resistance result from a series of interactions between insects and morphological and physiological and/or biochemical plant characters.

***Morphological characters*:** Many morphological characters contribute to the resistance of plants to insect pests. These include trichomes, surface waxes, and hardness of plant tissues; thickening of cell walls and cuticle, rapid proliferation of tissues, anatomical modifications of plant organs and colour and shape of plant parts.

***Chemical constituents*:** Insect resistance in plants may be due to the chemical constituents; qualitative or quantitative. These chemicals occur within certain parts of the plant or at specific stages of plant growth. These chemicals may act as physiological inhibitors and/or cause nutritional deficiencies.

***Plant nutrition*:** The nutrition and physiological status of the host plant as well as quantitative differences in the nutritional components can influence their resistance or susceptibility to insects.

Disease resistance: Disease resistance is often defined as reduction of pathogen growth on or in the plant. The resistance is usually pathogen species-specific or pathogen strain-specific. Plants are almost always resistant to certain pathogens but susceptible to other pathogens. Many plants respond by synthesizing pathogenic response proteins (PR) such as hydrolytic enzymes to defend against pathogenic attack. Two such enzymes are chitinase which degrades chitin, the major cell wall component of oidium and β-glucanase which degrades β-glucans, constituents of cell walls of downy mildew hyphae.

Resistance mechanism to diseases is of two types; vertical or horizontal:

***Vertical resistance*:** It is qualitative resistance or race specific resistance governed by major genes and is characterized by phenotype specificity. It is easily overcome by new races of the pathogen.

Horizontal resistance: It is quantitative or durable resistance controlled by polygenes and is host nonspecific. The genes provide defensive structures or toxic substances to the plants that slow down or stop the advance of pathogen into the host tissues and reduce the damage caused by non-biotrophic pathogens. The defenses develop slower in quantitative resistance and perhaps reach a lower level than those in race specific resistance. It is durable and never breaks down to new strains of the disease as in case of vertical resistance.

The resistance against insect pests is generally lacking in many fruit crops. The use of chemicals as control measure has proved to be hazardous to the consumers and is also not environmentally sustainable. Both classical and conventional breeding methods including biotechnological approaches are possible to deliver resistance to major pests and diseases in fruit crops. These are being explored. The genetic improvement could also reduce the cost of chemical application to combat pests and diseases. This would provide significant economic benefit to the fruit grower. The following breeding methods are being attempted.

1. **Introduction:** The sources of genes conferring resistance to biotic stresses are usually available in the related wild species or nearly related genera in many fruit crops.The land race/cultivar/wild species showing resistance/tolerance can be introduced for use in hybridization programme to develop resistant cultivars. The species/cultivars showing resistance and having satisfactory yield/quality characters can be recommended directly for cultivation (Table 8).

Table 8. Species/cultivars having biotic stress resistance.

Fruit	Disease/pest	Species/ cultivar
Apple	Fire blight Scab	*Malus robusta* *M. floribunda*
Peach	Root knot nematode	Flordaguard, Nemaguard, Nemared
Banana	Black Sigatoka leaf spot	*Musa acuminata* ssp. *burmannicoides.*
	Fusarium wilt (Race 1,2,4) Burrowing nematode (*Radopholus similis*)	*Musa acuminata* ssp.*burmannica, M. acuminata* spp. *malaccensis* Pisang Jari Buaya (PJB).
Citrus	Phytophthora Greening disease (HLB)	Trifoliate orange, Swingle citrumello, Bento citrange (hybrid) Australian finger lime (*Citrus australasica* syn. *Microcitrus australasica.*
	Tristeza virus	Rangpur lime, Rough lemon, Cleopatra mandarin, Trifoliate orange
	Root knot nematodes	*Poncirus trifoliata*
Grape	Downy mildew	*Muscadinia rotundifolia, V. amurensis*
Guava	Wilt	*Psidium friedrichsthalianum,* *Psidium molle, P. molle x P. guajava*
Mango	Anthracnose	*Mangifera laurina*, Parish, Fairchild,
	Malformation	Tommy Atkins, Bhadauran, Elaichi
	Powdery mildew	Janardhan Pasand, Neelum
Papaya	Papaya Ring spot virus	*Vasconcellea cauliflora, V. quercifolia, V. pubscence*
Pear	Fire blight	*Pyrus communis*

2. **Selection:** Most of the fruit crops exhibit considerable genetic variation for disease/insect pest resistance/tolerance. Thus, there is enough scope to select tolerant types from the available germplasm. The classical selection methods for insect-pest resistance may be by *field screening* or *greenhouse screening*. *Primary screening* for resistance is usually done in the natural field conditions, where the plants are exposed to mixed populations and there is always a possibility that the isolate or another pathogen that has not been tested previously may cause break down of resistance. When screening under natural field conditions,

it must be understood that infection intensity may vary with environmental factors as well as with microclimatic conditions and there is also possibility of disease escape. For example, in screening studies for resistance to fruit fly (*Anastrepha obliqua*) in Brazil, mango cultivars Alfa, Espada and Stahland IAC 111 exhibited significant field resistance but under artificial infestation, while Alfa maintained the resistance, the other two varieties became as susceptible as Tommy Atkins. This was attributed to the possible interfering factors in the field and higher pest population pressure in the artificial cage conditions used for screening. Similarly, detached leaf and berry screening under laboratory conditions against anthracnose, brown leaf spot, gray mold, and ripe rot in wild grapes indicated that *Vitis flexuosa, V. amurensis, V. coignetiae* to exhibit a lower infection rate than the conventional cultivars Campbell Early and Kyoho.

Greenhouse screening is the rapid method for evaluating a large number of breeding lines, but the resistant entries should also be screened in the field. Approaches such as field screening and controlled screening together can lead to reliable outcomes in scoring for resistance. To select banana cultivars with durable resistance to *Mycospharella* leaf spot diseases, Abadie *et al.* (2009) used initial field screening with varieties surrounded by susceptible check and followed it by detached leaf assays and the selections made were again screened to test the efficacy of resistance in field trials because under field conditions the variety may later succumb to the particular biotic stress.

To select appropriate isolates for employing them in resistance breeding programmes, the available varieties and closely related species are first screened for resistance attributes. Following points are taken into consideration while screening:-

(a) Scoring technique must be decided after careful consideration of the nature of the biotic stress in question. The procedure employed should take into account the population build up and extent of damage together.

(b) Test conditions such as temperature and inoculum density need to be standardized so that results from different locations are comparable. Disease reaction of the plants could be modified by external weather conditions and physiological and nutritive status of the tissues.

(c) The methodology used may have to be appropriately and carefully modified considering the nature of the propagule or growth conditions. For example, for screening root traits for resistance, modification may be necessary when *in vitro* derived plants are employed as compared to conventional propagules.

(d) Periodical and timely observations would enable the breeder to identify and quantify the reaction of genotypes to the imposed biological stress.

3. **Hybridization**: The base population should include wide genetic resources from which genes incorporate the desirable quantitative resistance. The sources of

exotic resistance must undergo a pre-breeding process to form a base population specifically synthesized with a wide genetic basis for transferring desirable and useful genetic traits from unadapted donor lines to well adapted cultivated varieties. Such pre-breeding attempts have been employed in banana to generate improved diploids that can be used as a parental material to incorporate resistance to nematodes, Sigatoka and *Fusarium* wilt in commercial cultivars by involving *Musa acuminata* ssp. *burmannica* and *Musa acuminata* ssp. *malaccensis.*

Commercial release of tolerant cultivars through selection and hybridization has been rather slow, which is attributable to the complex inheritance of tolerance and the problem of linkage drag in the conventional methods of improvement. Slow progress in this direction has spurred interest in alternative means. Development of easily propagable tolerant rootstocks is one of them. The rootstock breeding programmes were started at East Malling and Merton (UK). Northern Spy was used as a source of resistance to Woolly aphid along with Malling series. The Malling-Merton rootstock series, e.g. MM 104 (M.2 x Northern Spy), MM. 106 (M1 x Northern Spy) and MM. 111 (Merton 793 x Northern Spy) derived from Northern Spy were resistant to WAA. The hybrid rootstocks *Citrus reshni* x *Poncirus trifoliata* and *C. volkameriana* x *P. trifoliata* are highly resistant to nematodes.

A number of efforts have been made in different fruit crops to develop disease resistant/tolerant hybrids as listed in Table 9.

Table 9. Fruit cultivars having biotic stress resistance.

Fruit	Hybrids	Resistance
Apple	Borovinka, Hagloe Crab, Tita Zetei, Braeburn, Britegold, Delicious, Empire Red, Enterprise, Fuji, Goldstar, Jonafree, Nittany and Wealthy Red	Powdery mildew
	Prima, Primula, Sir Prize, Jonafree, Dayton	Scab
Banana	**Diploids-** SH2095, H109 **Triploid-** NARITA-1 **Tetraploids** - FHIA 01, 07 & 23, BITA 17	Fusarium wilt, Fusarium wilt Race 4. Sigatoka spot, Fusarium wilt Race 4.
	FHIA 01, 02, 17 & 18, 21, BITA 3 and BITA 17	Sigatoka leaf spot
Guava	Peipa, TS-G1, TS-G2	Wilt
	TS-G3	Wilt

4. Biotechnology

(a) ***In vitro derived somaclonal variants:*** The Pei-Chiao clone of banana with resistance to the disease and improved characteristics has been selected and was named Tai-Chiao No.1. Observations on sucker derived clones from TC plants of Giant Cavendish led to the identification of a promising FOC wilt resistant GCTCV 218 clone which was later named as Formosana. The cultivar Novaria

(AAA), derived from Grand Naine (AAA), shows early flowering. The AAB cultivar Pisang Mutiara, tolerant to black leaf streak, was selected in Malaysia as a somaclonal variant of Pisang Rasthali (Azhar and Heslop-Harrison, 2008). Somaclonal variants of banana cv. Rasthali (Silk, AAB) has been identified which showed efficient resistance reaction to FOC race 1 in bioassays.

(b) ***Protoplast culture and somatic hybridization:*** Plant somatic hybridization via protoplast fusion has become an important tool for ploidy manipulation in crop improvement programmes. This technique can be advantageously employed when barriers in crossing pose difficulties in conventional breeding. The method has been widely exploited in *Citrus* to derive a large array of somatic hybrids with potential resistance and tolerance to major disease problems and for scion or rootstock improvement. Standardization of the technique may pave way in generating resistant rootstocks or scions in other woody perennial fruit crops like avocado, mango, guava, grapes, etc.

Somatic hybridization can be exploited to manipulate polygenic traits without the requirement to isolate DNA, or to have knowledge of genes or their DNA-base sequences. Polygenic traits can be introgressed by nuclear and/or organelle transfer through symmetric and asymmetric protoplast fusion.

It is possible, by protoplast fusion, to generate somatic hybrid tetraploid parents for use in interploid crosses with other diploid lines, or for direct release of triploid somatic hybrids by haploid/diploid protoplast fusion. Procedures for protoplast fusion generally involve exposure of isolated protoplasts to chemical fusion agents, such as polyethylene glycol (PEG), exposure of mixtures of parental protoplasts to an alternating current, followed by high voltage direct current pulses (electrofusion), or a combination of these procedures.

(c) ***Marker assisted selection:*** Development of molecular markers for screening a population for specific pest or disease response will be of immense use to the breeder especially when large number of population are to be screened. Most of the markers so far developed or studied have been reported for distinguishing the genetic relationship among the genotypes. The reports on marker usage from limited trials are available in fruit crops but effective and reliable use of the identified markers in early screening in large scale population for pest or disease resistance is yet to be witnessed in India.

(d) ***Genetic transformation:*** Transgenic approaches would enable integrating the genes that may confer resistance without altering the genetic constitution of plants and thus avoiding the costs and time involved to purify or reselect a recombinant/mutant for favourable traits apart from resistance. So far only in two fruit crops, papaya and plum, transgenic technology has been commercialized for overcoming disease susceptibility although successful results have been obtained demonstrating the usefulness of the technique in enhancing resistance in fruit crops like citrus and grapes. Through transgenic approach, UH-Rainbow, the world's first transgenic papaya developed with coat-protein mediated virus resistance, was released for commercial adoption by the University of Hawaii (Gonsalves, 1998). As resistance of Rainbow has been shown to be specific to the virus strains only in Hawaii, other countries have initiated their own transgenic programmes. Strain specific coat protein mediated resistance may be overcome by breakdown of resistance by other PRSV strains or new unrelated viruses like the papaya leaf distortion mosaic virus. Concerted effort is needed in India towards accelerating papaya breeding for PRSV-P resistance. In plum, successful transformation against Plum pox virus (PPV), the causative agent of Sharka, was carried out by *Agrobacterium*-mediated transformation of pathogen derived coat protein. Transgenic resistance was further transferred to other seedlings through cross breeding to develop PPV-resistant cultivars. This resistance is transferred to the offspring as a single dominant gene. It is suggested that the transgene resistance technology and post-transcriptional gene silencing tested in "Honey Sweet" can be used to develop resistance in stone fruits such as peach, apricot, Japanese plum and cherry, which are susceptible to PPV. In citrus, both direct and indirect methods of gene transfer have been used. The potential of rice chitinase genes to enhance resistance against *M. fijiensis* in banana as well as the usefulness of the leaf disk bioassay for early disease screening in transgenic banana lines was demonstrated by transgenic approach.

Several alternative transformation systems are being explored by International research groups. Marker free transformants have been obtained in pineapple cv. of sweet orange using clean binary vector including only the gene of interest and without any associated markers. However, marker free technology may also lead to appearance of chimeric transformants. Cisgenic approaches are also now being explored to overcome the concerns expressed among public against the transgenic approach. In this method, recombinant DNA technology to introduce genes from crossable donor plants, isolated from within the existing genome or sexually compatible relative species is advocated thus avoiding radical alteration of plant genome. Through this Cisgenic approach in apple cv. Gala, scab resistance gene Rvi6 (formerly HcrVf2) has been integrated. Cisgenic approaches are yet to be fully explored in tropical and subtropical fruit crops. For genetic improvement of crop plants, RNAi has advantage in terms of its efficiency and stability over antisense-mediated gene silencing and co-suppression. RNAi approaches are now being researched upon to enhance resistance especially against viral pathogens. This method is now being utilized to generate BBTV resistance in banana, grapevine Fan leaf virus, PRSV in papaya and for gene silencing in *Diaphorina citri*, a phloem sap sucking insect vector of citrus greening. Transgenic development work in fruit crops for imparting biotic stress tolerance/resistance has been summarized in Table 10.

Table 10. Transgenic fruits for biotic stress resistance.

Fruit	Gene and genetic transformation method used	Mechanism of action	Target trait	Trait improvement (transgenic vs normal plant
Banana	Maize Cystatin gene (*OC-IΔD86*); *Agrobacterium tumefaciens*-mediated gene transfer	Binding of cystatins to the active sites of proteinases inhibit their activity, thus affecting their proteolytic digestion.	Resistance against *Radopholus similis.*	Transgenic banana plants exhibited 70% resistance to the migratory endoparasite, *R. similis.*
Strawberry (*Fragaria* × *ananassa*)	*CpTi* (*Cowpea protease trypsin inhibitor*) gene; *Agrobacterium tumefaciens*-mediated gene transfer	*CpTi* protein inhibits the internal proteases within the plant cell on pest attack and thus protect the integral protein functions.	Resistance against vine weevil (*Otiorhynchus sulcatus*).	*CpTi* transgenic lines reduced the frequency of survival of weevil larvae and pupae during insect bioassays.
Strawberry cv. Pajaro	Chitinase and a glucanse gene; *Agrobacterium tumefaciens*-mediated gene transfer	Chitinase and glucanase degrades chitin and glucan, cell wall composition of fungi, by hydrolysis and thus codes for disease resistance.	Resistance to *Botrytis cinerea.*	Two transgenic lines expressing chitinase gene showed enhanced tolerance to *Botrytis cinerea.*
Grapevine	Vv NPR1,1 gene (derived from *Vitis vinifera*); *Agrobacterium tumefaciens*-mediated gene transfer	NPR1 gene plays a key role in regulating salicylic acid (SA)-mediated systemic acquired resistance (SAR) in plants.	Resistance against powdery mildew.	Enhanced resistance to powdery mildew disease.
Grape vine rootstock of *V. berlandieri* × *V. rupestris* cv. Richter 110	*Agrobacterium oncogene-silencing gene; Agrobacterium tumefaciens*-mediated gene transfer	Oncogenes when silenced lead to no crown gall formation.	Resistance against crown gall disease.	Enhanced resistance to crown gall-diseas.
Apple (*Malus domestica)*	*Malus pumila* NPR1 (MpNPR1) gene; *Agrobacterium tumefaciens*-mediated gene transfer	NPR1 gene plays a key role in regulating salicylic acid (SA)-mediated systemic acquired resistance (SAR) in plants.	Fungal and bacterial resistance.	Resistance to two fungal pathogens, *Venturia inaequalis* and *Gymnos-porangium juniperi-verginianae*, and a bacteria, *Erwinia amylovora.*

Apple cv. Golden Delicious	DIPM-1, DIPM-2 and DIPM-4 gene; *Agrobacterium tumefaciens*-mediated gene transfer.	Genome editing to bring targeted mutagenesis.	Fire blight disease resistance.	Enhanced resistance to fire blight disease in transgenic plants.
Litchi cv. Bedana	*Chitinase* gene (rice derived); *Agrobacterium tumefaciens*-mediated gene transfer.	*Chitinase* degrades chitin, which constitutes 3–60% of cell wall composition of fungi, by hydrolysis and thus codes for disease resistance.	Fungal resistance against *Phomopsis* sp.	The transgenic lines exhibited higher chitinase activity and disease resistance than the non-transformed plants.
Mango (*Mangifera indica*) cv. Ataulfo	Defensin J1 gene from bell pepper; Biolistic-mediated DNA delivery method.	Defensin gene acts as proteinase inhibitor, also creates pores in cell membrane of fungal hyphae.	Resistance against anthracnose disease caused by *Colletotrichum gloeosporioides.*	Protein extract from processed embryos of transgenic mango cv. Ataulfo inhibited the growth of *C. gloeosporioides, Aspergillus niger* and *Fusarium* sps.

(e) ***Transcriptomics:*** Transcriptomic studies are now increasingly being employed to understand the molecular events due to infective process of pest or pathogen so as to find means to regulate the metabolic process to achieve tolerance or resistance. In banana, the transcriptome profiles of resistant and susceptible Cavendish banana roots infected with Foc TR4 were compared by generating more than 8 billion bases of high-quality DNA sequences using Illumina technology. Digital gene expression (DGE) analysis revealed large differences in the transcriptome profiles of the Foc TR4-resistant somaclonal variant and its susceptible wild type. Expression patterns of genes involved in pathogen-associated molecular pattern (PAMP) recognition, activation of effector-triggered immunity (ETI), ion influx, and biosynthesis of hormones as well as pathogenesis-related (PR) genes, transcription factors, signaling/regulatory genes, cell wall modification genes and genes with other functions were analyzed and compared. The results indicated that basal defense mechanisms are involved in the recognition of PAMPs, and that high levels of defense-related transcripts may contribute to Foc TR4 resistance in banana.

Trancriptome profiling of banana, *M. balbisiana* Attikol (BB), offers scope for selection of resistance genes as well as the development of trait-specific markers. Martinelli *et al*. (2012) reported the transcriptome profile of citrus fruit response to Huanglongbing Disease (greening) caused by *Candidatus* Liberibacter *asiaticus* (CaLas). HLB strongly affected pathways involved in source-sink communication, including sucrose and starch metabolism and hormone synthesis and signaling. Transcription of several genes involved in the synthesis and signal transduction of cytokinins and gibberellins was repressed while that of genes involved in ethylene pathways was induced. CaLas (*Candidatus* Liberibacter *asiaticus*) infection triggered technology, and analyzing expression profiles of 35 defense-related genes by qRT-PCR indicated that most of these defense-related genes were up-regulated after *C. gloeosporioides* infection. The study provides scope for identifying or discovering new candidate genes in relation to pathogen resistance in mango.

(f) ***CRISPR-Cas9 technology:*** The technology has been employed to obtain disease resistant plants. Since its successful application in 2013 for obtaining stable transgenic lines, CRISPR-Cas9 technology has been used against viral, fungal and bacterial infection, which causes severe losses in tomato.

(g) ***Genome editing:*** The *Xanthomonas citri subsp. citri* facilitated agroinfiltration to deliver Cas9, along with a synthetic sgRNA targeting the *CsPDS* gene, into sweet orange for enhancing transient protein expression in sweet orange leaves was employed by Jia and Wang (2014). DNA sequencing confirmed that the *CsPDS* gene was mutated at the target site in treated sweet orange leaves. This system or development of similar system holds significant promise for the study of citrus gene function and for targeted genetic modification including development of resistance or suppression of susceptibility, a response via both the salicylic acid and jasmonic acid pathways and increased the transcript abundance of several members of the WRKY family of transcription factors. Findings focussed on the fruit provide valuable insight to understanding the mechanisms of the HLB-induced fruit disorder and eventually developing methods based on small molecule applications to mitigate its devastating effects on fruit production.

BREEDING FOR ABIOTIC STRESS RESISTANCE

Abiotic stress caused by salinity, water logging, heat, cold and drought affects production and productivity of fruit crops. Plants experience water stress as a result of water deficit conditions under lack of uptake of water from the soil solution caused by high soil salinity, floods or low soil temperature. This is known as physiological drought. During drought, many plants die due to dehydration. Water potential of plant cell and turgor is reduced due to water stress and elevate the solute concentration in cytosol and extracellular matrices. Reduction of water content, turgor and total water disturbs the plant water relation manifested by stomatal closure, limited gaseous exchange, reduction in the transpiration and carbon assimilation (photosynthesis) rate.

Oxidative stress is caused due to high temperature, salinity and drought. It has serious secondary effect on cells. The formation of reactive oxygen species (ROSs) like oxides, water and hydroxides are due to the effect of oxidative stress due to which membranes and macromolecules of plant are damaged and affect the metabolism of cells. An imbalance is created between light capture and its utilization. This imbalance prevents the photosynthesis in leaves. Reactive Oxygen Species (ROSs) is generated by the degeneration of excessive light energy in photosynthesis apparatus. ROS results in denaturation of functional and structural macromolecules, DNA nicking, amino acids, protein and photosynthetic pigments oxidation and lipid peroxidation. Then, some responses are activated by cells against the ROS, such as an increase in the expression of genes for antioxidant functions and production of stress proteins, up-regulation of antioxidant systems, including antioxidant enzymes and accumulation of compatible solutes.

Salinity is a major stress that limits the increment in the demand of food crops. On the basis of adaptive evolution, plants are categorized as *halophytes* (plants that can withstand salt stress) and *glycophytes* (plants that cannot withstand salt stress and die). Majority of plants in the world are glycophytes and therefore, salinity causes the most dangerous environmental stress. It affects various physiological and metabolic processes. The changes depend upon the severity and duration of the stress and thus the extent of damage to production.

In the initial stage of salinity stress, osmotic stress causes water loss from the leaves and decrement in the water absorption capacity of root systems. This osmotic stress disrupts the cell ions homeostasis through the inhibition of uptake of essential elements such as potassium, calcium and nitrogen trioxide and high accumulation of sodium and chlorine. The accumulation of sodium, chlorine and boron in the tissue of transpiring leaves creates high ion toxicity. Accumulation of high salt in soil and plants is considered as hyperosmotic stress. Salinity also affects the photosynthesis reaction as it decreases the carbon dioxide availability due to which the diffusion in the plants is limited by which the contents of photosynthesis pigments are reduced. Total photosynthetic capacity of the plant is decreased by salinity resulting in decrement of leaf growth limiting the ability to grow.

Abiotic stress resistance mechanisms are of three types: i) *avoidance*, preventing exposure to stress, ii) *tolerance*, permiting the plant to withstand stress, and iii) *acclimation*, altering physiology of plants in response to stress. Abiotic stress resistant/ tolerant cultivars can be developed by the use of following breeding methods.

1. **Introduction:** The sources of resistance/tolerance to abiotic stress may exist in wild relatives, landraces, cultivated varieties, initial breeding materials and advance breeding materials. These can play an important role as donor for abiotic stress resistance in developing genetically superior varieties.
2. **Selection:** The use of stress tolerant rootstocks in grafting could safeguard the actively photosynthesizing leaves and shoots from stress injury. Drought tolerant species/hybrids/rootstocks have been recommended in fruit crops. Some of these are Rangpur lime, Desert Lime (*Citrus glauca* syn. *Eremocitrus glauca*) and Dragon trifoliate orange and Sunki mandarin in citrus, *V. rupestris* × *V. berlandieri,* 110 R (Richter), Ramsey (*Vitis champini*), 1103 Paulsen, 140 Ruggeri, and Kober 5BB (*Vitis berlandieri* x *V. riparia*) in grapes and M 9, M 26 and MM 111 in apple.

   The salt tolerant species/rootstocks in citrus are Rough lemon, *C. macrophylla*, Volkamer lemon (*Citrus volkameriana*), *C. amblycarpa*, and Cleopatra mandarin, Swingle Citrumelo, Troyer Citrange, Rusk Citrange, Gou Tou Chen, Us-897 x 639 and Carrizo Citrange. The salt tolerant *Vitis* species are Salt Creek, *V. champini* (syn.Ramsey), Freedom (Complex hybrid of *V. champinii, V. riparia, V. labrusca, V. vinifera, V. solonis*), 110R (*V. berlandieri* x *V. rupestris*), 1103P (*V. berlandieri* x *V. rupestris*), SO4 (*V. berlandieri* x *V. riparia*) and 41B (*V. vinifera* x *V. berlandieri*).

The species/rootstock that can be used for imparting salt tolerance in fruit crops are Tab-o-Larz in pomegranate; *P. atlantica*, UCB-1 in pistachio; *Prunus* spp., GF677, Myrobalan, Bright's Hybrid @ 5 in peach; *Z. rotundifolia, Z. nummularia* and *Z. spina-christi* in ber (Indian jujube); Crioula in guava, 13-1, Gomera-1, GPL-1; ML-2, Olour, Nekkare and *M. zeylanica* in mango and Hansen 536 and Brights Hybrid in almond.

The tolerant rootstocks identified for waterlogged conditions are *Pyrus betulifolia, P. calleryana, Cydonia oblonga* and *P. communis* in pear, M1, M 3, M 6, M 7 and M 26 in apple and *Prunus cerasifera*, *P. domestica*, Damas and GF-1 in *Prunus* species. The cold tolerant species/rootstocks identified in fruit crops are given in Table 11.

Table 11. Cold tolerant rootstocks in some fruits.

Fruit	Rootstocks
Apple	Ottawa 3,Ottawa 8, G 935, G 11, G 30, G 4 1, B 9, P 2, M 26, B 9, M 26, MM 106, MM 111 and M 7.
Pear	*Pyrus betulifolia* seedling, *Pyrus nivalis, Pyrus communis* cv. Barlett, Winter Nelis, OH x F, Chui Lum Tao (Peach).
Peach	Siberian C, St. Julien A, Marianna 2624.
Grapes	*Vitis labrusca, V. amurensis,* French and American grape hybrids.
Guava	*Psidium cattleyanum.*

3. **Hybridization:** The commercial cultivars could be improved by transfer of abiotic stress resistance genes from their wild relatives and their co-specific subspecies. These stress-tolerant traits can be introduced through wide distant (interspecific and intergeneric) hybridization.
4. **Biotechnology:**

(a) ***In vitro mutagenesis:*** It is an important technique which can induce stress tolerance and improve the yield and quality of crop plant. *In vitro* mutagenesis combines tissue culture technique and mutation induction strategy.

(b) ***Marker-Assisted Selection (MAS):*** The use of molecular markers gives support to the phenotypic evaluation for selection of new rootstock genotypes. Marker-Assisted Selection (MAS) can shorten the time required to obtain new cultivars and can make the process more cost-effective than selection based exclusively on phenotype. However, more basic research is needed to well understand the molecular and physiological mechanisms involved.

(c) ***Genetic engineering:*** Conventional breeding methods have not been much successful in mitigation of abiotic stress perhaps because the traits are controlled by a number of genes present at a quantitative trait locus (QTL). Therefore, it is a pre-requisite to identify potential candidate genes or QTLs (gene networks) associated with broad spectrum multiple abiotic stress tolerance.

Various genes controlling and signaling gene regulatory pathways offer certain key targets for genetic engineering for abiotic stress tolerance. Transcription factors (TFs) that regulate or switch on the expression of a number of genes involved in imparting abiotic stress tolerance in plants have been proposed as the most efficient targets for genetic transformation. These transcription factors include DREB1 gene family, Myb gene family, etc. Pasquali *et al.* (2008) reported improved tolerance to cold and drought stress in transgenic apple by the over expression of a cold-inducible Osmyb 4 gene from rice, which codes for a TF belonging to Myb family. The over-expression of DREB1b TF gene had also been reported to induce cold tolerance and drought tolerance in transgenic grapevine. The transgenic plants displayed increased drought tolerance in polyethylene glycol (PEG) stress as compared to control plants. Also, multiple abiotic stress tolerance in banana had been reported by the over-expression of *Musa* WRKY71 gene, which is a very potential abiotic stress responsive WRKY TF gene, cloned from *Musa* species cv. Karibale Monthan. Drought (polyethylene glycol in medium) and salinity (sodium chloride in medium) tolerance tests revealed that transgenic lines exhibited a higher tolerance for abiotic stresses than non transformed plants.

Tolerance to heat stress is straightway correlated with the increased capacity of plants to scavenge ROS. Plants have developed a variety of mechanisms to combat ROS by the production of various enzymatic systems like superoxide dismutase (SOD) to remove superoxide ions, glutathione reductase (GR) and peroxidase to scavenge peroxide ions (H_2O_2). Thus, over-expression of ROS scavenging enzymes in plants via genetic transformation offers a much potential strategy to overcome heat stress. Wisniewski *et al.* (2002) reported that over expression of cytosolic ascorbate peroxidase (cAPX) gene improved tolerance to heat stress in transgenic apple.

The non-enzymatic methods involve the production of a variety of chemical compounds including polyamines, carotenoids, ascorbic acid, tocopherol, etc., which directly react with ROS, scavenge them and thus provide protection to the plants against heat stress. Polyamines play an important role in imparting thermal stress tolerance in plants. S-adenosyl-I-methionine decarboxylase (SAMDC) is one of the key regulatory target enzymes in polyamines biosynthesis in plants for enhanced polyamines production. Transgenic lines produced 1.7–2.4-fold higher levels of spermidine and spermine with enhanced antioxidant enzyme activity and better protection of membrane lipid peroxidation as compared to wild-type plants, leading to enhanced tolerance to high temperature stress (38°C). Over-expression of heat shock proteins in plants has been proposed as one of the potential strategies to combat heat stress. HSPs function as molecular chaperons that are involved in correct protein folding, assembly, translocation, and degradation and provide stability to integral proteins and cell membranes under heat stress. The transgenic lines exhibited an increased peroxidase

(POD) activity, higher proline content and reduced malondialdehyde (MDA) content. Proline is an important osmoprotectant that protects cells from damage under heat stress and transgenic plants were better able to tolerate heat stress than wild-type plants. Transgenic citrumelo plants overexpressing the enzyme Δ1-pyrroline-5-carboxylate synthase, required for proline synthesis, showed a 2.5-fold increase in drought tolerance (Campos *et al*., 2011).

Cold tolerance is an important breeding target in breeding fruit improvement. Overexpression of the transcription factor MYB4, which regulates cold-induced dormancy and stress pathways, allowed the transgenic plants to tolerate cold temperatures for long periods in apple (Malnoy *et al*., 2007). Overexpression of the Arabidopsis dehydration response element-binding 1b protein in grape vine reduced cold induced wilting by 73% (Jin *et al*., 2009). Similarly, the expression of a *Poncirus trifoliata* basic helix-loop-helix protein in pummelo (*Citrus grandis*) enhanced cold tolerance, reduced electrolyte leakage by 13% and increased proline levels by up to 67% compared to wild-type plants (Geng *et al*., 2019).

Salinity tolerance is a complex mechanism governed by many genes. Plants which are exposed to abiotic stress conditions produce several pathogenesis-related proteins to compensate the adverse effect of stress conditions. Osmotin is one of the important pathogenesis-related proteins, which is produced by the plants to combat various biotic and abiotic stresses. Husaini and Abdin (2008) over-expressed tobacco osmotin gene in strawberry (*Fragaria* x *ananassa* Duch.) and found that the transgenic strawberry plants exhibited tolerance to salt stress. Transgenic development work carried out in fruit crops for imparting abiotic stress resistance has been summarized in Table 12a, b.

Table 12a. Transgenic fruits for abiotic stress tolerance.

Fruit species/ cultivar	Gene & genetic transformation method used	Mechanism of action	Target trait	Trait improvement
Apple	*MdNHX1* gene; *Agrobacterium tumefaciens*-mediated gene transfer.	Acts as Na^+/ H^+ antiporter.	Salinity tolerance.	Enhanced tolerance to salinity stress in transgenic plants
Banana	*MusaDHN-I gene*; *Agrobacterium tumefaciens*-mediated gene transfer.	Over-expression of dehydrin gene and a LEA protein.	Drought and salinity.	Enhanced tolerance to drought and salinity stress
Kiwi	*AtNHX1 gene*; *Agrobacterium tumefaciens*-mediated gene transfer.	Keeps K+/Na ratio high during salinity stress conditions.	Salinity.	Enhanced tolerance to salinity stress in transgenic plant
Pear	*SAMDC2 gene*; *Agrobacterium tumefaciens*-mediated gene transfer.	Encodes for the biosynthesis of polyamines.	Salinity.	Enhanced tolerance to salinity stress in transgenic plants

Citrus (Carrizo citrange root stock)	*Yeast HAL2 gene*; *Agrobacterium tumefaciens*-mediated gene transfer.	HAL-2 gene is involved in the pathway of methionine biosynthesis and confers tolerance to Lithium and Sodium.	Salinity tolerance.	Transgenic plants showed better tolerance to salt stress than non-transgenic plants
Mulberry (*Morus indica*) cv. K-2	Hval gene (from barley); *Agrobacterium tumefaciens*-mediated gene transfer.	*LEA* proteins (encoded by *HvaI* gene) protect cells against abiotic stresses by sequestering ions, stabilization of macro molecules and membranes, and act as antioxidants.	Salinity, drought and cold tolerance.	The transgenic lines displayed an enhanced level of tolerance to drought, salinity and cold conditions than normal plants as quantified by free proline, membrane stability index (MSI) and PSII activity

Table 12b. Drought resistance

Banana cv. Gonja manjaya	*Musa*WRKY71; *Agrobacterium tumefaciens*-mediated gene transfer.	Encodes a WRKY transcription factor.	Enhanced tolerance to drought, salinity and high temperature.
Strawberry (*Fragaria* x *ananassa*)	Osmotin; *Agrobacterium tumefaciens*-mediated gene transfer.	Increase the level of proline accumulation.	Enhanced tolerance to drought, heat and salinity stress in transgenic plants.

4 Tropical and Subtropical Fruits

MANGO

Origin: Indo-Mayanmar region

Taxonomy status:

Order: Sapindales

Family: Anacardiaceae

Subfamily: Anacardioideae

Section: *Mangifera*

Genus: *Mangifera*

Subgenus: *Mangifera*

Species: *indica*

Botanical name: *Mangifera indica* L.

Chromosome number: n = 20; 2n = 2x = 40; 80 (4x)

(Found in Gomera-I); Amphidiploidy

Genome size: 439 Mb

The related edible species are *Mangifera altissima* Blanco, *M. caesia* Jack., *M. foetida* Lour., *M. lagenifera* Griff., *M. odorata* Griff., *M. zeylanica* (Bl.) Hooker and *M. sylvatica* Roxb.

Botany

It is evergreen, branched tree with dome canopy and grows as high as 8-45m. Leaves are simple, alternate, lanceolate to oblong, petiolate, entire, coriacious or membranous. Flower is small, 5–10 mm in diameter, polygamous (functionally male and hermaphrodite) borne on terminal panicle. The male flower is essentially the same, except for the absence of an ovary. Perfect flower consists of 4-6 sepals, 4—5 (rarely 6) pinkish-white petals with several yellow ridges on the inner surface, free/adnate to the disc, stamens 5 (one functional and 4 staminodes), rarely 10—12 inserted within the disc or on it, free or connate at base; disc (nectary) is swollen, 4 to 5-lobed or pedicelate or absent. Ovary is superior, sessile, always one celled, oblique or globose. Style is lateral. Fruit is fleshy drupe, vary in size, shape and colour having compressed stone and fibrous seed. Embryo is monoembryonic or polyembryonic.

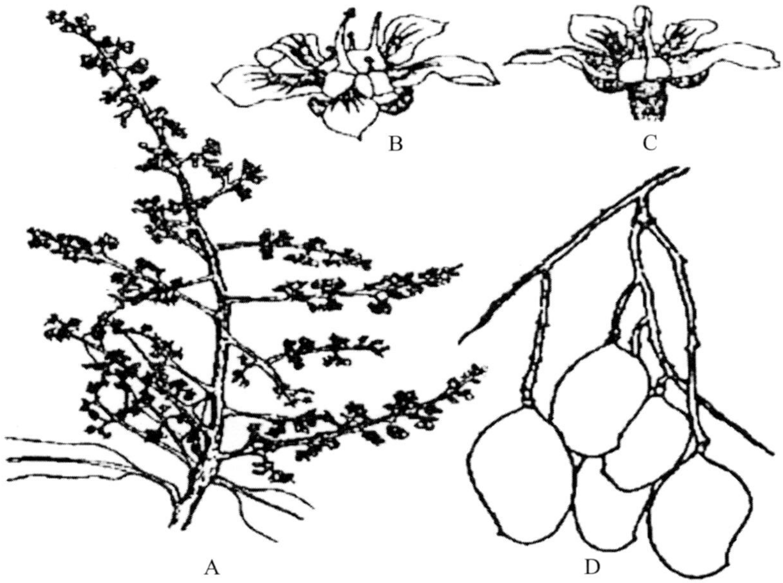

A—inflorescence; B—hermaphrodite flower;
C—male flower; D—shoot with fruits

Fig. 1. Mango inflorescence and fruit.

(Source: Purseglove, 1968)

Floral Biology

Flower bud differentiation in India takes place during October-December and flowering starts during Jan-Feb in North India. Primarily terminal but frequently axillary and multiple panicles may also arise from axillary buds. Flowers are andromonoecious (male and hermaphrodite). Total number of flowers varies from 300 to 6,000 per inflorescence. The percentage of perfect flowers on a panicle ranges from 1.25 to 81%, with a strong varietal difference in sex expression.The percentage of perfect flowers is higher in the apical zone of the panicle than in the basal and central portions; climate can also affect the percentage, as can the position of the panicle in the canopy. Flowers start opening in the morning and the anthesis is completed generally by noon. Maximum flowers open between 9 to 10.30 AM. Anther dehiscence takes place after the flowers have opened and completes at noon. Stigmas are receptive from 1 day before anthesis until 2 days after, being optimum on the day of anthesis. Although the receptivity of stigma continues up to 72 hrs after anthesis, it is most receptive about 6 hrs after anthesis. Mango is highly cross-pollinated. Pollination is entomophilous, done by houseflies and honey bees. Pollen viability is very high (>90%).

Breeding objectives for scion:

(i) Dwarf tree growth habit with precocity.

(ii) Regular bearing, off season cropping, high productivity and extended harvest season.

(iii) Attractive fruit shape, colour and good size (350-450 g) with good quality (tasty and firm flesh without fibres) with extended storage, processing quality and shelf life.
(iv) Resistance to major diseases (floral malformation, powdery mildew, anthracnose) and insects (mango hopper, mealy bugs, fruit fly and stone weevil).
(v) Freedom from sap burn injury, jelly seed and internal break down; and
(vi) Tolerant to abiotic stress (salinity, water logging and frost).

Breeding Objectives for Rootstock

1. Nucellar/polyembryonic rootstocks for uniformity.
2. Impart dwarfness on scion.
3. Scion compatibility.
4. Resistant to adverse soil conditions.
5. Tolerance to soil borne pathogens.

Limitations in Breeding

1. High level of heterozygosity and unpredictable outcomes of crossing.
2. Long juvenile phase.
3. Very small flowers, pointed stigma and poor fruit set.
4. Narrow genetic base.
5. Early post zygotic incompatibility and cross incompatibility.
6. Excessive fruit drop with loss of hybrid fruit resulting in poor success in hybridization.
7. Only one seed per fruit.
8. Recalcitrant seeds.
9. Prevalence of polyembryony.
10. Lack of adequate information on inheritance pattern.

Genetic Resources

In India, The ICAR-Central Institute for Subtropical Horticulture (CISH), Lucknow has a rich assemblage of 745 accessions. ICAR-IIHR, Bengaluru (767), IARI, New Delhi, RFRS, Vengurla, BAU, Sabour, BCKV, Mohanpur, FRS, Sangareddy, HARP, Ranchi, TNAU, Coimbatore and FRS, Rewa also maintains the germplasm of mango. NGR-Miami shares responsibility for maintaining the US clonal collections of mango with Mayaguez - Puerto Rico.

Sources of desirable resistance traits:

(i) **Hoppers, tip borers and seed borers:** *M. altissima.*
(ii) **Fruit fly *(Aanastrepha obliqua)*:** Alfa, IAC 111 and Espada Stahl.
(iii) **Bacterial black spot:** Heidi, Sensation and Bombay Green.

(iv) **Malformation:** Bhadauran, Ellaichi, Bhyadayam Dula, Samar Bahist Rampur and Miam Sahib.

(v) **Powdery mildew:** Neelum, Totapari, Zardalu.

(vi) **Anthracnose:** Arunika, Tommy Atkins, Parish, Fairchild, *M. laurina, M. minor,* Alphonso, Brasamasi, Carabao, Carrie, Early Gold, Keaw, Kent, Krishan Bhog, Rad, Saigon, Tommy Atkins and Van Dyke.

(vii) **Bacterial canker:** Bombay, Amarpali.

(viii) **High hills or high from mean sea level:** *M. orophila* and *M. dongnaiensis.*

(ix) **Wet climates:** *M. laurina.*

(x) **Cold:** *M. orophila,* Nimrod.

(xi) **Waterlogged condition:** *M. gedebe, M. decandra, M. inocarpoides, M. laurina.*

(xii) **Salinity:** 13-1, *M. zeylanica,* Babbaki,Olour and Kurukkan.

(xiii) **Calcareous soil:** 13/1, Kurukkan, Turpentine, Deorakhio Olour and Bappakkai.

Sources of desirable growth and fruit bearing traits

1. **Dwarfness:** Creeping, Neelum, Kalapad, Belkhas, Parikhas, Amrapali, Kerela Dwarf/ Kodur Dwarf, Hybrid-1, Rumani, Janardan Pasand, Ladavio, Manjeera, Imperial and Vellaikulamban, Latra, Ellachi, and Meghlonton.
2. **Polyembryony:** Turpentine, Deorakhio, Olour, EC-95862, Bappakkai, Vellai kolamban, Nekkare, Muvandan, Kurukkan, Kensington, Olour, Manipur, Deorakhio, Sabre, Mdu, Vattam, Mylepelian, Sabre and Kitchene.
3. **High sex ratio:** Dashehari, Neelum, Amrapali, Alphonso, Raspuri, Langra.
4. **Self incompatibility:** Dashehari, Langra, Chausa, Bombay Green.
5. **Pollinizers:** *M. pentandra* (5 stamens).
6. **Prolific bearing:** *M. casturi.*
7. **Double cropping:** *M. indica* var. *mekongensis.*
8. **Good fruit Qualty:** *M. odorata, M. foetida, M. caesia, M. casturi.*
9. **Banana like peel**: *M. pajang.*
10. **Big fruit size:** Banganapalli, Janardan Pasand, Kensington, Fazali, Langra, Mulgoa, Mallika.
11. **Red peel colour:** Ambika,, Janardan Pasand, Sinduria, Benazir, Fernandin, Suvarnarekha, Vanraj, Sandersha, Kensington, Tommy Atkins, Keitt, Julie, Sensation, Irwin, Arka Puneet, Nazuk Badan, Raspuri, Miyazaki.
12. **White skin and pulp:** *Mangifera caesia.*
13. **High pulp content:** Sindhu, Sahib Pasand, Amrapalli, Sunderja, Mallika, Bhutto, Bombay, Fazli, Khudadad, Chausa, Sora, Vanraj, Langra, Dashehari, Pairi, Arka Aruna, Jawahar.
14. **Fiberless pulp:** *M. mangifica.*
15. **Freestone:** *M. similis.*
16. **Seedlessness:** Sindhu.

17. **Freedom from spongy tissue:** Doophperda, Pairi, Kesar, Neelum and Dashehari.
18. **Good fruit quality:** *M. pajang* and *M. foetida.*
19. **Juicy fruit (sucking mango):** Peddarasam, Chinnarasam, Cherukurasam, and Panchadarakalasa.
20. **High TSS content:** Sahib Pasand, Doodh Peda, Dashehari, Langra, Arka Puneet, Mallika, Lazzat Baksh, Amrapalli, Sakkar China, Gopal Bhog, Rani Pasand, Sunderja, Neelphonso and Neeleshwari.
21. **Good keeping quality:** Pusa Surya, Alphonso, Totapuri, Sundershah, Dashehari, Banganpalli, Kesar, Rataul, Arka Anmol, Arka Puneet, Mallika, Fazli, Gulab Khas, Himsagar, Kishan Bhog, Mankurad and Neelum.
22. **Unique flavor:** Kensington Pride
23. **Off-season flowering:** *M rufocostat* and *M. swintonioides,* Baramasia, Royal Special, Amino-Do-Phasla and Teenphasla.
24. **Regular bearing:** Neelum, Kalapad, Bangalora (Totapuri), Totapuri Red Small, Arka Anmol, Arka Neelkiran, Fazli, Kesar, Rumani, Arka Aruna, PKM-1, PKM-2, Mahmud Bahar, Prabha Shankar, Sabori, Jawahar, Neeludin, Neeleshan, Mallika, Ruchi, Neeleshan Gujrat, Neeleshwari and Sai Sugandh.
25. **Early maturity:** Raspuri, Gaur Jeet, Mithua, Zardalu, Lazzat Baksh, Guruvarm, Chandrakaran, Panakalu, Olour, Vellaikolumban.
26. **Late maturity:** Neelum, Katki, Bathui, Fazli, Arka Anmol, Chausa, Fakhr-us-Samar, Puttu, Sora, Sultanus-Samar, Azam-us Samar and Totapuri.
27. **Processing for pulp/juice:** Alphonso, Rani Pasand, Sahib Pasand, Pairi, Baganapalli, Totapuri, Hyder Sahib, Pedda Rasam, Sukul, Allumpur Baneshan, Neelum, Chausa, Alif Laila, Prabha Shankar, Arka Anmol, Arka Puneet and Kesar.
28. **Processing for pickle and chutney:** Amlette, Chandrakaran, Karanjio and Ram Kela.

Desired traits in relation to its gene pattern

Upright growth of tree: Governed by dominant gene.

Long juvenile period: Governed by dominant gene.

Dwarfness: Recessive gene.

Precocious: Recessive gene.

Regular bearing: Recessive gene.

Biennial bearing: Dominant gene.

Bunch bearing habit: Dominant gene.

Fruit size/weight: Additive genetic control.

Fruit quality: Gene linkage.

Flesh colour: Light yellow dominat over orange yellow.

Orange flesh colour: Duplicate gene action.

Beta carotene and TSS: Additive gene action.

Fruit size: Polygenes.

Fruit beak: Dominant gene.

Red skin: Duplicate gene action; Broad sense heritability of 0.41–0.59 for colour in families with Tommy Atkins and Irwin as a female parent.

Polyembryony: Single dominant gene.

Bacterial canker: Cytoplasmic nature.

Floral malformation: Dominant gene.

Pre-selection Criteria

Pre-selection indices are defined as the characters showing any positive correlation with growth and fruit characters in the mature tree. The indices are valuable in discarding the seedlings showing traits correlated with undesirable characters in the mature tree. Also these help to a great extent in overcoming the problem of isolating worthless hybrids.

Pre-selection indices have been utilized to identify both wild and cultivated species having important traits for mango improvement. A higher phloem to xylem ratio of >1.0 (Kurian and Iyer, 1992), high phenolics content in the apical bud and lower density of stomata was found associated with dwarfness. Leaf flavour was directly correlated with fruit flavour (Majumder *et al.*, 1972). The emergence of new growth flushes simultaneously with fruiting or immediately after harvest was indicative of regular bearing.

Breeding methods

1. Introduction: A number of mango cultivars were introduced in India from different countries. These cultivars are Haden, Keitt, Kent, Tommy Atkins, Glenn, Lippens, Van Dyke, Parvin, Springfels and Zill from USA, Kensation from Australia, Julie and Carbao from Philippines and Amolie and Sweet from Thailand. There is need to introduce the newley developed cultivars Calypso™, Honey GoldA and R2R2 from Australia; Joa and Chene from South Africa; Maha Chanok from Thailand and Alpha, Lita and Roxa from Brazil. The cultivars introduced for use as donor for red skin colour are Tommy Atkins, Zilete, Haden and Sensation.

In Florida USA, mango varieties from many regions of the world were introduced from 1861, resulting in a wide diversity of mango germplasm in the area and thus it became secondary region of diversity of mango. Natural outcrossing in this diversity led to the production of hybrid seedlings with novel characteristics. In 1910, a high yielding monoembryonic seedling from variety Mulgoba, with an attractive red blush, was selected and named Haden. Following Haden, selections were made from the other seedling progeny Mulgoba. The success of the Florida varieties is due to their adaptability to many agroclimatic regions while retaining their fruit quality and regular bearing characteristics.

2. Selection: This technique has been used by both scientists and farmers. Most of the cultivars of mango are the result of open pollination leading to promising chance seedlings maintained asexually. Several selections for better fruit quality have also been made from commercial cultivars, e.g. Subash from Zardalu, Niranjan (off-season bearing) from Royal Special, Niranjan (thrice flowering in August, October and December) developed at Gujarat university, Menaka from Gulabkhas, Subhas from seedlings of Zardalu at BAU, Sabour, Pusa Surya from Eldon developed at IARI, New Delhi, Paiyur from Neelum, Alphonso 900 from Alphonso, developed at Konkan Krishi Vidyapeeth, Dapoli, Naomi from Palmer, Irwin from Irwin, Tongo from Naomi, Tommy Atkins from Haden, Keitt from Mulgoba and Golden Nuggets Selection and Brooks from Sandersha.

Clonal selection

Clonal selection have also resulted in identification of Few elite clones, e.g. Pant Chandra and Pant Sinduri from Dashehari developed at Govind Ballabh Pant University of Agriculture & Technology, Pantnagar, Chhattisgarh Nandiraj from Dashehari at Indira Gandhi Krishi Vishwavidyalaya, Raipur, Niranjan from Parbhani, Arka Neelachal Kesari from Gulabkhas released from Central Horticultural Experiment Station, Bhubaneswar and Cardozo Mankurad from Goa Mankurad. Clonal selections from Dashehari are Dashehari-35, C-24, 18, 35, 45 and 51.

3. Hybridization: The technique

(a) Select fewer panicles with lesser number flower/ flower buds.

(b) Emasculation and pollination using the polythene bag (24" x 21") technique.

(c) Use of self-incompatible parental lines with male parents in an insect proof cage.

(d) Use of gene marker i.e. purple colour of new leaves and panicles and beak characters of fruits to reduce contamination and achieve better fruit set.

(e) A new off- season crossing technique was suggested by Kulkarni (1986). It involves induction of flowering in the desired parents in off season by veneer grafting their defoliated shoots on to leafy shoots of off season flowering cv. Royal Special and allowing open pollination between them.

Hybrids: In India, first attempt was made at Pune by Burns and Prayag in 1911 to develop hybrids having regular bearing and good yield e.g. Cowasji Patel x Pairi. In the second phase, systematic work started in 1961 at IARI, New Delhi to develop regular bearing varieties Mallika and Amarpalli. Improvement work was started at IIHR, Bengaluru, RFRS, Vengrurule and Paria, Gujarat to develop varieties free from spongy tissue. Work was initiated to develop hybrids having red peel colour and dwarf tree stature using exotic and indigenous cultivars as donar parents at IARI, New Delhi and CISH, Lucknow. The other institutes where work was done to develop hybrids are Bihar Agricultural University, Sabour, Fruit Research Station, Kodur and Fruit Research Station, Sangareddy, AP.

Indian Agricultural Research Institute, New Delhi:

(i) **Mallika:** Neelum × Dashehari.

(ii) **Amrapali:** Dashehari × Neelum.

(iii) **Pusa Arunima:** Amrapali × Sensation.

(iv) **Pusa Pratibha:** Amrapali × Sensation.
(v) **Pusa Shreshth:** Amrapali × Sensation.
(vi) **Pusa Peetamber:** Amrapali × Lal Sundari.
(vii) **Pusa Lalima:** Dashehari × Sensation.
(viii) **Pusa Deepshikha:** Amrapali × Sensation.

Central Institute for Subtropical Horticulture, Lucknow:

(i) **Ambika:** This is a hybrid between the cross, Amrapali × Janaradhan Pasand.
(ii) **Arunika:** This is a hybrid between the cross, Amrapali × Vanraj.

Bihar Agriculture University, Sabour

(i) **Prabhashankar:** Bombai × Kalapady.
(ii) **Mahmood Bahar**: Bombai × Kalapady.
(iii) **Sabri:** This is a hybrid between the cross, Gulabkhas × Bombay Green.
(iv) **Al Fazli:** It is from the parentage Alphonso × Fazli.
(v) **Jawahar:** This is a hybrid between the cross, Gulabkhas × Mahmood Bahar.
(vi) **Sundar Langra:** This is from the cross, Sardar Pasand × Langra.

Indian Institute of Horticultural Research, Bengaluru:

(i) **Arka Aruna:** It is from the parentage, Banganapalli x Alphonso.
(ii) **Arka Anamol:** Alphonso × Janardan Pasand.
(iii) **Arka Neelkiran:** Alphonso × Neelum.
(iv) **Arka Puneeet:** Alphonso × Baganpalli.
(v) **Arka Udaya:** Amarpali × Arka Anamol.

Fruit Research Station, Kodur (Dr. Y.S.R. Horticultural University, West Godavari):

(i) **Au-Rumani:** This is a hybrid by the combination, Rumani × Mulgoa.
(ii) **KMH 1 (Kodur Mango Hybrid 1):** It is from the parentage, Cherukurasam X Khader.

Fruit Research Station, Sangareddy, AP:

(i) **Manjeera:** It is from the parentage, Rumani × Neelum.
(ii) **Swarna Jehangir:** This is a hybrid from the cross, China Suvarnarekha × Jehangir.
(iii) **Neelgoa:** This is a hybrid from the cross, Neelum × Mulgoa.
(iv) **Neeluddin:** This is from the parentage, Neelum × Himayuddin.
(v) **Neeleshan:** It is a hybrid between the cross Neelum × Baneshan.

Konkan Krishi Vidyapeeth, Dapoli:

(i) **Konkan Ruchi:** Neelum × Alphonso tree no. 900 (especially for pickle-making).

(ii) **Konkan Raja**: Bangalora × Himayuddin.

(iii) **Konkan Samrat:** Alphonso × Tommy Atkins.

Regional Fruit Research Station, Vengurla:

(i) **Ratna:** Neelum × Alphonso.

(ii) **Sindhu**: Ratna × Alphonso (Fruits are almost seedless, with a very thin stone).

Mahatma Phule Krishi Vidyapeeth (MPKV), Rahuri:

(i) **Sai Sugandha:** Totapuri × Kesar.

Agricultural Experimental Station, NAU, Paria, Gujarat:

(i) **Neeleshan Gujarat**: This is a hybrid from the parentage, Neelum × Baneshan.

(ii) **Neeleswari**: This is from the parentage, Neelum × Dashehari.

(ii) **Neelphonso:** This is from the parentage, Neelum × Alphonso.

(iii) **Sonpari:** This is a hybrid from the cross, Alphonso × Baneshan.

Horticultural Research Station, Periyakulam:

(i) **PKM 1:** It is a clonal selection from the hybrid, Chinnaswarnarekha and Neelum

(ii) **PKM 2:** Hybrid between Neelum and Mulgoa.

KAU, Thrissur:

Hybrid No 45: Bennet Alphonso × Himayuddin.

Hybrid No 87: Kalapady × Alampur Baneshan.

Hybrid No 151: Kalapady × Neelum.

New hybrids

India:

Arka Suprabath: It is a double-cross hybrid developed by using parents Amrapali (which is a cross of Dashehari and Neelam) and Arka Anmol (which is a cross of Alphonso and Janardhan Pasand). Released by IIHR, Bengaluru.

CARC, Brazil:

Embrapa Roxa 141: Amrapali × Tommy Atkins

Embrapa Alfa 142: Mallika × Van Dyke

BRS Beta: Amrapali × Winter

Embrapa Lita: Amrapali × Tommy Atkins

BRS Omega: Amrapali × Tommy Atkins

New Hybrids:

***Florida*, USA, Australia and Thailand:**

Bailey's Marvel: Haden × Bombay.

Anderson: Sandersha × Haden

Haden- JIRCAS: Mulgoba × Turpentine

Haden-OPARC: Mulgoba × Turpentine

Jacquelin-OPARC: Haden × Bombay

Jacquelin-JIRCAS: Haden × Bombay

NMBP-1243: It is a hybrid between Irwin and Kensington Pride. It is developed by the National Mango Breeding Program (NMBP), Western Australia

Mahachanok (Rainbow mango): It is a hybrid between Nang Klang Wan and Sunset and developed on Tiew-Thong Farm in Chiang Mai, Thailand.

Calypso: It is hybrid between Kensington Pride and Sensation developed in Australia.

Sensation: Haden × Brooks. Originated in South Florida

Haden: Mulgoba × Turpentine. Originated in South Florida

Interspecific Hybrids: Hybrids selected from segregating populations from crosses between heterozygous clones of different species. Modified back crossing may be adopted to transfer a desirable gene to the segregating population. *Mangifera odorata* and *Mangifera zeylanica* are used as parental sources.

4. Heterosis: Large population had bigger fruits in hybrid progenies obtained with Banganpalli as one of the parents. This has been reported in the progenies from crosses between Alphonso and Banganpalli. Transgressive segregation of this character was also observed.

5. Mutation: Repeated heading back after treatment with mutagen like gamma irradiation (2.9 kr), 1.5% EMS or 0.05% N-Nitroso Methyl Urea (NMU) resulted into a mutant with dwarf habit and small leaves. Davis Haden orignated as a mutant from Haden at Florida and the mutant Rosica from Rosado-de–lca in Peru.The other mutant cultivars are Hongmang 6 from Zill and IAC 100 Bourbon from Boubon.

6. Biotechnology:

Micropropagation: Micropropagation has not given commercial success in mango due to problems such as latent microbial infection, excessive polyphenol exudation and early explant necrosis. *In vitro* propagation of juvenile phase material has been reported in 2-year old seedlings of polyembryonic Gomera, Sabre, Turpentine and 13-1. Growth medium (G formulation) supplemented with 4.4 μM BA, 4.6 μM zeatin, 9.8 μM 2iP, 2.9 μM IAA, 2.5 μM IBA, 300 mg/liter glutamine and 300 mg/l casein hydrolysate was utilized to stimulate proliferation of axillary buds from the shoot tips.

Somatic Embryogenesis

Somatic embrygenesis involves stimulation of embryos from somatic tissues. Embryogenic cells can be present within a tissue at the time of explanting or they can be induced, although this can involve subculture to remove competent cells from the effect of repressors of embryogenesis that may be present in the explant. Litz described somatic embryogenesis from explanted polyembryonic mango ovules on MS medium containing various phytohormones.

Protoplast Culture

Protoplast technology may not impact mango improvement since the sexual compatibilities of *Mangifera* spp. with the common mango are unknown. Although many of the newly described *Mangifera* species have interesting environmental and pest resistance, it is uncertain if they are isolated genetically from mango. Somatic hybridization could provide a means for genetic recombination of the common mango with some of these species as a way to develop rootstocks. However, it is unlikely that this approach will have utility for scion development due to genetic heterogeneity caused by octoploidy.

Molecular Marker

Mango cultivars are currently identified on the basis of distinct morphological characters. However, morphological traits cannot serve as unambiguous markers since they may vary with environmental conditions. Reliable genetic markers have been developed and introduced for mango cultivar identification. These include isozymes, RAPDs (random amplified polymorphic DNAs) and VNTRs (variable number tandem repeats).

In mango, the feasibility of using isozymes to detect possible genetic variation among individuals of so-called clones was shown by Gan *et al.* (1981). Several polymorphic enzymes were employed for systematic characterization and parentage analysis of mango cultivars. They showed that morphologically very similar cultivars such as "Pico" and "Carabao" can be easily distinguished by isozyme analysis. Their isozyme banding patterns supported the origin of some cultivars such as "Haden" from "Mulgoba", "Zill" from "Haden" and "Tahar" from Irwin. In addition, the accepted hybrid origin of some other mango cultivars was refuted by the isozyme analysis. Edward is commonly accepted to be a hybrid of "Haden" and "Carabao". However, isozyme evidence showed that "Carabao" cannot be the male parent of "Edward". In addition it was shown that, contrary to common assumption, "Mulgoba" cannot be a parent of "Keitt" and "Givataim" cannot be an offspring of the Polyembryonic cultivar "13-1".

Schnell *et al.* (1995) studied the application of RAPDs to the identification of mango cultivars. Among 11 primers selected, no single primer gave a unique banding pattern for any of the 25 accessions examined; however several different combinations of two primer banding patterns produces unique finger prints for each accession. Adato *et al.* (1995) evaluated the usefulness of 10 different mini- and microsatellite probes for the identification of mango cultivars. Genetic markers can serve to detect linkages with agriculturally important traits. If the gene(s) to be transferred is marked by a tightly linked genetic marker, the segregating progeny can be screened at the seedling stage before the trait is expressed.

Genetic Transformation

Success has been achieved in obtaining genetically transformed mango embryogenic cultures using *Agrobacterium tumefaciens*. Engineered Agrobacterium strains, viz., (i) C 58CI with plasmid pG3850::1103 with selectable marker NPT II and (ii) A208 contained pTiT37-SE:: pMON9749, a co-integrate vector with genes encoding for NPT-II and GUS were used for transformation of 'Keitt' and 'Hindi' mango, respectively (Mathews *et al*., 1993).

Mango fruits have short shelf life. A major goal in the genetic engineering of fruit quality is through control of ethylene production and action. Two different strategies have been employed to achieve this aim: inhibition of expression of genes encoding ethylene biosynthetic enzymes by transformation with the respective antisense genes and lowering the level of the ethylene precursor ACC by deamination or hydrolysis. Ethylene biosynthesis is catalyzed by three major enzymes, Sadenosyl methionine (SAM) synthase, ACC synthase and ACC oxidase. Two cDNA clones have been identified as ACC synthase and ACC oxidase from a mango cDNA library. These two clones are currently being used in antisense experiments of genetic transformation of mango to try and block ethylene production. The resultant transgenic fruits should present an extended shelf life and an enhanced fruit quality.

7. Conservation of germplasm: *In vitro* based storage techniques are the only satisfactory way of assuring constant supply of germplasm. Alternatives to *ex situ* collections include the use of slow growing *in vitro* cultures for short and medium term storage and the use of seed banks and cryopreservation of *in vitro* cultures for long term storage.

BANANA

Origin: Indo-Malaysian region

Taxonomy status:

Order: Zingiberales

Family: Musaceae

Section: *Eumusa*

Genus: *Musa*

Botanical name: *Musa* spp.

Chromosome number: x = 11; 2n = 2*x* = 22; 2n = 3x = 33; 2n = 4x = 44

Genome size: 523 mb (*Musa acuminata*)

Musa acuminata and *Musa balbisiana* are the ancestors of the edible cultivars, with the exception of the Fei group (Fei group is composed of cultigens characterized by reddish sap of the plant, shiny bright green bracts and carotenoid-rich orange fleshed fruit borne on erect bunches and not on hanging bunches typical of *Eumusa*) of the *Australimusa* section originating from intra- and interspecific hybridizations. Bananas are also classified into genomic groups based on their genome constitution, which are designated by the letters A, B, S and T to represent *M. acuminata* (AA), *M. balbisiana* (BB), *M. schizocarpa* (SS) and *M. textilis* (TT), respectively.

Edibility of the fruits of diploid *Musa acuminata* developed through selection. The crossing of the seeded edible diploids (AA), involving several of the many subspecies of *M. acuminata* in diverse regions, resulted in the formation of generally sterile hybrid progeny with the genome AAA, propagated vegetatively. Diverse hybridizations with *M. balbisiana* resulted in mostly sterile hybrid progeny mainly with the genomes AB (dessert bananas), AAB (plantains) and ABB (cooking bananas). Other combinations such as bispecific tetraploid AABB cultigens arose rarely, and quite infrequently a few additional species were involved.

Botany

It is a fast growing, herbaceous perennial plant, arising from underground stem or corm. The aerial shoot is the above ground part of the banana plant, consisting of the pseudostem, leaves and the inflorescence. The pseudostem is made of large overlapping leaf sheaths that are tightly rolled on each other to form a firm cylindrical structure. Leaves are oblong or elliptic, becoming as much as 15 m long. Approximately 44 leaves appear before the inflorescence. The inflorescence is a spike originating from the tip of the corm. The lower bracts of inflorescence axis enclose female flushes and are composed of one spathe and single or double rows of flowers at its axils. The middle few bracts may enclose neutral or hermaphrodite flowers, while at the tip of inflorescence are male flowers. Female flowers, with inferior trilocular ovaries with three carpels, occupy the lower 5—15 rows on stalk having 6 tepals (5 united and 1 free) and five or six fleshy and nonfunctional stamens (staminodes) or hermaphrodite flowers consisting ovary that does not develop into fruit in the centre, and males at the top. Male flowers and bracts are shed one day after opening. The flowers consist of three each of sepals and petals, two whorls of outer and inner stamen, and a tri-locular, inferior ovary having axile placentation. The adaxial petal is not opposed by an inner whorl of stamen, resulting in only two stamens in this whorl. Fruits are epigynous berry borne in hands of up to 20 fruits with 5—13 hands per spike. Fruits during growth appear as angled, slender and green fingers, reaching harvest maturity 90—120 days after flower opening. Cultivated varieties are typically seedless. When seeds are present, they vary among species in shape and morphology. Seeds of Musa balbisiana, parent of many commercial edible banana varieties, are dark brown, ovoid, about 4 mm long with a conspicuous white powdery endosperm.

Plate no. 1. Banana inflorescence and flower types.

(Source: ProMusa)

Floral Biology

Banana blooms year round, but most flowering is concentrated from February to August. The longevity of female flowers is more than that of male flowers and opening of both types of flowers occurs at any time during the day, but mostly in the morning. The natural volume and rate of emergence of female flowers are significantly higher than male flowers. Male flowers open in the evening and are ready for pollination early in the morning. The flowers remain open for 7-8 days and anthers begain dehiscing shortly after flower opening and continue the second day. Anthesis occurs from 6.30 AM to 8.00 AM. Stigma is receptive at the opening of flowers. Pollination is generally carried out between 7.00 and 10.00 AM. Undehisced anthers are collected from male flowers and are twisted gently to force them to dehisce. Using a soft sable (fur used for making artists brushes) hair brush, pollen grains are prised out and smeared gently over the stigmatic surface of those female flowers that open on the day of pollination. Pollinated flowers are covered with a soft cloth bag.

Floral morphology suggests that wild bananas are bat pollinated in their native range. Insects such as bumble bees (*Bombus eximius* and *B. montivolans*), honey bees (*Apis cerana* and *A. florea*) and wasps (*Vespa mandarinia*) are primary floral visitors, showing a preference for female flowers. It can take up to several months to obtain seeds from a desired cross, and production of seeds is generally poor, reported in the range of 0.3–21.7 seeds per bunch. Seeds obtained from crosses also germinate poorly. It is a standard practice in most programs to recover hybrids through *in vitro* culture.

Breeding Objectives

1. Development of dwarf statured cultivars with high suckering ability.
2. Production of early flowering/maturity, large bunch size, high yield and superior quality fruits with longer finger size suitable for export.
3. To develop varieties with wider adaptibility with enhanced productivity.
4. Resistance or tolerance to biotic stress; Resistance to sigatoka (*Mycosphaerella* spp.) complex, all known races of *Fusarium* wilt, Xanthomonas or bacterial wilt disease and banana Bunchy Top Virus (BBTV). Resistance to weevil (*Cosmopolites sordidus*) and *Radopholus similis, Pratylenchus* spp. *and Helicotylenchus* nematodes.
5. Resistance or tolerance to abiotic stress (drought and cold).
6. Development of male fertile parthenocarpic diploids with resistance to major diseases.

Limitations in Breeding

1. Polonged life cycle.
2. Narrow range of genetic variability.
3. High level of heterozygosity.

4. Low reproductive fertility.
5. Polyploidy.
6. Irregular meiotic behavior.
7. Parthencarpy.
8. Female and male sterility.
9. Low seed setting in crosses (average 1 to 1.5).
10. Diverse genomic configurations.
11. Little knowledge on the genetics of important agronomic traits.

Germplasm Resources

The International Transit Centre (ITC) holds the world's largest collection of bananas. The International Musa Germplasm Collection is managed by Bioversity International. Hosted by the Katholieke Universiteit Leuven in Belgium, the gene bank has more than 1,500 accessions most of them cultivars.

In India, ICAR National Research Centre on Banana, Tiruchirapalli is the national repository for Banana. It has the largest collection of banana with 310 indigenous accession and 30 exotic accessions. IIHR, Bengaluru, TNAU, Coimbatore, Banana Research Station, Kannara, KAU, Trichur and Banana Research Station, Hajpur, Rajender Prasad Central Agricultural University, Pusa are also maintaing the germplasm of banana.

Sources of desirable resistance traits:

Burrowing nematode (*Radopholus similis*): Pisang Lilin Buaya, Kadali, Octoman, Thaen Kunnan, Ney Poovan, Kunnan, Tongat, Pisang Lilin, Dudh Sagar, Chini Champa, Anil Vazhai, Manik Champa, H-1, Calcutta 4 (*M. acuminata* ssp. *burmannicoides*).

Black weevil (*Cosmopolites sordidus*): Calcutta 4 (*M. acuminata* ssp. *burmannicoides*).

Panama wilt: Basrai, Poovan, Champa, Cavendish group, Mongil, Peyladen, Rajabale, Vamanakeli, Giant Governor, Robusta, Gold Finger, Pisang Lilin, Pisang Raja, Calcutta 4, BITA-1, BITA-2, *Musa acuminata* ssp. *malaccensis*, *Musa acuminata* ssp. *burmannica,* Calcutta 4 (*M. acuminata* ssp. *burmannicoides*), *Ensete perrieri* and SH-3142.

Black sigatoka (*Mycosphaerella fijiensis*): Calcutta 4 (*M. acuminata* ssp. *burmannicoides*), *Musa acuminata* ssp. *malaccensis,* ssp. *burmannica* and ssp. *truncata*, TMP2x.

Yellow sigatoka (*Mycosphaerella musicola*): Calcutta 4, (*M. acuminata* ssp. *burmannicoides*).

Banana bunchy top virus: Poovan, Gold Finger.

Leaf spot: H1, H2.

Cold: *Musa basjoo,* Java Blue.

Heat: Sugar ABB Pisang Awak, *Musa balbisiana.*

Drought: Karpuravalli, FHIA-2, *Musa balbisiana.*

Salt: Karpuravalli.

Sources of desirable growth and fruiting traits

Parthenocarpy: *Musa acuminata* ssp. *banksii.*

High yield: FHIA-6.

Keeping quality: Hill Banana, Ney Poovan.

Suitability for long distance transport: Karpuravalli.

Export value: Grande Naine, Poyo and Williams.

High sugar content: H1.

Good pulp quality: Goldfinger.

Fibre yield: *Musa textilis* (abaca or Manila fibre).

Ornamental purpose: *M. ornata* and *M. veluntina.*

Desired traits in relation to its gene pattern

Male and female fertility: Recessive genes interacting with cytoplasm sensitive.

Persistence of male bracts: Two loci with complementary dominant gene.

Persistance of hermphrodite flowers and male bracts: Two independent loci with complementary dominant gene.

Seed sterility: Cytogenetic factors.

Vegetative parthenocarpy: Three independent complementary dominant genes.

Sterile edible diploids: Genetic female sterility.

Apical dominance: Recessive gene.

Suckering habit: Dominant gene.

Dwarf growth habit: Single recessive gene.

Bunching habit: Oligogenic.

Bunch orientation: Oligogenic (three loci with thresholds effects of dominant gene).

Fruit size: Governed by epistatic.

Resistance to sigatoka and panama wilt: Several dominant genes.

Resistance to Bacterial wilt (Moko disease): Several recessive genes.

Breeding Methods

1. **Introduction:** Several introductions of banana cultivars have been made. These are Lady Finger (EC 160160) resistant to bunchy top virus introduced from Australia and being evaluated at NRC on Banana, Tiruchirapalli, TNAU, Coimbatore, IIHR, Bengaluru; Namwa Khom introduced from Philippines, Saba (of ABB genome group and Bontha subgroup) introduced from Philippines, being evaluated at NRC on Banana, Tiruchirapalli, IIHR, Bengaluru and TNAU, Coimbatore, Grande Nain (G-9) MS (EC 27237) from France and Velery from West Indies. Exotic introductions FHIA-1, FHIA-3, FHIA-23, FHIA – I7,

FHIA –18 FHIA – 21 FHIA – 23 FHIA-25, SH-3640, and TMB x 5295-1 were identified for high productivity and resistance to Sigatoka leaf spot.

2. **Selection:** The majority of the varieties of banana cultivars in the world were developed by selection. A tall high yielding (70 kg) mutant Gandevi has been selected from Giant Cavendish. Udhayam is a single plant selection from "Kenthali belonging to Pisang Awak subgroup. It is suitable for table and processing purpose and tolerant to Black Sigatoka and leaf spot diseases and nematodes. Kovvur Bontha (Plantain) is a selection done at Kovvur, Andhra Pradesh. Bangrier of genomic group ABB and Bluggoe subgroup is a single plant selection from the local land race Bangrier. Manjeri Nendran II is a clonal Selection from Nendran. Kozhikodu and Bangladesh Malbhog are selections of Silk. Pisang Ceylon is a selection of Mysore subgroup. At TNAU, Coimbatore a shade tolerant and high yielding diploid accession of Sannachenkadali, christened as AA-1 with field tolerance to nematodes, infectious chlorosis, bract mosaic virus, resistant to *Eumusae* leaf spot and bacterial head rot diseases has been developed. Ambala Kadali, Erachivazhai, Thattilla Kunnan, Veneetu Kunnan and Rasa Kadali are the selections from natural mutant which showed significant variations from their parental genotypes.
3. **Hybridization:** Two major breeding approaches are followed in banana breeding.

Pragmatic Breeding

The origin of this approach is at the dawn of banana breeding, when breeders tried to develop Gros Michel hybrids resistant to Fusarium wilt. Triploid cultivars show a residual fertility and when crossed with diploid cultivars, may produce a few seeds. The primary products issued from these crosses show a considerable genotypic (and phenotypic) variation, ranging from diploid to highly polyploid hybrids. Among them, tetraploid hybrids, arising from unreduced triploid egg cells (2n=3x=33) were selected. The value of these hybrids is that the genes from the mother plant do not segregate and their main characteristics are maintained in the tetraploid, while the haploid genome added from the diploid confers disease resistance. This strategy was taken up for other dessert and cooking bananas to confer resistance to black leaf streak and nematodes.

Dessert tetraploid hybrids were developed at FHIA from crosses between dwarf mutants of Gros Michel and Prata with improved diploids resistant to Sigatoka diseases and nematodes. Cooking bananas were developed from crosses between plantains and resistant diploids at IITA, CARBAP and FHIA to confer resistance to black leaf streak. Some outstanding hybrids were obtained by this strategy. FHIA21, a cooking tetraploid hybrid released by FHIA, is now being cultivated for local markets in some countries in West Africa, in Central and South America and in the Caribbean, as a substitute to black leaf streak susceptible plantains. A major outbreak of this approach is the discovery of endogenous integrated sequences of the endogenous banana streak virus (eBSV) in the plantain genome, releasing infectious viral particle in the progenies issued from crosses. However, these viral sequences have been shown to behave as pseudo-genes and in some cases can segregate as heterozygous locus (Gayral *et al.*, 2008), paving the way for the elimination of infectious eBSV sequences. Primary tetraploid hybrids, AAAB, AAAA and AABB are often much more fertile than the triploid parent, and can produce seeds when crossed with a diploid accession. The descendants obtained from these 4x x 2x crosses are predominantly triploids (secondary triploid hybrids). Redistributions and recombinations between A and B chromosome during meiosis can produce triploid hybrids free from infectious eBSV from these crosses.

As mentioned above, the initial 3x x 2x cross also produces diploid hybrids, whose genetic background comes from the triploid mother plant. These primary diploid hybrids are eventually used in the 4x x 2x cross to bring additional mother plant background to the secondary triploid hybrids. In this breeding scheme, selected resistant primary tetraploid hybrids can be either released as new improved cultivars or subsequently crossed with other improved diploid selections to obtain secondary triploid hybrids. In this approach, the genetic diversity used on the triploid side is very limited due to low fertility of triploid cultivars. The choice of the diploid parents to be crossed in the first or second step is therefore crucial. Considering that the unreduced triploid eggs of the maternal parent are genetically homogenous, the breeding effort is only based on the improvement of the diploid parent. Within the great diversity of the diploid pool, wild and edible diploids have been selected to introgress pest and disease resistance in triploid cultivars.

However, wild relatives are highly fertile but have very few of the outward appearances of cultivated bananas, while edible diploids, even if they are more attractive in terms of bunch appearance and fruit qualities, are at most moderately fertile and often not resistant to diseases. Consequently, pre-breeding at the diploid level appears as a preliminary step. The most notable outcome of genetic improvement are the AA hybrid M53 (resistant to Sigatoka diseases and to Fusarium wilt), bred in the 1950s by the breeding programme of the former Jamaican Banana Board, and several outstanding diploids created at FHIA, some of which have multiple resistance (to Sigatoka diseases, nematodes and Fusarium wilt) and a huge number of hands.

Pragmatic breeding steps

↓

Triploid parent (cultivar to be improved) 2n = 3x = 33 x Diploid parent 1″ (source of resistance)

↓

Primary hybrid progeny

↓

Primary diploid or "diploid parent 2" Primary tetraploid

↓

Secondary progeny 2n = 22 or 33

↓

Secondary triploid (3x)

Reconstructive Breeding

CIRAD developed a new breeding strategy aimed at the creation of triploid hybrids directly from diploid germplasm. The advantages of this approach are manifold: the genetic combinations selected at the diploid level are totally or partially preserved through the doubles-diploid. This breeding strategy is based on a search of specific combining ability between the two diploids of which one is donor of diplo-gametes. As the production of 2n gametes is uncontrolled and fairly rare in diploid clones, the regular production of diplo-gametes has been achieved through chromosome doubling of diploid varieties.

The ability to set progeny is strongly linked to gamete diploid parent 2" 2n = 2x = 22 chromosome doubling mainly triploid progeny 2n = 33 selected triploid hybrids (3x). The ability to set progeny is strongly linked to gamete fertility, which is highly variable from one to another clone. An intensive programme of hybridization combining various parents and a larger number of doubled genotypes has indicated that the studied doubled diploid clones are almost exclusively producing polyploid gametes, for a major part, containing exactly 22 chromosomes. Then, the progenies obtained from doubled-diploids x diploids are essentially triploid, which is the expected objective.

This strategy has several advantages, such as; the genetic combinations selected at the diploid level are totally or partially preserved in the eventual triploids; the hybrids display the desired sterility because of their triploid nature; utilization of very fertile wild parents sometimes enabling production of large triploid progenies in which selection is very effective and that this strategy enables breeders to tap the wide diversity in diploids. It was used for dessert and cooking bananas to confer resistance to black leaf streak and to nematodes.

Reconstructive Breeding

↓

Diploid parent 1" 2n = 2x = 22

↓ Chromosome doubling

Doubled diploid 1 2n = 4x = 44 X Diploid 2n = 2x = 22

↓

Mainly triploid progeny 2n = 33

↓

Selected triploid hybrids (3x)

Diploids

Diploids are usually less productive and less vigorous than triploids, with smaller bunches and sometimes containing seeds in fruits and are thus not suitable as final breeding products. However, some commercial trades of significant importance, high value niche markets far from mass consumption, rely on diploids like Kunnan (ABcv) in India or Pisang Mas/Sucrier (AAcv) in Colombia.

The narrow genetic variability and limited fertility among cultivated triploid bananas make diploid bananas vital to genetic improvement. A number of fertile improved diploids with varying degree of disease resistance have been released by IITA and FHIA. Diploid improvement has almost exclusively been done through the use of *M. acuminata* cultivars such as "Calcutta 4" (*M. acuminata*), source of resistance to the Sigatoka complex, yellow Sigatoka, *Fusarium* wilt, banana weevil and burrowing nematodes. Decades of breeding effort utilizing this material has resulted in the production of improved diploid lines which combine disease/pest resistance, short stature, and interesting bunch characteristics. Developing improved diploids that possess multiple sources of resistance, preserving the quality characteristics of the preferred triploid, would greatly increase the efficiency of breeding efforts.

M. balbisiana diploid cultivars have made limited contribution to breeding due to the presence of endogenous banana streak virus (eBSV) sequences originating from the B genome that are activated under appropriate conditions. Progenies of interspecific *acuminata-balbisiana* hybridizations have often been associated with the occurrence of banana streak disease. This has resulted in underutilization of the B genome which could be an important source of drought tolerance and of resistance alleles not found in *M. acuminata*. Diploid breeding at TNAU, Coimbatore, primarily involved the use of cultivars Anaikomban, Namarai, Erachi vazhai, Pisang lilin and Tongat (all AA) as male parents with Matti (AA) as the female parent led to development of potential synthetic diploid hybrids H.59, H 65 (Matti x Anaikomban) and H109 (Matti x Tangat). Many synthetic diploids have been found to have good resistance to burrowing nematode, Sigatoka leaf spot and *Fusarium* wilt. Cultivar Matti showed higher female fertility than the other diploids which were generally female sterile. A plausible explanation for high seed fertility of Matti could be attributed to its localised cultivation in Kanyakumari district of Tamil Nadu near the foot hills of Southern Travancore where its wild ancestor *M. acuminata* subsp. *burmanica* could be traced even now. This wild species is highly female fertile and is self propagated through seeds. The parthenocarpic type Matti would have arisen as a clonal selection from this wild species retaining the seed fertility trait.

Triploids

Triploid hybrids are thought to be the best objective for banana breeding as triploidy gives a selective advantage over the other ploidy levels. Crosses Diploid x diploid, triploid x diploid and tetraploid x diploid results in the development of commercial triploids. TNAU, Coimbatore attempting to improve the Pome (Subgroup AAB) cultivar Kallar Ladan developed the promising hybrid CO 1, which is the result of three way sequential crosses i.e., Kallar Ladan (AAB) x *M. balbisiana* clone Sawai (BB) and F1 (AB) x Kadali (AA). Subsequent triploid x diploid crosses like Karpuravalli x H.201 led to the development of a promising ABB hybrid 96/7 having bright yellow fruits without any ashy coating. Likewise Karpuravalli x Pisang Lilin also resulted in a parthenocarpic triploid AAB (H-212). Banana Research Station, Kannara released two triploid hybrids namely BRS-1 (Agnishwar x Pisang Lilin) and BRS-2 (Vannan x Pisang Lilin) through triploid x diploid breeding.

4x × 2x schemes should be initially improved through recurrent selection to accumulate the favourable alleles. Crossing H-02-02 (AABB) with Pisang Lilin resulted in the development of a primary triploid H-03-15 (AAB) resistant to wilt and nematodes giving good bunch weight (14.5 kg.) TNAU, Coimbatore. This indicates that tetraploid x diploid cross has to be done in large numbers to develop triploids having novel genomic constitution.

The National Agricultural Research Organization, Uganda (NARO) and the International Institute of Tropical Agriculture have developed NARITA hybrids (related to a group of cooking and juice banana) which are high yielding and disease resistant. The NARITAs are secondary triploid (3x) hybrids. These were obtained by crossing a female fertile EAHB (3x) cultivar with diploid NARITA 1 (2x), a source of resistance.

Tetraploids

Tetraploids are produced by crossing a triploid female (ABB or AAA) with improved diploid male (AB or AA). Autótetraploids (AAAA) are used for breeding export quality AAA clones and AABB tetraploids are used to breed AAB/ABB types. Triploid x diploid crosses (Karpuravalli x Pisang Lilin) at TNAU, Coimbatore gave three parthenocarpic tetraploids with AABB genome. Triploid x triploid crosses resulted in the production of tetraploids e.g., Karpooravalli (ABB) x Red banana/Robusta (AAA), having male and female fertility and resistance to nematodes and Sigatoka leaf spot, produced tetraploid (AABB) progenies H-02-21 and H-02-34. Tetraploids derived from triploids are generally too big, so care should be taken to select a dwarf triploid as a parent.

Breeding of banana at the Brazilian Enterprise for Agricultural Research, National Centre for Research on Cassava and Fruit Crops (EMBRAPA-CNPMF) has been targeted to the production of tetraploids with AAAB genomes resistant to *Fusarium* wilt (*F. oxysporum* f. sp. *cubense* Race 1), nematodes, Yellow Sigatoka caused by *Mycosphaerella musicola*, bacterial wilt (*Xanthomonas campestris* pv. *musacearum*), weevils and black leaf streak (*M. fijiensis*). BITA-2 is a synthetic hybrid bred by IITA. It is a cross between Fougamou (AAB, Pisang Awak) and Musa balbisiana 1-63. The starchy tetraploid cultivars such as BITA 03 and PITA 16 were produced at the International Institute of Tropical Agriculture, Uganda (IITA), which are being used for the production of wine and beer, respectively. At the Banana Research Station (BRS), India, the tetraploid hybrid BRS-01 of the Pome subgroup (*Musa* spp. AAA cv. Agniswar × *M. acuminata* A (*Musa* spp. AAA cv. Agniswar × *M. acuminata* AA cv. Pisang Lilin) and the hybrid BRS-02 (*Musa* spp. ABB cv. Vannan × *M. acuminata* AA cv. Pisang Lilin) of the Mysore subgroup exhibited immunity to *M. eumusae* and *M. musicola*. Fundación Hondureña de Investigación Agrícola (FHIA) is known for being the first to have delivered disease resistant hybrids to farmers. Important released tetraploid cultivars are FHIA-01 (AAAB; Gold Finger), FHIA-02, FHIA-03 (Sweet Finger), FHIA-17, FHIA-18, FHIA-18, FHIA-21 (resistant to black Sigatoka), FHIA-23, FHIA-25 (resistant to black Sigatoka), SH-3640, BITA-2 and BITA-3. These dessert/cooking hybrids are resistant to bunchy top and *Fusarium* wilt. Diploids could be made into tetraploid either through conventional breeding or through *in vitro* polyploidization of diploids using antimitotic agents like colchicine/oryzalin.

4. Polyploidy: The approach involves manipulation of ploidy by *in vitro* mutation technology for production of tetraploids from diploids by colchicines/oryzalin treatment. Subsequently improved tetraploid lines are selected and the selected tetraploids are crossed with diploids to produce triploids suitable for final evaluation. PITA-9 is a selected tetraploid hybrid derived from the female fertile French somaclonal variant of False Horn Plantain. It is resistant to black Sigatoka, produces medium to high yield of big fruits in the humid forest zones of Nigeria and Cameroon. Increased fertility status allows using PITA -1 as a parent for further ploidy manipulations via 4x x 2x crosses and for population improvement within the False Horn plantains.

5. Mutation: Spontaneous mutants appear in banana at a higher rate than in other crops. Several natural sports have been identified. Highgate from Gros Michel, Motta Poovan from Poovan, Sombrani Monthan from Monthan, Krishna Vazhai from Virupakshi, Ayiranka Rasthali from Rasthali and Barhari Malbhog from Malbhhog are spontaneous mutation.

Physical mutagenesis is an attractive technique for banana and plantain improvement. For example, aluminium tolerant mutants of the AAA banana cv. Nanicão, early-flowering mutants GN-60 Gy/A and FATOM-1 (GN-60/A) from cv. Grande Naine, and improved agronomic characteristics in Dwarf Parfitt, a Cavendish cultivar resistant to *Fusarium oxysporum* f. sp. *cubense* race 4 could be obtained by irradiation with co-operation of FAO and IAEA (International Atomic Energy Agency). "Novaria" (early flowering) is a promising mutant of Grand Naine developed by gamma irradiation and released in Malaysia. An improved mutant cultivar Klue Hom Thong KU1 bearing large fruit bunches was released in Thailand through gamma irradiation.

6. Biotechnology:

Micropropagation: Micropropagation systems are well developed for bananas and can be used to multiply elite genotypes. Tissue culture Shoot tip culture was first applied to *M. acuminata* AAA cv. Cavendish and subsequently to other cultivars. Similarly tissue explants including meristems, rhizomes, inflorescences and immature male flowers, immature zygotic embryos and leaf baseswere exploited. Now micropropagation has become a standard practice for the production of tissue culture derived banana plants. Such plants generally perform better than the plants derived from conventional planting materials with respect to yield, finger size, cycle time, number of suckers, efficiency of nutrient uptake and emergence and uniformity even in ratoon crops.

Organogenesis and somatic embryogenesis: Plant regeneration via somatic embryogenesis has been reported in bananas from embryogenic cell suspensions established from embryogenic callus induced from apical meristems, corm like tissues, pseudostems, leaf bases and rhizome fragments, proliferating scalps, immature zygotic embryos, immature male flowers and immature female flowers. In general, embryogenic cell suspensions of banana are usually established from immature male flowers and scalps. However, somatic embryogenesis in banana is constrained by limited choice of explants, restricted and often variable embryogenic response of cells and tissues *in vitro*, labour intensive and time consuming establishment of embryogenic cell suspensions, loss of embryogenic capability and high incidence of somaclonal variation associated with long term culture.

Embryo Rescue: This technique is considered crucial for breeding of bananas because only a few viable seeds normally result from sexual crosses involving edible cultivars. It is used to circumvent post-zygotic incompatibility and to facilitate the recovery of progeny from new parental combinations following sexual hybridization.

In banana, it is crucial to excise embryos before they reach maturity to recover the maximum number of hybrids. It is reported that in the wild banana *M. acuminata* cv Pisang Jeje (AA), for embryos at 95% and 100% maturity Murashige and Skoog medium lacking growth regulators was adequate for direct shoot regeneration, whereas less mature embryos had to be excised from parental tissues and regenerated indirectly through a callus stage on MS-based medium supplemented with 4.4 μM benzyl adenine and 2.8 μM indole acetic acid.

Somaclonal variation: Mutant and somaclonal variant banana plants exhibiting traits such as tolerance to aluminum, dwarfism and resistance to Panama and Sigatoka diseases have been released for commercial production. Tai-Chiao No.1, a variant from the banana cv. Pei-Chiao that resulted from multiplication of *in vitro* material, showed improved agronomic characteristics including resistance to *Fusarium* wilt Tropical Race 4. An improved somaclonal variant of Cavendish type banana was selected and released in 2007 as banana cultivar Tai-Chiao No. 5. It inherited the semi-dwarf status, resistance to Foc race 4 and moderate resistance to flower thrips (*Thrip hawaiiensis)* with higher yield potential and good eating quality. It is now well accepted in the local and Japanese market.

Observations on sucker derived clones from TC plants of Giant Cavendish led to the identification of a promising FOC wilt resistant GCTCV 218 clone which was later named as Formosana 4. The cultivar Novaria (AAA), derived from Grand Naine (AAA), shows early flowering The AAB cultivar Pisang Mutiara, tolerant to black leaf streak, was selected in Malaysia as a somaclonal variant of Pisang Rasthali.

Somatic hybridization: Somatic hybridization is the only way to generate banana hybrids between highly sterile cultivars, especially in the triploid Cavendish group. Protoplast fusion can accelerate and facilitate the crossing of bananas which is difficult to achieve by conventional breeding methods. Generally, electrofusion is the preferred procedure to fuse banana protoplasts and is the most efficient procedure to generate somatic hybrid plants. Matsumoto *et al.* (2002) generated pentaploid somatic hybrid cells resulting from the fusion of banana protoplasts.

Molecular markers: Most frequent use of molecular marker methods in Musa has been limited to germplasm characterization and diversity analysis. RFLP analyses of diverse germplasm have been exploited to study the taxonomy and phylogeny of *Musa* spp. and variation in the chloroplast genome within the genus. The use of RFLP and RAPD markers can increase the efficiency of identification of promising new banana genotypes by early detection of desirable genome combination. RAPD and PCR-RFLP markers specific for the A and B genomes have been identified, whereas RAPD has been used to distinguish diverse *Musa* germplasm.

AFLP is powerful tool in the molecular breeding of bananas relative to its ability to identify large numbers of polymorphic bands within germplasm without any prior knowledge of genome markers for parthenocarpy, dwarfism and apical dominance. Sequence tagged microsatellite site (STMS) markers based on VNTR polymorphism of microsatellites has also been developed, a technique based on PCR. STMS markers can be used to overcome time constraints when dealing with large number of genotypes because they are SSRs. The STMS and AFLP markers are more efficient than isozyme, RFLP and RAPD markers for genetic mapping and marker-assisted breeding in banana. Retrotransposon-derived markers have been used to identify and characterize banana cultivars and to classify banana genomes. All marker systems have different advantages and disadvantages with respect to specific applications. Consequently, it is important to develop the capacity to perform several assays on targeted germplasms to choose the most suitable systems/procedures for molecular breeding of bananas.

DNA fingerprinting has been useful to characterize *Musa* diversity. Thus, AFLP analysis of total DNA revealed that the *M. acuminata* complex of about 50 morphological subspecies can be ascribed to only three genetic subspecies (*burmannica*, *malaccensis* and *microcarpa*) while *M. balbisiana* was shown to contain two forms (Singapuri and I-63), suggesting that there might be at least three A genomes and two B genomes. Furthermore, it seems that A genome in the plantains came from the subspecies *microcarpa* while B genome came from I-63 form of *M. balbisiana*. Molecular markers are not only useful for the detection of introgression of genes but also useful to screen the tissue cultured banana plantlets at an early stage for various biotic and abiotic stress resistance. Many reports are available on the development of markers for biotic stresses especially for *Fusarium* wilt.

Genetic transformation: *Agrobacterium*-mediated DNA delivery resulted in low genetic transformation rates with the induction of chimeras. As a consequence, this technique is not commonly used for this crop compared with biolistics-mediated gene delivery. However, the synthesis of vaccines, antibodies, and other therapeutic proteins in transgenic bananas has several advantages as it eliminates costly and time consuming processing, such as extraction and purification.

Direct genetic transformation of *Musa* is likely to have a major impact on banana and plantain improvement. The first genetic modification by microprojectile bombardment was achieved using embryogenic cell suspensions from meristematic buds of the ABB cooking banana Bluggoe. The efficiency of one to four transgenic plants regenerated per bombardment plate made this technique practical. The first *Agrobacterium tumefaciens*-mediated transformation was achieved using meristems from tissue cultured plants of the AAA Cavendish clone Grande Naine. Transformation by electroporation has been achieved using protoplasts of the ABB Bluggoe and meristems from the AAB plantain Harton.

The International Institute of Tropical Agriculture (IITA), in a partnership with Uganda's National Agricultural Research Organisation (NARO) and other institutions have developed transgenic bananas constitutively expressing the hypersensitivity response assisting protein (Hrap) and the plant ferredoxin-like protein (Pflp) genes from sweet pepper (*Capsicum annuum*). These transgenic bananas have exhibited enhanced resistance to *X. campestris* pv. *musacearum*. Transgenic bananas did not show significant changes in plant morphology versus the non-transgenic banana cultivar. Most transgenic bananas had significantly higher host plant resistance than the non-transgenic control. Improved resistance to Sigatoka was obtained when banana was transformed with the endochitinase gene ThEn-42 from *Trichoderma harzianum* and the grape stilbene synthase (StSy) gene (Vishnevetsky *et al.*, 2011).

CRISPR/Cas9 system: A successful strategy to eliminate banana streak virus in plantains is reported in a study published in Communications Biology. This strategy, which uses CRISPR technology, could help improve plantain growth and productivity. Jaindra Tripathi and colleagues used the CRISPR/Cas9 system to inactivate viral DNA from the B genome of the false horn plantain Gonja Manjaya. Gonja Manjaya is a false horn plantain variety of the Musa genus commonly grown in East and Central Africa. The authors found that, when exposed to drought stress, 75% of the edited plants did not show any symptoms of banana streak virus compared to non-edited plants, which confirmed that the viral DNA was deactivated (Tripathi et al., 2019).

***In vitro* conservation:** *In vitro* conservation has been established and practiced worldwide to complement field and seed bank conservation. Presently, culture based technologies are the most suitable and preferred approaches to conserve and distribute banana germplasm. Two conservation methods are followed.

Short to medium term storage (usually for 2 to 10 months): It is done in defined culture media under controlled environmental conditions to reduce growth to limit growth by increasing the osmotic potential of the culture medium using sugars and sugar alcohols combined with reduced temperature. Medium-term storage of banana plantlets *in vitro* is best achieved at 16°C under low-intensity illumination. Cultured plant material may also exhibit somaclonal variation. Such variation may be incited by the intrinsic nature and genetic stability of the plant material and by external factors, including the culture system, particularly growth regulators in the medium, the environmental conditions, and duration of storage.

Long term storage by cryopreservation: Cryopreservation is considered to be the only valid alternative for long term conservation of *Musa* germplasm. It is done in liquid nitrogen at –196 ^{0}C where all metabolic processes are inactivated keeping the samples intact, e.g. the materials are frozen and preserved safely without alteration for an indefinite period of time. Cryopreservation complements other conservation approaches and offers an alternative or additional tool for germplasm collections. It is currently being exploited to overcome some of the serious limitations encountered in the traditional germplasm conservation methods such as field, seed and *in vitro* banks.

CITRUS

Origin: Southeast foothills of the Himalayas, in a region that includes the eastern area of Assam, northern Myanmar and western Yunnan (Wu *et al.*, 2018).

Taxonomy status:

Order: Sapindales

Family: Rutaceae

Subfamily: Aurantioideae

Tribe: Citreae

Subtribe: Citrinae

Botanical name: *Citrus* spp.

Chromosome number: n=9; 2x=2n=18; 3x=27 (Tahiti or Bears lime)

Genome Size: 380 Mb (sweet orange)

Cultivated citrus arose from natural interspecific hybridizations between four ancestors: citron (*Citrus medica* L.), mandarin (*C. reticulata* Blanco), pummelo (*C. maxima* (L.) Osb.) and wild papeda citrus (*C. micrantha* Wester).

Botany

Plants are small to medium sized, spreading, evergreen (deciduous in *Poncirus trifoliata*), trees with thorny shoots, growing to about 2–15m tall with most species being single-trunked having very hard wood but with a relatively thin (<5 mm) bark. The stem is green, with unifoliate, alternate leaves. Leaves are thick, leathery, dotted with transluscent oil cells. Leaf shape varies from ovate to lanceolate; varies from 4 to 10 cm in length and may possess more or less broadly winged petioles. Flowers are perfect or staminate, fragrant, borne solitary or in short cymes in the axils of the leaves or in small lateral or terminal cymes. The sepals are four to five lobed. The petals are five, scented, white or whitish purple with pink tinge on outer surface. The stamens are polyadelphous, cohering toward the bases in a few bundles and are usually four times as many as petals. The yellow, four-lobed anthers surround the pistil at or near the level of the stigma. The ovary is superior and is composed of 6–14 carpels joined to each other and to a central axis. The style is prominent but usually deciduous. The fruit is a hesperidium berry whose form and size vary from globose to oblong and oblate. The outer rind is known as flavedo (exocarp and endocarp). It is covered with tiny pockets containing aromatic oil. The albedo (mesocarp) covers the endocarp. The pulp (endocarp) is divided into 6–14 sections join to each other and to a central axis containing the juice vesicles (sacs). They are separated by thin septa. Seed content is variable among and within species. Seeds are few to many, monoembryonic or polyembryonic.

Floral Biology

Citrus species flower once in a year during early spring (Feb- March) while flower induction is initiated in early January in subtropical zones having cool winter as in north India. In the tropics and coastal areas, flowering takes place several times a year. Anthesis takes place between 9 AM to 12 PM. Dehiscence of anthers takes

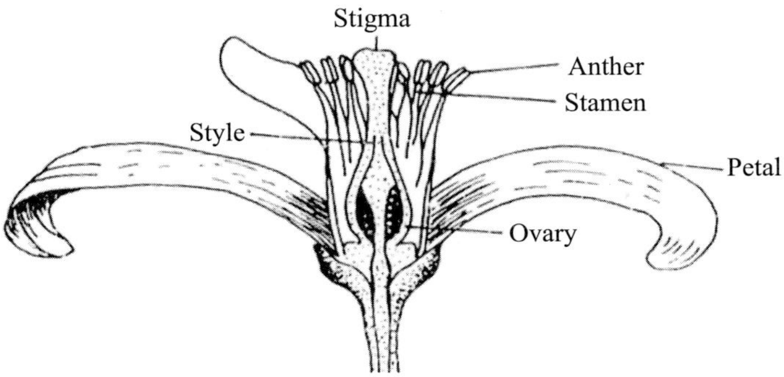

Fig. 2. L.S. of citrus flower

(Source: Hume, 1957)

place in morning hours. It occurs before anthesis in sweet ornage, sweet lime, etc. Normal anthers are bright yellow at maturity. Stigma is receptive for one to a few days before anthesis and for several days afterwards with a maximum on the day of anthesis. Pollen is sticky, adherent type, entomophilous. Pollination is by honeybees, mites and thrips. Self-pollination can occur because of nearness of anthers to stigma.

Breeding Objectives for Scion

(i) Varieties having dwarf tree size.

(ii) Varieties suitable for different agro-ecological regions.

(iii) Varieties bearing fruits of good size, shape, exterior appearance, rind colour, ease of peeling and high quality, good flavour and seedlessness.

(iv) Varieties with high juice and sugar content in fruit and good for processing.

(v) Expanding the harvesting period and enhancement of keeping quality.

(vi) Regular bearing.

(vii) Tolerance to biotic (citrus canker, *Alternaria* brown spot (*Alternaria alternata*), *Phytophthora*, HLB (greening disease), viruses (Tristeza, citrus variegated chlorosis) and fruit fly (*Ceratitis capitata)*) and abiotic stresses (freezing, waterlogging, salinity and drought).

Breeding Objectives for Rootstock

(i) Impart dwarfness in scion.

(ii) High degree of apomixis or nucellar embryony.

(iii) Impart high yield, superior fruit quality and size and other essential traits to the scion.

(iv) Compatibility with scion.

(v) Adaptatability to drought, cold, salinity, waterlogging and tolerance to iron chlorosis in calcareous soils.

(vi) Tolerance to citrus canker (*Xanthomonas axonopodis* pv. *citri,* Citrus tristeza virus (CTV), *Phytophthora* sp. and Huanglongbing (greening disease) caused by *Candidatus* Liberibacter *asiaticus,* nematodes (mainly *Tylenchulus semipenetrans* Cobb) and resistance to exocortis and tatter leaf virus.

Limitations in Breeding

(a) Long juvenile phase.
(b) Highly heterozygous.
(c) Inbreeding depression.
(d) Apomixis and nucellar polyembryony.
(e) Pollen and/or ovule sterility.
(f) Self or cross incompatibility.
(g) Little information on inheritance pattern.

Germplasm Resources

In US National Clonal Germplasm Repository for Citrus and Dates, 19 of >1300 accessions listed in the genus *Citrus* appear to be wild collections, 17 of the 83 *Poncirus* accessions are listed as donated or collected in Asia and most other genera in the Aurantioideae are represented by 6 or fewer accessions.

The Citrus germplasm collections being maintained are: 27 at Abu Dhabi Food Control Authority (ADFCA), Abu Dhabi, UAE; 1250 at Chinese Academy of Agricultural Science, Citrus Research Institute (CAAS-CRI), Beibei, China; 400 at National Citrus Breeding Centre, Huazhong Agricultural University (NCBC), Huazhong, Wuhan, China; 1718 at Institute of Fruit Tree Science, National Agricultural and Food Research Organization (IFTS-NARO), Shizuoka, Japan; 219 at Research Farm of Kinki University (RFKU), Nishimitani, Kinokawa-City, Japan; 447 at Plant Resources Centre (PRC), An-Khanh-Hoai Duc-Ha Noi, Vietnam and 37 at Royal Botanic Gardens & Domain Trust (RBGDT), Sydney, NWS, Australia.

In India, ICAR- Central Citrus Research Institute, Nagpur is the national repository for citrus. It has the largest collection of 614 accessions. The other centres are MPKV, Rahuri with 250 accessions, APHU, Triputi with 150 accessions, IIHR CHES, Chethali with 280 accessions, PAU RRS, Abohar with 150 accessions and CAU, Pasighat with 400 accessions.

Sources of Desirable Traits

(a) For resistance to biotic stress

Root rot/foot rot (Phytophthora): *Severnia buxifolia*, Rangpur lime, Marmalade orange or Sour orange, Troyer citrange, Carrizo citrange, Trifoliate orange, Volkamer lemon, *Citrus australasica* syn. *Microcitrus australasica* and *C. amblycarpa.*

Nematode: Marmalade orange, Sweet lime, Trifoliate orange, Troyer citrange, Carrizo citrange, Karna Khatta, Nasnaran, *Microcitrus* spp., *Severnia bauxifolia, S. disticha*, CRH-3, CRH-5, CRH-41 (IIHR, Bengaluru).

Tristeza virus (CTV): Rangpur lime, Marmalade orange, Rough lemon, Cleopatra mandarin, Trifoliate orange, Troyer citrange, Carrizo citrange, Nasnaran, Sunki mandarin, *C. amblycarpa* and Volkamer Lemon.

Xyloporosis: Rough lemon, Trifoliate orange, Sweet orange, Sour orange, Cleopatra mandarin, Grapefruit, Troyer citrange.

Exocortis: Rough lemon, Sweet orange, Sour orange, Cleopatra mandarin, grapefruit, *Fortunella, Citrus glauca* syn. *Eremocitrus glauca, C. depressa.*

Gummosis: Trifoliate orange, Sour orange, Rangpur lime, Cleopatra mandarin, Troyer citrange.

Citrus canker: *C. junos,* Satsuma and Dancy.

Greening disease or Huanglongbing (HLB): Amilbed, Gajanima, Florida rough lemon, Karna Khatta, Desert lime (*Citrus glauca*), Australian round lime, Finger lime (*Citrus australasica*) and *Balsamocitrus daweii.*

(b) For tolerance/resistance to abiotic stress

Cold: Trifoliate orange, Swingle Citrumelos, *C. junos, C. ichangensis,* Jiouyuezao (*C. reticulata*), Jiouyuezao (*C. reticulata*) Changsha mandarin Satsuma mandarin, Owari Satsuma, Robinson, Page, Nova, *Citrus glauca* syn. *Eremocitrus glauca,* Kumquat.

Iron chlorosis: Rough lemon (*C. jambhiri), C. macrophylla*, Volkamer lemon and *C. amblycarpa*

Drought: Rangpur lime, Marmalade orange, Rough lemon, Sour orange and *Citrus glauca* syn. *Eremocitrus glauca.*

Salinity: *Severinia buxifolia,* Rangpur lime, Marmalade orange, Sour orange, Nasnaran, Cleopatra mandarin, Sunki mandarin, *Citrus glauca* syn. *Eremocitrus glauca.*

Boron toxicity: *Severinia buxifolia.*

Aluminium toxicity: Cleopatra mandarin, Rough lemon.

(c) For growth and fruit yield and quality traits

Dwarf tree stature: Flying Dragon trifoliate orange, *Citrus indica* (Indian wild orange).

Female sterility: Mukaku Kishu (*Citrus kinokuni*).

High yield: Rangpur lime, Marmalade orange, Rough lemon, Cleopatra mandarin, Sour orange, Sweet orange, Karna Khatta.

Good quality: Sour orange, Troyer Citrange, Carrizo citrange, Sweet orange, Nasnaran.

Zygotic citrus species: Clementine mandarin, Temple mandarin and *C. maxima.*

Desired traits in relation to its gene pattern:

Bitter component (flavanoids neohesperisidosire): Two dominant multiple genes.

Trifoliate leaf/anthocyanin colouration in leaf: Single dominant gene.

Nucellar embryony: One or two dominant genes.

Seedlessness: Recessive oligogenic character.

Fruit peeling: Crosses between sweet orange and mandarin yielded "loose skin" progeny at high frequency of 87%.

Thorniness, pubescence and presence of oil glands: Dominant genes.

Cold tolerance: Multigenes.

Tristeza (CTV): Dominant single gene.

Pre Selection Indices

Since stomata density in leaves and bark percentage in relation to thickness of twig (BPHT) had significant correlation with the height and crown volume of grafted trees, these can be used as the main indices for dwarfing effect. The vigorous rough lemon (*C. jambhiri*) rootstock possessed lower proportions of bark (phloem) in the stems and roots and larger vessel elements in xylem compared to that in the least vigorous flying dragon *(P. trifoliata)* rootstock.

Citrus FT (CiFT) cDNA, which promoted transition from vegetative to reproductive phase in *Arabidopsis thaliana* when constitutively expressed, was introduced into the transgenic trifoliate orange plants, showed early flowering, fruiting and characteristic morphological changes. Most of the transgenic flowers developed on leafy inflorescences, apparently in place of thorns. However, wild type adult trifoliate orange usually develops solitary flowers in the axils of leaves. The transgenic lines showed variation in phenotypes, such as time to first flowering and tree shape. In F_1 progeny obtained by crossing Kiyomi tangor (*C. unshiu x C. sinensis*) with the pollen of one transgenic line, very early flowering was observed immediately after germination. The transgene segregated in a Mendelian fashion in the F_1 progeny, with complete co-segregation of the transgene and early flowering phenotype. These results showed that constitutive expression of CiFT can reduce generation time in trifoliate orange (Endo *et al*., 2005).

Breeding Methods

1. Introduction: It is usually attempted merely to enrich the existing germplasm or when it lacks some desirable characters. The sweet orange cultivars Washington Navel, Valencia, Cutter Valencia, Jaffa, Blood Red, Moro, Tarocco and Sanguinello; mandarin cultivars kinnow, Cleopatra, Daisy, Fortune, Murcott and Clementine; Pummelo US -145; grapefruit cultivars Marsh Seedless, Duncan, Foster, Thompson, Hudson, Flame, Star Ruby, Ray Ruby, Rio Red and lemon cvs. Ponderosa, Lisbon, Eureka and Villafranca have been introduced from USA and other countries.

2. Selection: All commonly grown citrus cultivars have originated as a result of selection of chance seedlings or branch mutations found by observant growers. Promising selections in sweet orange are Yuvraj Blood Red, a clonal selection from Blood Red, Phule Mosambi (at Rahuri), Pusa Sharad and Pusa Round (at IARI, New Delhi). Valencia clones have been selected at University of California, Riverside.

These are Pehrson #3 Valencia Orange and Pehrson #4 Valencia Orange (VI 749 and VI 750). New selections made from Tarocco and Moro sweet oranges by the Italian program show differences in ripening time, anthocyanin content in pulp and peel and fruit shape, size and firmness. These have been registered.

In Nagpur mandarin, National Research Centre for Citrus, Nagpur, India, selected 7 promising clones (N_4, N_2, N_{28}, N_{34}, N_{38}, N_{43} and N_{51}) of mandarin. Of these, clone N4 is a seedless cultivar, N2 is early maturing clone and Mudkhed Seedless is a bud mutant.

In Satsuma mandarin, varieties "Mihu" and "Okitsu" are nucellar selections from "Miyagawa". In Italy, the Istituto Sperimentale per l Agrumicoltura of Acireale (ISA) obtained several nucellar selections from different citrus species or clones. In mandarin, the most diffused (spread) originated from "Mandarino tardivo di Ciaculli" (nuc.60-22A-2, nuc.60-22A-7, 19 CN. L, 18 CN. L, 3DN6 N L) and "Mandarino Avana apireno" (nuc. 62- Ap-9, nuc.-62-Ap-14).

Hickson Mandarin: It is reported to have originated on Ellendale tangor near the town of Roma, Queensland, Australia, as a sporting limb.

Alborea: It is a seedless cultivar obtained from an open pollinated Fortune mandarin.

Seedless Kinnow: This has been developed by using selection by Niaz Ahmad Chaudhry in Pakistan.

NRCC, Nagpur selected five pummelo types: NRCC Pummelo 1, NRCC Pummelo 2, NRCC Pummelo 3, NRCC Pummelo 4 and NRCC Pummelo 5. NRCC Grapefruit-6 have also released from local material.

Selection made in Mexican or West Indian lime are Jai Devi (at FRS, Periakulam), Balaji (at Tirupati), PDKV (at Akola), Phule Sharbati and Sai Sharbati at Rahuri, Chakradhar, Pramalini and Vikram at MAU, Parbhani, PKM-1 at Periyakulam, Pusa Abhinav and Pusa Udit at IARI, New Delhi, NRCC Acid Lime-7 and NRCC Acid Lime-7 and ARL-l at CCRI, Nagpur. PAU Selection is from Baramasi lemon.

Flame: Originated from open pollinated seedlings of Henderson grapefruit.

3. Hybridization:

Sweet orange hybrids: US 7650, US 7651, and US 7652; Ambersweet × FF-1-30-52 (Wilking × Valencia) developed by C. J. Hearn at USDA-ARS, Orlando, Florida.

US SunDragon: Crossing between FF-5-14-96 and FF-1-11-7, developed by J. Chaparro at USDA-ARS, Leesburg, Florida. It is sweet orange like hybrid having *Poncirus* in its pedigree, with remarkable tolerance to Huanglongbing.

Mandarin Hybrids

Tangor: It is a hybrid between mandarin orange (*Citrus reticulata*) and sweet orange (*Citrus sinensis*), e.g., Natural hybrids Clementine mandarin (mediterranean area) Ponkan. Ellendale discovered in the nineteenth century in Australia.

Murcott (Honey): It is a mandarin and sweet orange hybrid. The Murcott arose out of Walter Tennyson Swingle's pioneer attempts to produce novel citrus hybrids. Its seed parent has been identified as the King tangelo; the pollen parent remains to be identified.

Imperial Mandarin: It is a chance hybrid of Mediterranean or Willowleaf and another mandarin, possibly Emperor, in approximately 1890.

Kinnow: It is a hybrid between King orange and Willow leaf mandarin developed in 1935 by Howard B. Frost at University of California Citrus Experiment Station.

Afourer or Nadorcott mandarin: Probably a hybrid between Murcott and unknown pollen parents, discovered in Morocco.

Ortanique: It was **o**riginally found in Jamaica. The name comes from the words "orange", tangerine" and "unique". It is a late maturing tangor (*Citrus sinensis* × *C. reticulata*) with high carotenoid content.

Temple: *Citrus sinensis* × *C. reticulata*. Originated in Jamaica.

Umatilla: Hybrid of satsuma mandarin and Ruby orange.

Setoka: It is a hybrid of Murcott and Kuchinotsu No.37, from Japan.

Daisy: Fortune x Fremont, developed by the US Department of Agriculture, California, USA.

Sugar Belle: Clementine x Honeybelle. It is tolerant to HLB.

Dekopon or Suma mandarin: It is a hybrid between Kiyomi and Ponkan (Nakano no.3), developed in Japan in 1972. It is a seedless and sweet variety of mandarin orange.

Bingo: It arose from a seedling tree selected among a family of hybrids from a cross of the seed parent LB7-11 (Clementine mandarin × Valencia sweet orange) and pollen parent Seedless Kishu. It produces annual crops of well coloured fruits of excellent eating quality that are essentially seedless, very easy to peel and mature early in the season.

Roe Tangerine: It is a produce of W. G. Roe & Sons from private breeding program and is very typical of a traditional Florida tangerine in appearance and flavour.

RBB7-34: It is a navel-like seedless orange/mandarin hybrid with a round shape and vastly superior in colour. It has higher brix and higher acid than in navel orange and has very good flavour.

C4-15-19: It is early maturing (Oct – Nov), seedless tangerine having good exterior and internal colour.

UF 950: It is easy to peel, seedless, convenient to eat having crisp texture, better color and flavour than clementine.

Satsuma tangors (*Citrus unshiu* Marcovitch):

- Iyokan (Miyagawa × unknown sweet orange).
- Miyauchi Iyo (early ripening Iyo mutant).
- Othani Iyo (late ripening Miyauchi mutant).
- Kiyomi (Miyagawa × Trovita navel orange).

- Seto (Sugiyama Unshiu × Trovita navel orange).
- Reikou ((Kiyomi x Encore) x (No. 5 x Marcott)).

Tangelo: It is a hybrid of mandarin orange or tangerine) and grapefruit e.g. Orlando, Minneola, K- Early, Nova,Thornton and Ugli.

Jamaican tangelo or Ugli: A natural hybrid is thought to be a grapefruit x mandarin. Large knobbly fruit, the variety was discovered in Jamaica in 1914.

Chironja orange: *Citrus paradisi* x *Citrus sinensis*. An believed to have originated in Puerto Rico.

Osceola: Clementine mandarin x Orlando tangelo made by Gardner and Bellows in 1942 at the United States Department of Agriculture Laboratory in Orlando.

Robinson: Cross between Clementine mandarin and Orlando tangelo. This is very early maturing, rather large fruited variety, a sister to Lee and Osceola, all three resulting from a Clementine mandarin-Orlando tangelo cross made by Gardner and Bellows of the US Department of Agriculture in Florida in 1942 and released in 1959.

Nova: Clementine tangerine crossed with Orlando tangelo released in 1964 by the US Department of Agriculture.

Minneolo: It is hybrid between *Citrus paradisi* cv. Duncan and Dancy mandarin.

Orlando: It is a sibling of Minneola tangelo having Duncan grapefruit and Dancy mandarin as its parent.

Primosole: *C. unshiu* Miho (Carvalhais mandarin) x *C. reticulata* (Miho wase), developed in 1986 in Italy.

Desiderio: *C. unshiu* Miho x *C. clementina*, developed in 1986 in Italy.

Bellezza: *C. unshiu* Okitsu x *C. reticulata*, developed in 1986 in Italy.

Sirio: *C. unshiu* x *C. clementina*, developed in 1986 in Italy.

Simeto: *C. unshiu Miho* x *C. deliciosa* Avana apireno, released in 1993 in Italy.

Cami: (*C. clementina* Comune x *C. deliciosa* Avana) x Mapo tangelo, released in 1992 in Italy.

Tacle: Triploid hybrid mandarin (*C. clementina* Monreal x *C. sinensis* Tarocco 4x), released in 1999 in Italy.

Clara: *C. clementina* Monreal x *C. sinensis* Tarocco 4x, released in 1994 in Italy.

Gold nugget: Wilking (*C. deliciosa* Willow leaf x *C. nobilis* King) x Kincy (King x Dancy tangerine), released in 2000 in University of California, Riverside, California.

Winola: Wilking (*C. deliciosa* Willow leaf x *C. nobilis* King) x Minneola tangelo, released in 1992 in Israel.

Amaka: Mandarin hybrid (Kiyomi tangor (Miyagawa satsuma x Trovita orange) x Encore (King x Willow leaf), released in 1996 in Japan.

Youkou: Kiyomi tangor (Miyagawa satsuma x Trovita orange) x *C. reticulata* Nakano 3 ponkan, released in 1995 in Japan.

Afourer: Murcott tangor x unknown pollen parent, released in 1997 in Morocco.

Hayaka: Mandarin hybrid (Imamura satsuma x *C. reticulata* Nakano 3), released in 1991 in Japan.

Ariake: Tangor (*C. sinensis* Seike navel x *C. clementina*), released in 1992 in Japan.

Miho-core: Mandarin hybrid (Miho-wase satsuma x Encore (*C. nobilis* King x *C. deliciosa* Willowleaf), released in 1994 in Japan.

Nankou: Miho-wase satsuma x *C. clementina*, released in 1991 in Japan.

Amakusa: Mandarin hybrid, (Kiyomi tangor (Miyagawa satsuma x Trovita orange) x Encore (King x Willowleaf), released in 1993 in Japan.

Kiyomi: It is a tangor hybrid and has a parentage of Miyagawa-wase (Satsuma mandarin) crossed with a Trovita orange at the Okitsu branch of Fruit Research station, Japan.

Seihou: It originated from the cross of "Kiyomi" tangor x Minneola tangelo and was released by the Kuchinotsu Branch, Fruit Tree Research Station, Japan in 1988.

Hareyaka: Encore (*C. nobilis* King x *C. deliciosa* Willowleaf) x *C. reticulata* Nakano 3 ponkan, released in 1994 in Japan.

Tsunokaori: Kiyomi tangor x Okitsu Wase Satsuma, released in 1991 in Japan.

Sunset: *C. reticulata* Imperial x Ellendale tangor, released in 1991 in Australia.

Rishon: Temple tangor x *C. reticulata* Mikhal, released in 1993 in Israel.

Orah: Temple tangor x *C. reticulata* Dancy, released in 1992 in Israel.

Fairchild: A hybrid of Clementine mandarin and Orlando tangelo.

Fremont: Hybrid of Clementine mandarin and Orlando tangelo.

Fortune: Hybrid of Clementine mandarin and Orlando tangelo.

Encore: Originated from a cross between King and Willowleaf mandarins.

Page: Clementine mandarin x Minneolo. The cultivar was released in 1963 from a cross made in 1942 by Gardner and Bellows of the United States Department of Agriculture facility in Orlando.

Lee: Arising from a cross Clementine mandarin with Orlando tangelo made by Gardner and Bellows in 1942 at the United States Department of Agriculture Laboratory, Orlando, USA.

Ambersweet: (Clementine × Orlando) × Sweet Orange, developed by Dr. C. J. Hearn of the United States Department of Agriculture Horticultural Research Station, Orlando.

Sunburst: Robinson x Osceolo. This cross was made in 1961 and was released in 1979 by C. J. Hearn of the United States Department of Agriculture in Orlando.

Albir: Seedless mandarin was obtained from a cross between Fortune and Kara mandarins by Centro de Protección Vegetal y Biotecnología, Instituto Valenciano de Investigaciones Agrarias (IVIA), Spain.

Fallglo: Bower x Temple, made by P. C. Reese in 1962.

USDA 88-2: Cross between Lee and Nova mandarins, is very juicy and easy-to-peel, early season, seedless mandarin with small to medium size fruit.

USDA 88-3: Cross between Robinson and Lee mandarins, an early season mandarin, but the fruit is very seedy.

Tacle and Clara: Cross between Monreal Clementine and Tarocco orange.

Daisy: Hybrid between Fortune and Fremont mandarins.

Camel: A seedless hybrid of Nules Clementine and Willowleaf mandarin.

Shasta Gold1orTDE2: (Temple tangor x 4n Dancy mandarin) and Encore mandarin, A late maturing hybrid.

Tahoe Gold1 or TDE3: (Temple tangor x 4n Dancy mandarin) and Encore mandarin Seedless, mid-season maturing.

Yosemite Gold1or TDE4: (Temple tangor x 4n Dancy mandarin) and Encore mandarin.

Winola: A spontaneous triploid hybrid selected from a population of diploid hybrids between Wilking mandarin and Minneola tangelo.

Primosole, Simeto, Desiderio, Bellezza, Sirio and Cami: These are some other seedless hybrids.

Gold Nugget mandarin: Wilking mandarin x Kincy mandarin, fruit is sweet and seedless.

Pixie: It is result of open pollination of Kincy (King x Dancy).

Encore mandarin: It is a cross between King and Willow Leaf mandarin.

C54-4-4: Cross of Clementine X Murcott, made at the USDA Horticultural Research Laboratory in Orlando, Florida, by Dr Phil Reece.

Marathon mandarin: Fruit is seedless and easy to peel. Its name is because of its exceptional ability to hold fruit on the tree for long.

950: It is a clementine-sized easy to peel seedless mandarin from UF.

Tango: It is a low-seeded mandarin that has excellent internal colour. The fruit does not degreen and therefore must achieve maximum colour in the field.

US Superna: Lee × Nova, developed by C.J. Hearn at USDA-ARS, Leesburg, FL.

Mandared: A late, seedless, red flesh mandarin variety obtained by crossing Nules clementines (female parent, 2x) and Tarocco oranges (male parent, 4x). It is the patented result of research carried out by Giuseppe Reforgiato Recupero, Giuseppe Russo and Santo Recupero for the CRA-ACM of Acireale, Italy.

UF Glow: It is a seedless, easy peeling, early mandarin from UF with a mild flavour. It is self fertile and cold tolerant. Trees appear to do better than HLB.

Hybrids of Pummelo

Melogold or **Melogold grapefruit**: *C. grandis* Osbeck × *C. paradisi* Macf.

Oroblanco graprfruit or Sweetie: *C. grandis* Osbeck × *C. paradisi* Macf. It is a sweet seedless hybrid similar to grapefruit.

UF Citrus Selection 914: It is a grapefruit-sized grapefruit/pumelo hybrid with great internal and external colour.

Valentine Pummelo Hybrid (VI 597): Siamese Sweet pummelo x (Ruby blood orange x Dancy mandarin). It was selected by Drs. Soost and Cameron in 1986.

Hybrids with Australian limes:

a. **Australian Desert Lime (*Citrus glauca*** (Lindl.) Burkill syn. *Eremocitrus glauca* (Lindl.) Swingle.

   Citrus eremorange: Natural hybrid between *Citrus glauca* syn. *Eremocitrus glauca* and *Citrus sinensis.*

   Citrus eremolemon: Hybrid between *Citrus glauca* syn. *Eremocitrus glauca* and *Citrus limon* cv. Meyer lemon.

   Citrus eremoradia: Hybrid with *Citrus glauca* syn. *Eremocitrus glauca* x *Citrus aurantium.*

b. **Finger lime or caviar (***Citrus australasica* F. Muell. syn., *Microcitrus australasica* (F.Muell.) Swingle.

CSIRO citrus breeder Dr Steve Sykes took on the challenge of selecting and crossing native citrus with modern cultivars. The result is three high-yielding citrus varieties with larger than normal fruit, consistent yields and the unique flavour and texture of Australian native limes.

Australian Blood: from a cross between an acid mandarin and a native finger lime. It has blood-red rind, flesh and juice.

Australian Sunrise: It is a hybrid selection from a cross between a calamondin (mandarin crossed with cumquat) and a native finger lime. It produces a pear-shaped, orange fruit that makes an excellent marmalade.

Australian Outback (Australian Desert): is a variety selected and developed from a collection of different native desert lime trees. It produces small green, juicy fruits which ripen at Christmas time. Its fruits can be used for preparing sauces.

Hybrids of Papeda

Sudachi Ichandarin (VI 693): Hybrid of a papeda and a mandarin, Sudachiarose as a chance seedling in the Tokushima Prefecture of Japan, on Shikoku Island.

Hybrids of lemon

Lemonange: Lemon x Grapefruit.

Lemonime: Lemon x Lime. e.g., Perrine.

Hybrids of lime

Lemonime: Lime x Lemon.

Limequat: Key lime x Kumquat. e.g., Lakeland

Hybrids of citron

Meyer lemon: (*Citrus* × *meyeri*), cross between a citron and a mandarin/pomelo hybrid distinct from the common or bitter oranges.

Intergeneric hybrids

(a) **With *Poncirus trifoliata:*** No edible fruit producing seedling was obtained among the progeny consisting of the numerous inter-generic hybrids, but these were selected for use as rootstocks. The hybrids developed with *Poncirus trifoliata* are listed below.

- **Citrange:** Sweet orange × *P. trifoliata.* cvs. Troyer, Carrizo, C-35, Bitters C-22 Citrange (VI 792), Carpenter C-54 Citrange.
- **Citradia:** Sour orange x *P. trifoliata.*
- **Cicitrange:** Back cross hybrid between Citrange and *Poncirus trifoliata* x *C. paradisi.*
- **Citrandarin:** *P. trifoliata* x *Citrus reticulata.*
- **Citrumelos:** Grapefruit × *P. trifoliata.* cvs. Swingle, Sacaton, 4475.
- **Citrangequat**: It is a trigeneric hybrid of a citrange and a kumquat, developed by Dr. Walter Swingle at Eustis, Florida in 1909.
- **Citrandarin**: *Citrus reticulata* x *Poncirus trifoliata.*
- **Citremon**: An inter-generic hybrid between *Poncirus* and Meyer Lemon. This is one of the hardier citrus types.
- **US-852**: [Changsha mandarin (*C. reticulata*) x English Large Flowered Trifoliate Orange (*Poncirus trifoliata*)], resistant to citrus nematode and good tolerance to calcareous soils and flooding.
- **US-812**: (Sunki mandarin x Benecke trifoliate orange), which makes healthy trees, excellent fruit quality and are very productive.
- **F&A 5**: (Cleopatra mandarin x *P. trifoliata*) is resistant to nematode and has good tolerance to calcareous soils and flooding.
- **F&A 13**: (Cleopatra mandarin x *P. trifoliata*) is susceptible to nematode but is highly tolerant to salinity and flooding.
- **F&A 418**: (Troyer citrange x common mandarin) is dwarfing rootstock with good tolerance to calcareous soils, susceptible to nematode and induces big fruits on the scion.
- **F&A 517**: (King mandarin x *P. trifoliata*) is also a dwarfing rootstock and has good tolerance to calcareous soils and salinity.

(b) With *Fortunella:*

- **Procimequat**: *Citrus aurantifolia* Mexican x *Fortunella japonica* x *Fortunella hindsii.*
- **Limequat**: It is a citrofortunella hybrid that is the result of a cross between key lime and kumquat, hybridized by Walter Tennyson Swingle in 1909.
- **Orangequat**: It is a cross between Satsuma mandarin and Meiwa kumquat. e.g. Nippon orangequat.
- **Citrangequat:** Kumquat x citrange.
- **Calamonsi (calamondin):** Cross between mandarin and *Fortunella.*

(c) **Trigeneric hybrids:**

West Indian lime or Key lime: Hybrid between Citron (*Citrus medica*) x pummelo (*Citrus grandis*) and a *citrus australasica* species, *Citrus micrantha.*

Rootstock Breeding: Many of the newer rootstock cultivars are of partially trifoliate origin, thereby inheriting some degree of tolerance to *Phytophthora*. To *Phytophthora/Diaprepes* root weevil complex, US-802, US-897, US-942, and UFR-1 are more tolerant in comparison with other rootstocks. The trees grown on these rootstock cultivars produce good yields under high HLB pressure and exhibit lower fruit drop. In 2015, five rootstocks with improved tolerance to HLB were released by the USDA: US-1279, US-1281, US-1282, US-1283, and US-1284. All the five are hybrids of mandarin and trifoliate orange, produce medium sized trees, and appear adapted to Florida's flatwood soils. In 2018, the USDA released three rootstocks, US SuperSour 1 (SS1), US SuperSour 2 (SS2) and US SuperSour 3 (SS3).

4. Polyploidy: Seedlessness is one of the most important characteristics for citrus in fresh fruit market, because consumers do not accept seedy fruits. Parthenocarpy is an essential trait for seedless fruit production and this is present in citrus germplasm. Triploid plants have very low fertility and are generally sterile and do not produce seeds by cross pollination (Frost and Soost, 1968). Consequently, triploid breeding is becoming the most important strategy to produce seedless cultivars. Production of triploid hybrids is the most promising approach to obtain cultivars that do not produce seeds even with substantial cross pollination. Recovery of sexual triploid hybrids (3x = 27) has been reported from 2n x 4n, 4n x 2n and 2n x 2n crosses. When tetraploid was used as pistillate parent, 85% triploid progeny were recovered (Aleza *et al.,* 2012a,b). Most important triploid cultivars are Gold Nugget, Tacle, Clara, Winnola mandarin, Oroblanco, Melogold or Melogold grapefruit. The Hong Kong wild Kumquat and *Fortunella hindsii* are reported as tetraploids. The tetraploids are generally compact, slow in growth and less fruitful and bear fruits of thicker rind, larger oil glands and lesser juice than the diploids.

Triploids are more vigorous than tetraploids. Triploid mandarin cultivars Shasta Gold®, Tahoe Gold® and Yosemite Gold® were released in USA. The high triploid recovery arises from normal sexual fertilization of the diploid female gamete with a haploid male gamete. Shasta Gold1 or TDE2 (late maturing), Tahoe Gold1 or TDE3 (mid-season) and Yosemite Gold1 or TDE4 (mid-late season) are triploid mandarin hybrids from tetraploid female parent (Temple tangor x 4n Dancy mandarin) and diploid male parent Encore mandarin that combine large fruit size, attractive deep orange rind colour, rich fruit flavour and the virtual absence of seeds. Unlike 4n x 2n crosses, 2n x 4n crosses gave rise to only 9% triploid hybrids. Tacle and Clara Seedless triploid mandarins were obtained by crossing diploid female parent Monreal Clementine and tetraploid male parent Tarocco orange, while Camel mandarin was a triploid hybrid from diploid female parent Nules Clementine and tetraploid male parent Willow leaf mandarin. Reale isolated from a cross of Monreal Clementine and *Fortunella hindisii* (4x) is an interesting ornamental potted tree which is everblooming and early fruiting.

5. Mutation: In citrus, induced mutation breeding has been done since 1935, when the first report on the use of irradiation was published. In sweet orange, cultivar Washington Navel gave At Wood, Fisher, Newhall, Naveline and Navelate mutants. Low seeded selection Murcott, called Mor mandarin, was produced by budwood mutation, and the seedless mandarin Tango was produced by irradiation of budwood of Afourer (Nadorcott). Owari cultivars gave mutant Clausellina. Roose and Williams developed low seeded "KinnowLS" cv. from irradiated bud at Riverside, Calif. USA. Similarly PAU Kinnow-1 was developed through mutation by Dr. H.S.Rattanpal and workers at PAU, Ludhiana, India.

Fairchild LS Mandarin (VI 762): It is a selection of Fairchild mandarin developed by irradiation at University of California, Riverside. The distinctive trait of Fairchild LS is less seeds.

Tango Mandarin (VI 765): It is a selection of W. Murcott mandarin developed by irradiation at University of California, Riverside. Fruits of Tango are similar to W. Murcott in all characteristics including appearance, quality and production with the exception of seed number.

US Early Pride: It is a very low seeded selection by induced mutation of Fallglo.

Satsuma and Clementine originated either as bud sport or as apomictic seedling mutants.The mutants of the satsuma are Ueno and Miyagawa, the mandarin varieties 'Tardivo di Ciaculli' and 'Avana seedless' and the clementine varieties 'Arrufatina', 'Clemenpons', 'Hernadina', 'Marisol', 'Nour', 'Nules', 'Oronules', 'Comune ISA', 'Fedele', 'Spinoso', 'Rubino' and 'Tardivo.'

Orri variety was obtained by Vardi *et al.* (2003) from irradiated budwood of Orah mandarin. It produces zero or few seeds. Tango mandarin was recovered by Roose and Williams (2007) from irradiated budwood of Nadorcott and it also produces very few seeds.

Other examples of low seed varieties obtained by irradiation are Mor and Murina derived from Murcott tangor and Nulessin and Nero derived from the Clemenules clementine variety. Budwoods of the clementine "Monreal", treated by gamma rays, produced several types of mutants. Among these almost seedless varieties "Monreal verde" and "Monreal rosso" grow well even in the areas where the clementine "Comune" grows with difficulty.

Similar programmes were carried out in China at the Cash Crop Station, Xiangxi, where "Fuyu 28" selection was obtained from irradiated (gamma rays) buds of

"Ponggan". Guandong Fruit Research Institute identified some promising seedless selections which were released as "Huaqing Seedless" and "Huaqing Few Seeded" varieties.

Grapefruit Cultivar Walters (white fleshed) gave budsport Foster (pink fleshed). Thompson (light pink flesh) gave budsport Redblush or Ruby. The grapefruit cultivars Rio Red, Foster, Hudson, Red Blush, Flame and Star Ruby were developed by treatments of mutagenic chemicals or ionising rays. A low seeded grapefruit like hybrid (USDA 77-19) was developed by USDA citrus breeding program by irradiating the hybrid USDA 75-8 selected from a population of Pearl tangelo x grapefruit.

Irradiated buds of Kutdiken lemon gave rise to plants showing variations in branching habit, thorniness and flowering, fruit maturity time.

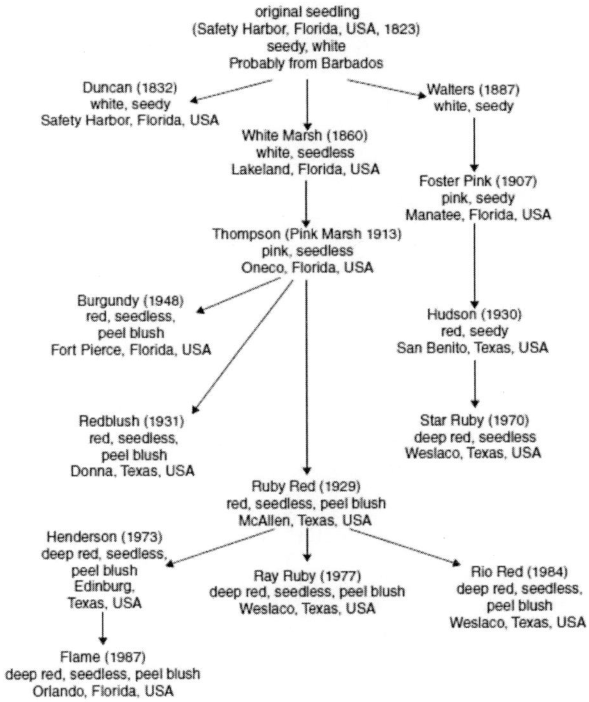

Fig. 3. Origin of the important grapefruit cultivars.

(Source: Albrigo *et. al.*,___2019))

6. Biotechnology:

Regeneration and micropropagation

The direction for genetic improvement of citrus is greatly impacted by the advances in plant cell and tissue culture. Plant regeneration systems are potentially useful for obtaining genetic change through cell transformation or mutagenesis. The amenability of citrus to be regenerated via organogenesis and somatic embryogenesis is the fundamental basis that makes possible much of the potential for these genetic advances.

It is possible to create new genotypes in citrus *in vitro* by organogenesis from explants of cotyledons, epicotyls, buds, roots and juice vesicles. Organogenesis has been induced *in vitro* from explants such as shoot meristems of seedling and mature trees, nodal explants, stem internode, leaf sections, and root tissues. *In vitro* culture of excised, fully developed embryos, early heart shaped embryos, globular embryos within undeveloped ovules of mature fruits, and immature embryos has also been used to recover plants. *In vitro* seedling explants have also been used for multiple shoot formation and/or regeneration. Regeneration has also been achieved by culturing thin sections of mature stem segments.

In case of polyembryonic citrus, it is possible to regenerate plants by somatic embryogenesis from the nucellar tissue through *in vitro* culture of undeveloped ovules extracted from ripe fruit embryonic callus, embryos and plantlets from polyembryonic citrus species and varteties producing seedless genotypes can be obtained.

Somatic embryogenesis is particularly attractive in citrus because many genotypes have the capacity for nucellar embryony. Somatic embryogenesis has been induced directly in cultured nucelli and undeveloped ovules or indirectly via callus formation. Embryogenesis has also been induced from endosperm derived callus, juice vesicles, anthers, styles and pistil thin cell layers. The rootstocks Citrange, Troyer, Carrizo and Flying Dragon are usually multiplied from axillary buds.

Somaclonal variation

Sweet orange has been well suited for studies of somaclonal variation due to its reproductive biology (nucellar embryony) and excellent performance in tissue cultures. Somaclones of Hamlin and Valencia have been identified. Somaclones of Valencia and Hamlin with elevated soluble solids and improved colour and early and late maturity having fresh market potential are being tried in USA.

Valquarius™: It is a somaclone of cv. Valencia. It is early maturing and seedless, suitable for fresh and processing market. Similar Valencia clones N7-3, B9-65, N13-32, OLL-8 and OLL-4, bearing large fruits precociously have been released by IFAS, Florida.

OLL-8: It is precocious bearing and produces high yields of high quality fruit with exceptional flavour and juice colour. It also has potential as a fresh fruit cultivar making it a dual use cultivar.

Anther culture

Anther culture may provide the ability to recover plants having reduced ploidy level which may facilitate the recovery of useful recessive alleles or mutations by haploid plant regeneration. Citrus and *Poncirus* anthers have been cultured in attempts to produce haploid plants. Colchicine treatment can be used for doubling of chromosome number to yield homozygous diploids. Such plants would facilitate considerable understanding of citrus genetics and would offer much greater control over phenotypic expression. However, anthers from sour orange, sweet orange, and lime yielded only diploid regenerants. Nuclear fusions that were sometimes observed in the routes of microspore development were suggested to be the cause of diploidy in regenerants.

The production of diploid plants by culturing anthers of tetraploid somatic hybrids may also provide breeders with greater access to these unique genetic combinations. In *C. natsudaidai*, haploid seedlings have been obtained by irradiation. Haploid plants and embryogenic calli of Clementine have been obtained after *in situ* parthenogenesis induced by irradiated pollen. Haploid plantlet regeneration through gynogenesis of Clementine has been induced by *in vitro* pollination with pollen from a triploid plant. Tri-haploids have been obtained by regeneration of anthers of *C. clementina.*

Somatic hybridization

Somatic hybridization consists of fusion of two protoplasts, one originating from embryonic callus and the other from leaf mesophyll, through polyethylene glycol (PEG) to develop allotetraploids. The first successful protoplast isolations were reported as early as in 1982 and the first citrus somatic hybrid was made between *C. sinensis* and *P. trifoliata*. These results allowed the establishment of citrus breeding programs for scion and rootstock improvement in several countries.

Somatic hybridization has been applied to overcome self and cross incompatibility barriers and generate interspecific and intergeneric hybrids. In citrus, this technology has been extensively used and has many important implications. It has provided a means of producing heterozygous tetraploid hybrids, which have incorporated complementary traits from donor parents. It has made production of hybrids possible from sexually incompatible or difficult to hybridize citrus relatives that possess valuable attributes and thus broadening the germplasm base available for rootstock improvement.

Somatic hybrids have been developed at the Citrus Research and Education Center, Florida, USA. The somatic hybrids have combined desirable disease resistance and stress tolerance traits, and confer varying degrees of tree size control and precocity. Somatic hybridization has also been used to create new tetraploid somatic hybrids that combine elite diploid scion material to be used as tetraploid breeding parents in interploid hybridization schemes to develop seedless and easy-to-peel new mandarin varieties Another approach to seedlessness is the transfer of cytoplasmic male sterility from Satsuma oranges to other elite but seedy scions via cybridization. This approach has the potential to make existing popular cultivars seedless, without altering the cultivar integrity in any other way. Creation of triploid citrus hybrids by electrofusion of haploid and diploid protoplasts is also promising. Progress has also been made in the development of improved acid fruits (lemons and limes) and ornamental citrus, through somatic hybridization and the subsequent use of the hybrids in sexual crosses. A hexaploid somatic hybrid has been obtained from electrofusion between diploids of *Citrus sinensis* and *Clausena lansium*.

Somatic hybridisation by protoplast fusion techniques is a way to create improved *citrus* rootstocks which can overcome the barriers imposed by its reproductive peculiarities. Grosser *et al*. (2003) have coined the term 'tetrazygs' to identify tetraploid progeny from such crosses. Somatic hybridization is very important to develop new rootstocks resistant to Blight, *Phytophthora*, nematodes, etc. Most of the institutions involved in breeding for rootstock improvement utilize this technology. It is developed by IFAS and CREC in University of Florida, by CIRAD in France, by IVIA in Spain, and by ISA and Centro di Studio per il Miglioramento Genetico degli Agrumi, CNR in Italy.

Sixty somatic hybrids (SH) have been produced by protoplast fusion at CREC, Florida. SH containing Flying Dragon germplasm were extremely dwarf and precocious, those with sweet orange parentage were consistently resistant to *Phytophtora*, while hybrids of *Fortunella*, *Citropsis*, *Atalantia*, *Citrus australasica* and *Citrus ichangensis* parentage was resistant to *Radopholus citrophilus* nematode. Moreover seeds from SH germinated well and produced vigorous seedlings. CIRAD in France selected a vigorous and tristeza resistant intergeneric somatic hybrid between *Poncirus trifoliata* and *Citrus deliciosa.*

The somatic hybrids Cleopatra' mandarin + sour orange, 'Rangpur' lime + 'Sunki' mandarin, 'Cleopatra' mandarin + 'Volkamer' lemon, 'Ruby Blood'sweet orange + 'Volkamer' lemon, 'Rohde Red'sweet orange + 'Volkamer' lemon, and 'Caipira' sweet orange + 'Volkamer' lemon had less trunk rot occurrence. The somatic hybrids 'Cleopatra' mandarin + 'Volkamer' lemon, 'Cleopatra' mandarin + sour orange, 'Caipira' sweet orange + 'Volkamer' lemon, and 'Caipira' sweet orange + 'Rangpur' lime were tolerant to root rot.

Molecular Markers

Isozyme analysis is effective to distinguish hybrids from nucellar seedlings. RFLP (Restriction Fragment Length Polymorphism), useful to develop genomic map, further locates the genes having specific functions. Genetic modification through Agrobacterium mediated transformation has emerged as an important tool for introducing agronomically important genes into *Citrus* species.

Gene constructs have been created for various types of CTV derived genes and other genes from the naturally resistant *Poncirus trifoliata*. These have been inserted into citrus genomes to induce resistance to CTV virus. A citrus blight associated gene has also been introduced into Carrizo citrange. Genetic transformation and regeneration of mature tissues of citrus, which could bypass the juvenile phase, has also been attempted. A gene for tolerance to salinity (HAL2) from yeast has been introduced into citrus. Genes that regulate vegetative and other behaviour in *Arabidopsis* have been engineered into citrus resulting in altered growth habits and greatly reduced juvenility. Genes involved in metabolic pathway regulation, such as those in the flavonoid pathway, have also been introduced in citrus.

The growing interest in manipulating carotenoid biosynthesis in plants is mainly related to human nutrition as precursors of vitamin A and natural antioxidants. Also, this kind of pathway manipulation holds promise of altering colour and flavour development in citrus. CS-ACS1 gene that controls the ethylene biosynthesis has also been introduced into citrus, and the transgenic lines that produced higher level of antisense ACS RNA repressed the increase in ACC content following chilling treatment. Attempts have also been made to introduce juice quality related pectin methylesterase gene into citrus. This exploitation of enzymes associated with cloud separation may also offer a great promise of targeted trait modification and improvement of the responsible genes by genetic manipulation. The cDNA of the Xa21, a *Xanthomonas* resistance gene, has been introduced into citrus via protoplast cotransformation.

Genetic transformation

Genetic transformation may be an efficient alternative for citrus improvement, opening the way for the introduction of specific traits into known genotypes without altering their elite genetic background. The characters that could potentially be manipulated by genetic transformation of Citrus include pest and disease resistance, growth habit and fruit quality. In order to use this technology, it is essential to develop efficient genetic transformation systems for Citrus. Genetic modification through Agrobacterium-mediated transformation has emerged as an important tool for introducing agronomically important genes into Citrus species.

Genetic transformation has been used for incorporation of transgenes of potential interest into Citrus. Wong *et al.* (2001) showed that citrus plants transformed with *ACC* synthase gene in antisense does not display the same ethylene synthesis as untransformed plants after chilling treatments.To develop a better understanding and reduce the juvenile phase, Carrizo seedlings constitutively overexpressing the *Arabidopsis* floral regulatory genes *leafy* (*LFY*) or *apetala1* (*AP1*) were generated. Both produced fertile flowers and fruits in the first year, considerably shortening the juvenile phase. Furthermore, sexual and nucellar seedlings derived from the transgenic plants had a very short juvenile period and flowered in the first spring. Transgenic *Poncirus* with ectopic expression of the *CiFT* gene (homolog to *FLOWERING LOCUS T*), another flowering time gene, also exhibited early flowering. These results open the way to very innovative approaches for citrus breeding over several recombining cycles by combining short juvenile phase and marker assisted selection.

Tolerance or resistance to *P. citrophthora*, the most widely distributed fungal disease in citrus growing areas, was evaluated by introducing the *p23* gene that codes for a pathogenesis related protein induced in tomato. The results provided evidence for the antifungal activity *in vivo* of the p23 pathogenesis related protein against *P. citrophthora*. Research effort is in progress to understand the basis of the resistance to *Citrus tristeza virus* (CTV). CTV resistant transformants have been obtained by genetically engineering the *p25* and *p23* genes from CTV, although the stability of the pathogen derived resistances over the large diversity of CTV strains still has to be demonstrated. An alternative strategy is to look for plant derived resistance genes. The general resistance gene to CTV (*Ctv*) in *Poncirus trifoliata* has been characterized and localized to a region comprised of 10 predicted genes. Each of the 10 genes has been individually introduced in grapefruit and although results of the CTV challenge are still preliminary, transgenic lines individually expressing each one of these 10 genes were susceptible to CTV, suggesting that more than one gene in the locus is involved in resistance to CTV. Work on other pivotal diseases such as mosaic virus, citrus canker and citrus blight, and for abiotic stress such as salt stress and osmotic stress is in progress. Regarding quality traits, a pectin methylesterase gene (*Cs-PME4*) isolated from sweet orange to prevent cloud separation in juice was also introduced via protoplasts and subsequent regeneration through somatic embryogenesis.

GUAVA

Origin: Tropical America

Taxonomy status:

Order: Myrtales

Family: Myrtaceae

Subfamily: Myrtoideae

Tribe: Myrteae

Sub-tribe: Pimentoid

Botanical name: *Psidium guajava* L

Chromosome number: 2n = 2x = 22 ; 3x = 33 (Seedless, Indonesian Seedless); Aneuploidy

Genome size: 443.8 Mbp (*P. guajava*)

987.78 Mbp (*P. guineense*); 515.48 Mbp (*P. cattleyanum*); 572.32 Mbp (*P. acutangulum*).

The related species are *P. littorale, P. acutangulum, P. cattleyanum, P. friedrichsthalianum, P. cujavillus, P. guineense and P. molle.*

Botany

A shrub or small tree, with smooth, light reddish brown bark, with pubescent 4-angled young branches, its trunk normally attains a diameter of about 25 cm. Leaves are opposite, ovate elliptic or oblong elliptic, acute acuminate, pubescent beneath, often rather brittle, prominently nerved, lateral nerves 10-20 pairs, blades mostly 10-15 cm long and 3-5 cm wide. Flowers are hermaphrodite, solitary or 2-4 together in leaf axils, peduncle about 1-2 cm long, pubescent, calyx 4-5 lobed, about 6-8 mm long, petals 5 white 10-15 cm long, fugacious, usually 4/5, obovate, slightly concave, stamens numerous, white, stigma peltate, style 10-12 mm long with inferior ovary. Fruits globose, ovoid or pyriform, many seeded, pulp granular, juicy, seeds yellowish reniform.

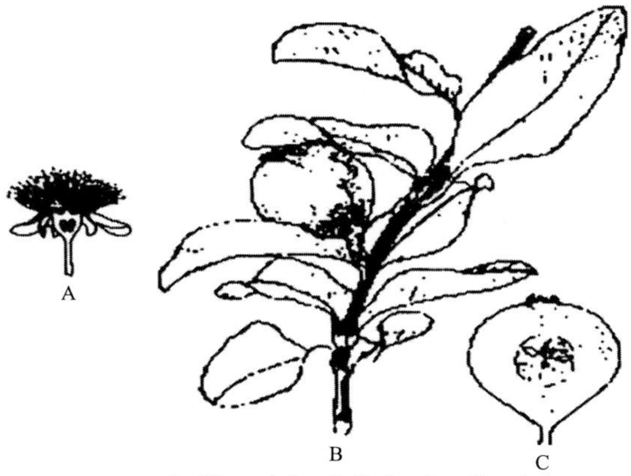

A—Flower in longitudinal section; B—a twig baring fruit; and C—fruit in longitudinal section

Fig. 4. Guava shoot twig, flower with fruit.

(Source: Purseglove, 1968)

Floral Biology

The cracking of calyx occurs nearly 24 hrs before flower opening. The peak anthesis is between 5.00 and 6.30 AM in most of the varieties. In north India, anthesis takes place between 5.00 and 7.30 AM. The dehiscence of anthers starts shortly before or after anthesis in all the varieties and continues up to 2 hrs. The stigma is receptive 2-5 hrs after opening and receptivity is maximum on the day of anthesis. Guava is primarily self fruitful, although some strains seem to produce more fruits when cross pollinated (35%) with another variety. Bees and other insects visit its flowers.

Breeding Objectives

1. Development of cultivars with dwarf and spreading growth habit.
2. Development of cultivars for dessert or processing purpose having better fruit quality in terms of shape, size, colour, firmness, flavour thick pulp, pleasing aroma, high Vitamin C content, higher lycopene content, good pectin content, better keeping quality and processing, etc.
3. Development of cultivars having pink or red coloured skin and flesh, free from muskiness.
4. Development of cultivars with least number of soft seeds.
5. Development of cultivars resistant to biotic (fruit fly, whitefly, mealy bug, nematode (*Meloidogyne spp.*); wilt, fruit rot and rust) and abiotic stress (waterlogging and cold).

Limitations in Breeding

1. Long juvenile phase/period.
2. Heterozygosity.
3. Self and cross incompatibility.
4. Triplody/aneuploidy leads to seedless fruit.
5. Lack of information on inheritance pattern.

Germplasm Resources

Guava germplasm is maintained at several institutes in Cuba, India, Colombia, Venezuela, USA, Malaysia and Nigeria.

The ICAR-Central Insitute for Subtropical Horticulture, Lucknow is the national repository for guava. It has the largest collection of 120 accessions. The IIHR, Bengaluru (67), JNKVV, Fruit Research Station, Rewa, Horticulture Experiment and Training Centre (HETC), Basti, UP and CCS HAU, Hisar etc., are also maintaing the germplasm of guava.

Sources of desirable resistance triats:

Biotic and abiotic stress

Wilt: *P. cattleyanum*, *P. friedrichsthalianum, P. molle* and *P. guineense.*

Root knot nematodes: *P. littorale*, *P. friedrichsthalianum*, *P. guajava* x *P. guineense*.

Drought: L-49, Allahabad Safeda, Chittidar, Safed Jam, Kohir Safeda.

Cold: Allahabad Safeda, *P.cattleyanum.*

Fruitfly: EC 147037, EC 147039, Kamsari, Red Flesh, Superior Sour lucidum.

Tea mosquito bug: EC 147037, EC 147039, Kamsari, Red Flesh, Superior Sour lucidum.

Spiraling whitefly: Arka Amulya, Benaras, Spear acid, *P. chinensis, P. friedrichsthalianum,* EC 147039.

Sources of desirable growth and fruiting traits

Dwarfness: *P. chinensis*, Aneuploid-82 and other Aneuploids, Hybrid-1, Apple Colour, Pusa Srijan (Tetrasomic aneuploid), *P. molle* and *P. friedrichsthalianum.*

Suckering habit: *P. chinensis.*

Earliness: Riverside Vermelah.

Globose fruit shape: Smooth Green, Allahabad Safeda, Apple Colour, Arka Amulya, Arka Mridula, Benaras, Behat Coconut, Hafsi, Sindh, Mirzapur seedling, Dharwad.

Big size fruit: One kg guava, Behat Coconut, Benaras, Kamsari, Dharwad, Chaikaiya Ruthmanagar.

Pink skin colour: Philippine guava**,** Apple colour, CISH-G-1, CISH-G-2, Hybrid 16-1 (Apple colour x Allahabad Safeda), Allahabad Surkha and Phillippine guava.

Red pulp color: Arka Rashmi, Kamsari, Tim, Purple Local, Gill, Rica and Paluma.

Pink pulp colour: Ruby, Butch, Anakapalli, Red Fleshed, Super Acid, Florida Seedling, Hybrid-1, Ruby Supreme, Purtagal, China Surkha, Kothred, Patillo, Beaumont.

Pink skin with pink flesh colour: Lal Sahiba, Sangam.

Processing: 7-12, EC 147036, 7-39 EC 147034**,** Safed Jam, Kohir Safeda.

Seedlessness: Seedless

Soft seed: Hybrid-1 (Allahabad x Seedless), Hybrid-16-1 (Apple colour x Allahabad Safeda), Safed Jam, Kohir Safeda, Chittidar.

High yield: Benaras, 7-38 EC147034, EC 162904, L-49, Allahabad Safeda, Lalit, Rewa-72, Chittidar, Dhareedar, Arka Mridula, CISH-G-4, Benaras, 7-39 EC147034, EC 162904, Behat Coconut.

High TSS: Allahabad Safeda, Chittidar, CISH-G-1, China White, Barampur, Kothred, Burafkhana, Dhareedar, Allahabad Safeda, Arka Mridula, Seedless, Sindh, Hafsi, Bangalore Local, Surka Chitti, Behat Coconut.

High Vitamin C: Mirzapur seedling, *P. chinensis*, EC 162904, G-6, Chakaiya Ruthmanagar, Dhareedar.

Good keeping quality: Harijha, Arka Mridula, Arka Amulya, Hybrid 16-1.

Inheritance pattern of desired traits

Bold seed: Dominant.

Skin colour and acidity: Polygene.

Red pulp: Dominant over white pulp colour; governed monogenically.

Red flesh and bold seed: Linkage between red flesh and seed size.

Obovoid shape: Dominant gene.

Triploidy: Female sterility.

Disease resistance and quality characteristics (Vitamin C, acidity, pectin): Low heritability.

Breeding methods

1. Introduction: Several cultivars were introduced from different countries for specific purposes, e.g. Beaumont, Khu Kula from Hawaii; Indonesian Seedless, Indonesian White from Indonesia; Acerapera from Brazil; Verdie, Patillo and Pink Acid from Florida, USA and Taiwanese Guava XXI, Century Guava and Pearl Guava and Paluma from Brazil and Taiwan. Thai guava has been introduced in India.

The cultivars viz., Hawaiian, Para guava, Mexican cream, Strawberry, Thai white, Pink supreme and Yellow Cherry of Indonesia need to be introduced in India. The Philippine Guava (Purple type) has shown some promise as a wilt-resistant rootstock but needs extensive experimentation.

2. Selection: Guava is an open pollinated crop and seedling selections are done to create new cultivars. Dinesh and Yadv (1998) carried out half sib analysis in progenies of Apple Colour. They observed that genotypic variance was lower than phenotypic variance and heritability was moderately high for all characters implying that selection may be practised for improvement of fruit characters. L-49 or Sardar (Cheema and Deshmukh (1927) is selection from Allahabad Safeda. Banarasi, Surkha, CISHG-1, CISHG-2, Bangalore Local, Arka Mridula, Sind-63, Dharwar-34, Kothrud- 19, Supreme and Red Indian are other varieties evolved by selection.

Gujarat:

Exotica and Matchless, Dhareedar, Dhawal (selection from Allahabad Safeda), **Anand Selection, Bhavnagar Selection**.

IIHR, Bengaluru: **Arka Mirdula (**selection from open pollinated seedling of Allahabad Safeda).

***CISH, Lucknow*:**

Shweta, **Lalit** and **Lalima** (selections from half sib population of apple cultivar), **CISHG**-1 (deep fruit colour with soft seed), **CISHG-5 (**selection from open pollinated seedling).

***Fruit Resarch Station, Saharanpur*:**

S-1 (superior qualty of fruit) was selected.

***Narendra Dev University of Agriculture and Technology, Faizabad (UP)*:**

Seedling selections were made from Allahabad Safeda' and many selections were made under "**Faizabad Selection**".

UAS, Dharwar:

Selection-1, Selection-3, Selection-5, Selection-6, Selection-7, Selection -9 and Selection 11(selection from Navalur cultivar).

GBPUAT, Pantnagar:

Pant Prabhat: A seedling selection from open pollinated seedlings.

MPUAT, Udaipur:

MPUAT S-1: A seedling selection.

MPUAT S-2: A seedling selection.

VNR Nursery, Raipur:

VNR Bihi: Selected by VNR Nursery. The fruit is big (ranging from 300 gm to 850 g), fewer seeds, and long shelf life.

Subtropical Expt. Stat., Homestead, Fl:

Supreme and Ruby: Selected from open pollinated seedling.

Hawaii Agricultural Experiment Station, University of Hawaii:

Ka Hua Kula: Selected from openpollinated seedling of Beaumont.

UNESP, Sao Paulo, Brazil:

Jaboticabal: Selections from UNESP started in 1976, originated from open pollinated American, Indian and Brazilian varieties of different provenance besides local selections, led to the cultivars Red Ogawa N° 1 Rica, Paluma and Pedro Sato.

8502-01: Productive plants, with ramifications predominantly horizontal, medium vigour and precocious maturation.

8503-08: Productive and vigorous plants with precocious maturation.

Sassaoka: Seedling selection in Brazil from Common Red.

3. Hybridization: Improvement work in guava was initiated at several Institue/ Reserch Stations in India. Ganeshkhind Fruit Research Station, Pune started improvement work in 1907.

IIHR, Bengaluru: Crosses among Allahabad Safeda, Red Flesh, Chittidar, Apple Colour, L-49 and Banarasi produced 600 F_1 hybrids. Hybrid Arka Amulya (Allahabad Safeda x Triploid) has been released and two hybrids (Hybrid-16-1 and Hybrid-31-1) have been found promising. A cross between Arka Mridula x Red Flesh has shown fruit with deep red colour and good TSS. Some crosses like Arka Amulya x Apple Colour and Apple Colour x Red Flesh have yielded spectacular hybrids.

Arka Rashmi: Kamsari x Puple Local.

Arka Kiran: Kamsari × Purple Local.

Arka Poorna: Purple local × Allahabad Safeda.

IARI, New Delhi: Crosses between a triploid (Seedless) and diploid (Allahabad Safeda) produced 26 diploid, 9 trisomic (2n +1), 5 double trisomic (2n +1+1), 14 tetrasomic (2n +2), one triple trisomic (2n +1+1+1) and one double tetrasomic hybrids. Amongst these, some tetrasomic and double tetrasomic hybrids had dwarf stature. In another case, two aneuploids were obtained, one with 21 and the other with 30 chromosomes, from progeny of triploid guava.

***HETC, Basti*:** A number of cross combinations of 'Seedless' x Allahabad Safeda, Seedless x L-49, Allahabad Safeda x Patillo, Apple Colour x Red Fleshed and Apple colour x Kothrud were made. None of the 55 hybrids obtained from these crosses were promising.

***Fruit Research Station, Sangareddy, AP*:**

Two hybrids, **Safed Jam** (Allahabad Safeda x Kohir) and **Kohir Safeda** (Kohir x Allahabad Safeda) were selected and released.

***Dr. Rajendra Prasad Central Agricultural University Pusa, Bihar*:**

F_1 hybrid seedlings were raised from various intervarietal crosses. A hybrid from the cross Apple Colour x Sardar had the maximum fruit weight, Vitamin C and pectin content.

***Fruit Research Station, Anantharajupet, Andhra Pradesh*:** Out of 6 hybirds, two (H1 and H6) were found promising with regard to fruit quality and precocious bearing.

***CCS HAU, Hisar*:**

Hisar Safeda: Allahabad x Seedless.

Hisar Surkha: Apple Colour x Banarsi Surkha.

***KVK, Patiala, PAU, Ludhaina*:**

Punjab Pink: It is a hybrid between (Portugal x L 49) x Apple colour.

Punjab Safeda: It is a hybrid between Shweta x 1716.

Punjab Kiran: It is a hybrid between Apple guava x 1716.

Other hybrids are:

Hybrid 16-1: Apple Colour x Allahabad Safeda.

RCGH -1: Sour type x Red fleshed local.

RCGH-4: Red fleshed x Allahabad Safeda.

RCGH -7: Lucknow-49 x Pear Shaped.

Bangkok Golden apple: Thai guava x Indonesia Seedless.

In Brazil, Pereira *et al*. (2003) released **XXI Century guava** cultivar after 10 years of evaluation of 219 plants obtained from several crossings. The cross Supreme-2 × Paluma gave very productive plant with a short cycle (130 days from bloom to harvest), big fruits (236 g in average), thick, rosy red pulp, great flavour and little and small seeds.

***Interspecific hybrids*:** Interspecific hybridization was carried out using the wild species *Psidium chinensis* with *P. guajava* cv. Beaumont to produce rootstocks resistant to wilt disease. At IIHR, Bengaluru, *P. molle, P.cattleyanum and P. guineense* were found wilt resistant. Interspecific crosses were made between *P. guajava, P. molle, P.cattleyanum* and *P. guineense* to incorporate dwarf habit and prolific pattern. Crosses, *P. chinensis* x *P. guajava, P. molle* x *P. guajava* and *P. chinensis* x *P. molle* were compatible.

At CISH, Lucknow, Cross between *P. guajava* x *Psidium molle* was incompatible and that between *Psidium molle* x *P. guajava* was compatible. Hybrid population (F_1) of these parent combinations was found free from wilt even after artificial inoculation by *F. oxysporum*, *F. solani* and *Gliocladium roseum* and were graft compatible with commercial guava varieties. *P. chinensis* when used as female parent with *P. molle,* the cross was compatible.

Rootstock breeding: Efforts were made to develop dwarf and wilt resistant rootstocks by using interspecific hybridization. Intensive work is needed to develop useful interspecific hybrids resistant to wilt using the resistant species in breeding programmes. The Philippine Guava (Purple type) has shown some promise as a wilt-resistant rootstock but needs extensive trials.

4. Polyploidy: Diploid, triploid, tetraploid and aneuploid forms have been reported. Aneuploids had promising qualities and may prove useful in developing plants with reduced seediness and, possibly, in providing dwarfing rootstocks.

Pusa Srijan: (Aneuploid. No. 82), a tetrasomic, has been found to have strong potentiality for use as dwarfing rootstock.

5. Mutation: In Sind and Deccan types of guava, bud variations are common. Seedless variety of guava is a selection from bud variation.

6. Biotechnology

Micropropagation: *In vitro* clonal propagation of guava was achieved by culturing nodal explants of mature trees on Murashige and Skoog (MS) medium supplemented with 1 mg l^{-1} benzyladenine (Liu and Ya, 2011). Nodal segments taken from *in vitro* proliferated shoots gave rise to 2–4 shoots by precocious axillary branching without initial lag period. Shoots obtained from the proliferation stage were rooted on half strength MS medium containing indole-3-butyric acid and α-naphthaleneacetic acid. Regenerated plantlets were successfully established in soil under field conditions.

Somatic embryogenesis: Efforts were made for *in vitro* regeneration via somatic embryogenesis. Immature zygotic embryos have been used as primary explants for induction of somatic embryogenesis by most workers. Of the other explants such as leaf, node, internode, petal and mesocarp only mesocarp gave some success.

Molecular breeding: DNA markers have been used to quantify genetic diversity and determine phylogenetic relationships. Different PCR-based techniques have been used to verify DNA quality, to establish fingerprint of individual accessions, to assess the genetic diversity, to construct a genetic linkage map and for tagging economically important genes for marker assisted selection. These are Random Amplified Polymorphic DNA (RAPD), Amplified Fragment Length Polymorphism (AFLP), Inverse Sequence Tagged Repeat (ISTR) and Simple Sequence Repeat (SSR) (Table 13).

Table 13. Achievements in guava through molecular approaches.

RAPD	Assessment of genetic relationship among four Mexican guava cultivars to estimate chemical (quercetin) diversity
AFLP	Genetic characterization of Mexican native 48 guava cultivars
ISSR	Developed molecular marker for pulp colour in guava.
	Assessed genetic variability among eleven *Psidium* sp.
SSR	Characterized 69 cultivars of guava for higher nutritional content

SSR	Characterization of 23 nuclear simple sequence repeat (SSR) loci in three guava species for cultivars identification and linkage mapping
	Studied the genetic distance among guava genotypes.

Biotechnological interventions may bring in desired traits in commercial guava variety. Recombinant DNA technology and marker-aided breeding could facilitate development of desired guava cultivars. There is a need to develop genetic map of guava. Work is being done on transforming guava with cold hardiness gene (CBF1, CBF2 and CBF3). There is a need to engineer genes controlling ethylene biosynthesis and ethylene sensitivity in guava for better shelf life. Insertion of genes encoding hydrolytic enzymes (which can degrade fungal cell wall) such as chitinase and glucanase could also be helpful for controlling diseases in guava. Many groups have reported *in vitro* regeneration protocol in guava using somatic embryogenesis and organogenesis pathway. Shoot bud culture is the most preferred pathway for clonal multiplication. There have been reports of successful micropropagation and plantlet production from apical and axillary bud explants. Callus has been produced from fruit tissue, however regeneration of organs or embryos or plantlets from callus appears to be more difficult, although embryogenic callus has been developed from anthers. So far there are no applications of molecular biology to *Psidium* species.

Genetic transformation: Genetic transformation of guava with CBF1, CBF2 and CBF3 cold hardy genes has been demonstrated at Fort Valley State University, USA. Genetic transformation system has been developed in guava using *endochitinase* gene. At CISH, Lucknow guava nodal explants were co-cultivated with *Agrobacterium tumefaciens* (LBA 4404) harboring a binary vector *p*BI121 having selectable markers (*nptll* and GUS) with CaMV 35S promoter gene. The resultant plants showed kanamycin resistance (Mishra *et al.*, 2014).

PINEAPPLE

Origin: South America (areas of north western and eastern Brazil, Colombia, Guyana and Venezuela).

Taxonomy status:

Order: Poales

Family: Bromeliaceae

Subfamily: Bromelioideae.

Botanical name: *Ananas comosus* (L.) Merr.

Chromosome number: 2n = 2x = 50; 2n = 3x = 75 (Cabezona,); 2n = 4x = 100 (*Ananas macrodontes* syn. *Pseudananas sagenarius*)

Genome size: 526 Mb

Other species of *Ananas* are *Ananas macrodontes* Morren, *Ananas comosus* var. *microstachys* (Mez) Smith, *Ananas comosus* var. *parguazens*is (Camargo & L.B.Sm.) Coppens & Leal, *Ananas comosus* var. *erectifolius* (L. B. Smith) Coppens & Leal and *Ananas comosus* var. *bracteatus* (Lindley) Coppens & Leal.

Botany

Pineapple is herbaceous, perennial plant with short stem containing a rosette of long narrow leaves that are relatively wide (more than 5 cm), spiny (antrorse spines), partially spiny or smooth with terminal inflorescence. The individual flowers are actinomorphic, hermaphrodite, each sitting on an axil of a bract, which secrets nectar, trimerous with three sepals, three petals, six stamens in two whorls of two and three carpellate, inferior ovary with numerous ovules. The plant produces a coenocarpium (a multiple fruit derived from ovaries, floral parts and receptacles of many coalesced flowers). Each fruitlet has small naked, winged or plumose seeds. Their endosperm is reduced with a small embryo. Cayenne cultivars produce seedless fruits because of self incompatibility. Even then they produce germ cells. The germ cells of Queen and Cayenne cultivars are mutually compatible and gametes are produced.

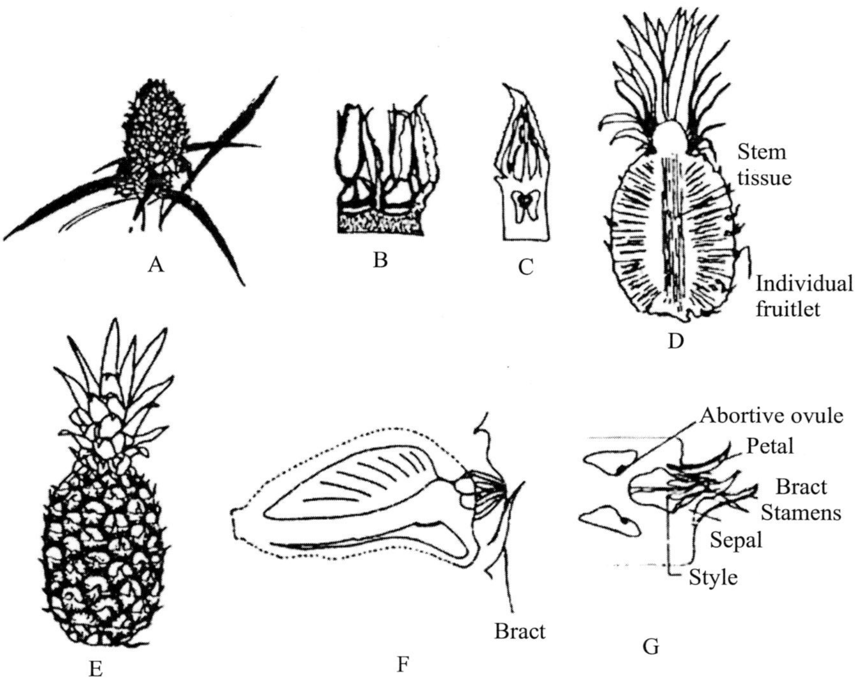

A—inflorescence; B—flowers; C—flower in longitudinal section; D—fruit in longitudinal section; E—fruit; F—individual fruitlet in longitudinal section; G—individual flower in cross section

Fig. 5. Pineapple inflorescence, flower and fruit.

(Source: Purseglove, 1986; Simpson and Conner-Ogorzaly, 1986)

Floral Biology

Pineapple flower open in early morning (6-9 AM) and fades in the evening. The maximum time of anthesis occurs during 8-10 AM. About 50-100 flowers open over

10-20 days daily from base to the top. The dehiscence of anthers and stigma receptivity occurs on the same day. Pineapple is self incompatible. The self sterility is mainly due to the inhibition of pollen tube growth in the upper third of the style. Certain varieties of the *Ananas* genus show natural cross pollination as in *A. macrodontes*. This is called as pseudo-self compatibility. They are self fertile and the offsprings are very homogenous and are autogamous. It is cross pollinated mostly by humming birds, bees and ants.

Breeding Objectives

1. To develop early maturing, good ratooning ability spineless, high yielding cultivars with wider geographic adaptability.
2. To develop cultivars suitable for table purposes, having medium sized (1-2 kg), cylindrical and sweet fruits.
3. To develop cultivars suitable for canning purposes, big sized (≥2 kg), cylindrical and sweet fruits having high juice content.
4. To develop varieties tolerant to physiological disorder (multiple crown, fasciation, collar of slips, knobs, malformed fruit).
5. Resistance to pest (mealy bugs) and diseases (fruitlet core rot caused by *Fusarium ananatum* and marbling disease caused by *Pantoea ananas*, heart rot and root rot and nematode, and abiotic stress (low temperatures during winter, and high winds during typhoons).

Breeding Limitations

1. Poor genetic base.
2. Heterozygosity.
3. Self incompatibility (gametophytically type).
4. Parthenocarpy.
5. Polyploidy.
6. Lack of inbreeding.
7. Less number of seed.
8. Cumbersome and labour intensive hybridization.

Germplasm Resources

The pineapple collection is managed by USDA's Agricultural Research Service (ARS) at Hilo as part of the agency's National Clonal Germplasm Repository for Tropical Fruit and Nut Crops. CIRAD pineapple gene bank was established in Martinique in 1985 and it has over 600 accessions.

In India, much attention has not been given in the development of field gene banks. However, commercial cultivars are maintained at IIHR, Bengaluru, BCKV, Kalyani, Agriculture Research Station, Kovvur (APAU), Department of Horticulture,

College of Agriculture, Jorhat, Regional Research Station, Diphu (Assam Agricultural University), Department of Horticulture, College of Horticulture, Navsari (GAU), College of Horticulture. At Vellanikara, Trichur, 33 accessions of pineapple are being maintained in field gene bank.

Sources of resistance traits:

Mealy bug wilt: Spanish group cultivars PR-1-67, Red Spanish, Singapore, Green Selangor, Castilla, Cabezona, PRI-56, Nangka, Gandol, Masmerah, Pernambuco, Queen, *Ananas ananassoides, Ananas bracteatus* and *Pseudananas sagenarius.*

Nematodes: Lot 520, *A. comosus* var. *ananassoides*, *A. comosus* var. *erectifolius,* and *A. macrodontes.*

Fusarium: *Ananas ananassoides***,** Perolera, Primavera and Sao Bento.

Gummosis: Smooth Cayenne, PR 1-67, Puerto Rico and *Ananas ananassoides.*

Root rot: Parnambuco, Red Spanish, hybrid No. 59-656, *Ananas ananassoides, A. comosus* var. *bracteatus.*

Dry climate: *Ananas bracteatus.*

Drought: *A. comosus* var. *ananassoides.*

Water logging: *A. macrodontes.*

Sources of quality traits

Ornamental value: *A. comosus* var. *bracteatus* (bright red coloration to the syncarp); *A. macrodontes* (pink colour); *A. comosus* var. *erectifolious* "Selvagem 6" (smooth reddish leaves).

Big fruit size: Giant Kew (Cayenne group).

Canning: Cayenne group cultivars Smooth Cayenne, Cayenne Lisse, Hilo Cayenne, Boron, Rothschild, Smooth, Gautenialan, Typhone, St. Michael, Emeralda, Kew, and Giant Kew, Champaka.

Export value: Monte Lirio, Smooth Cayenne, Cayenne Lisse, Hilo Cayenne, Boron, Rothschild, Smooth, Gautemalan, Typhone, St. Michael, Emeralada, Kew, Giant Kew and Champaka.

Good flavour: Parnambuco, *Ananas ananassoides.*

Early fruiting: Parnambuco.

Early ripening: Queen.

Fibre: *A. macrodontes*, Queen.

Inheritance pattern of desirable traits:

Fruit weight: Fruit weight is well correlated with fruitlet number which is highly heritable. However, there can be a moderate negative correlation between fruitlet number and fruitlet height × width. Fruit weight is highly correlated with fruitlet length [(fruit diameter-core diameter)/2] than fruitlet height × width. Fruitlet length is moderately positively correlated with fruitlet number.

Fruit size and quality: Polygene.

Total soluble solids (TSS): It is a quantitative trait controlled by multiple additive genes. It displays a moderately high heritability but with weak negative correlation with fruit weight. TSS can therefore be difficult to increase, substantially and quickly, simultaneously with fruit weight.

Peduncle Length: Peduncle length appears quantitative and displays a maternal effect. Collins (1960) indicates it to be controlled by several genes with a long peduncle being dominant. It is moderately heritable and not highly correlated with any other traits.

Slip presence: Slip number can be bimodal with populations segregating into zero or several slips. Slip presence has a low heritability with a strong environmental influence.

Plant sucker presence: The number of plant suckers in segregating seedling populations is a qualitative character with maternal effect. Sucker presence has a low heritability being strongly influenced by environment.

Spineless forms: Dominant gene.

Piping and non-piping: Controlled by another non linked pair alleles with gene P (piping) being epistatic to S and s. The homozygous PP genotypes produce pronounced piping than Pp genotypes.

Collar-of-slips: Dominant character and occurs in a heterozygous state in Smooth Cayenne.

Breeding Methods

1. Introduction: The important cultivars Smooth Cayenne or Cayenne, Kew, Queen and Mauritius have been introduced. A miniature pineapple Saigon Red, brought to Hawaii from Vietnam in 1938, is too tart to eat. However, its compact size, red skin and long, shapely leaves could prove ideal as a decorative plant for landscape plantings or indoor use.

2. Selection: Most of the Improvement in pineapple has taken place mostly by simple selection of mutant clones of the cultivars. During the second half of the 20th century crop improvement *via* clonal selection was carried out in Australia, Brazil, Cuba, India, Guinea and Cote dIvoire, Japan, Malaysia, Mexico, Puerto Rico, Soth Africa, Taiwan and Venezuela. Important clones selected from important cultivars are:-

Cayenne: Important clones are Smooth Cayenne, Cayenne Lissti, St. Michael, Typhone, Smooth Guatemalan, Esmeralda and Sarawak.

Queen: Important clones are Mac Gregor, Natal, Ripley and Alexandria.

Spanish: Important clones are Red Spanish, Singapore Spanish, Green Selangor, Cabezone, Masmerah, Nanas Jahor and Nanas Merah.

Abacaxi: Important clones are Pernambuco, Perola, Sugar Loaf, Papelon, Abbakka, Venezolana, Amarella, Perola, Eleuthera and Abacaxi.

Maipure: Important clones are Maipure, Perolera, Lerija, Monte Liro and Rondon.

Selection from Smooth Cayenne variety is known as Sarawak in Malaysia and Champaka in India. Cultivar **N67-10,** widely grown in Hawaii, is a vegetative selection from Hawaiian Smooth Cayenne released by Okinawa Prefectural Agricultural Research Center.

3. Hybridization: Emasculation is not needed in pineapple because of auto-incompatibility. The female parent inflorescence is covered with a mosquito net before flowering. Flower induction is done by using ethephon for synchronizing of flowering time. Normal breeding strategy in pineapple is intercrossing two heterozygous clones followed by 3 to 4 cycles of field observation on the progenies for selection of the best clone.

Important Hybrids

Amritha: Parents are Kew and Ripley Queen. Developed at Pineapple Research Centre, Kerala Agricultural University, Vellanikkara. Fruit yield is 85 tonnes per hectare.

Nanas Jahor: It is a Cayenne - Spanish hybrid and is susceptible to fruit marbling and cork spot.

FLHORAN 41: Cross between Smooth Cayenne cv. clone HA 10 and Manzana cv. clone CO 24, developed by CIRAD FLHOR, Montpellier, France.

PR-1-67 (Red Spanish x Cayenne): Good plant vigour and is resistant to wilt.

H-7 (Valera Monendi x Kew): Large size fruit and yield is 3-3.5 kg.

MD-2 (Gold Extra Sweet, Super Sweet pineapple, hybrid 73-114): It is cross between 58-1184 and 59-44314. The PRI hybrid 73-114 is named after Millie wife of Frank Dillard manager of DM. MD-2 pineapples are trade marked Del Monte Gold.

CO-2: It is interspecific cross between hybrid clones 58-1184 and 59-443 from Pineapple Research Institute. CO-2 achieved some success at MPC as Maui Gold is also grown by Dole (Castle Cooke was renamed Dole Food Company, Inc. in 1991) on Oahu and in the Philippines. The hybrid is also one of the main fresh fruit cultivars in Australia.

Hybrid 73-50 (MD-1): The PRI hybrid 73-50, named MD-1 by DM, was later patented by DM as CO-2. It was developed by the Hawaiian Pineapple Research Institute, Hawaii.

New cultivars have been produced in many other countries; Tainung 17 and Tainung 21 in Taiwan; Josapine in Malaysia; Aus Jubilee, Aus Carnival and Aus Festival in Australia; and Imperial in Brazil.

Okinawa Prefectural Agricultural Research Center (OPARC), Japan:

Breeding programs started in 1989 and released following cultivars:-

(i) **Soft Touch** (Hawaiian Smooth Cayenne × I-43-880) was released in 1999. This cultivar has a favorable peach-like aroma.

(ii) **Gold Barrel** (Cream Pineapple × McGregor ST-1), released in 2009. Eating quality is excellent.

(iii) **Julio Star** (N67-10 × Cream Pineapple), released in 2009. This cultivar has resistance to fruitlet core rot and a long shelf life.

(iv) **Okinou P17** (Yugafu × Summer Gold) was released in 2015, and has favourable characteristics such as excellent eating quality, resistance to fruitlet core rot and tolerance of typhoon.

Interspecific Hybrids

A. comosus x *A. ananassoides, A. comosus* x *A.bracteatus, A. comosus* x *A. erectifolius* are diploids, cross compatible and produce fertile F_1 progeny. A cross between diploid species and tetraploid *Pseudoananas sagenerius* produces triploids. Other important hybrids released are Nanas Johore, Josapine, etc.

4. Mutation: Natural mutation is common in pineapple, e.g. Selangor Green is natural mutant of Spanish group. Z-Queen or James Queen is reported to be a mutant of Natal Queen. Somatic mutation is common in pineapple. At least 30 clones have been identified in Smooth Cayenne. Pineapple is therefore considered to display a high level of genome instability.

Attempts to produce improved cultivars through gamma radiation on *in vitro* plantlets have been made. Gamma radiation causes severe and random changes. Irradiation sometimes shows negative response like growth retardation and premature suckering. Ibrahim *et al.* (2009) reported a beneficial mutation on resistance to bacterial heart rot (*Erwinia chrysanthemi*). Mutation with chemicals sometimes produces spineless plants in cv. Queen.

5. Polyploidy: The manipulation of ploidy was first explored in the PRI program in the 1940s. Cabezona, a natural triploid has large plants and fruit weighing 4.5–6.5 kg. *Ananas macrodontes* is tetraploid (2n=4x=100) in nature. Tetraploid Smooth Cayenne was found to be vigorous but produced poor quality fruit. Therefore, it was believed that tetraploids of elite cultivars, when used in combination with diploid parents, would produce triploid progeny very similar to the tetraploid parent with only a few traits inherited from the diploid. Consequently, considerable promise was placed in the production of tetraploids of elite cultivars using colchicine on growing points. After many efforts few useful triploids were produced. But triploids were mostly sterile. Hence this approach was abandoned in further breeding programme.

6. Biotechnology:

Micropropagation: Micropropagation has been adopted for commercial production of pineapple. Attempts have been made for rapid multiplication of plants through micropragation by using different kinds of explants i.e. leaf base, shoot base, excised lateral buds and meristem tips from crown.

The study was conducted at the Tissue Culture Laboratory, Horticulture Research Institute, Agricultural Research Center (ARC), Egypt during December 2013 to March 2016 to investigate the effect of different type Murashige and Skooge (MS), Gamborge (B5) and Woody plant media (WPM) at four salt concentrations (Full, ¾, ½ and ¼) of culture media on micropropagability of Smooth Cayenne during establishment stage. MS media at full strength was the best culture medium giving 1.75 shootlet/explant, shootlet length of 2.25 cm and 9.25 leaf per shootlet. BAP at 2.0 mg/l showed the highest shoot proliferation of 56 and 42 shoots per explant at the first and second subculture, respectively. The longest shoot (5.5 and 5.5 cm) was produced in the two subcultures by 0.25 mg/l BAP treatment. The highest number (10.67) of roots/shootlet was produced by 1.0 mg/l IAA and the tallest length of roots was obtained from explants cultured on MS media containing 3 mg/l IAA. The individual rooted explants derived from plantlets were transferred to poly bags in the green house.

In vitro regeneration of Josapine pineapple in tissue culture system was carried out using crown tip meristem. Optimum sterilization protocol was established. Plant growth regulators were applied in MS medium supplemented with 8.0 g/l agar and 30 g/l sucrose along with different concentrations of BAP alone and in combination with NAA. Optimum regeneration was obtained in MS medium supplemented with 3.0 mg/l BAP and that containing 1.0 mg/l BAP/0.5 mg/l NAA. The *in vitro* plantlet of Josapine pineapple was successfully acclimatized into the soil.

Somaclonal variation with certain mutations is often seen where plants are multiplied *in vitro*. Variations in spiny leaves are common and predictable to a certain extent. Most somaclonal variations are negative.

Protoplast Culture

Protoplast culture and somatic hybridization as a tool for introgression of genes have had no impact on pineapple improvement.

Molecular Markers

Molecular markers have become powerful tools in pineapple taxonomy and can assist in more precise identification and classification and use in hybridization programme for selection of parents containing useful traits.

The genetic diversity in *Ananas* germplasm was initially investigated by using isozyme markers. Fifteen out of 27 *A. comosus* cultivars were identified by five enzymatic systems, two peroxidases and three phosphoglucomutases. Pineapple accessions (161) from the Hawaiian collection, including four different species of *Ananas* and one species of *Pseudananas*, were identified by six isozyme systems involving seven putative loci.

DNA-based markers have been used to study the phylogenetic relationships between *Ananas* and related genera. Restriction fragment length polymorphism (RFLP) markers were analyzed among 301 accessions of *Ananas* and related genera including 168 *A. comosus* accessions, suggesting that *A. comosus* has lower level of polymorphism than the wild *Ananas* species. Similarly, based on AFLP markers pattern of Mexican germplasm collections, *A. comosus* accessions were reported to have a low level of diversity.

Three commercial cultivar groups, Cayenne, Queen, and Spanish were investigated by random amplified polymorphic DNA (RAPD). Cultivars Cayenne and Queen were grouped into two separate clusters, whereas Spanish failed to form amonophyletic group. However, major cultivar groups of the 148 *A. comosus* accessions such as Cayenne, Spanish and Queen could not be dinstinctively separated by AFLP. The genetic diversity and genetic relationships among 15 accessions of pineapple was studued by using Inter simple sequence repeats (ISSR) markers. ISSR analysis was found to be a rapid and suitable method for studying genetic diversity among indigenous Intrachit

(Thailand pineaple cultivars) and others. Chloroplast DNA (cp DNA) diversity of *Ananas* and related genera were evaluated by PCR-RFLP suggesting that the genetic diversity of *Ananas* was relative to the geographic origin of the accessions but not the species. The level of genetic diversity among commercial cultivars is still unclear.

Leaf margin phenotype-specific restriction-site-associated DNA-derived marker analysis was employed to analyze three bulked DNAs of F_1 progeny from a cross between a piping-leaf-type cultivar, Yugafu, and a spiny-tip-leaf-type variety, Yonekura (cultivars of Japan). The parents were both *Ananas comosus var. comosus*. From the analysis, piping-leaf and spiny tip-leaf gene-specific restriction-site-associated DNA sequencing tags were obtained and designated as PLSTs and STLSTs, respectively. SSR and CAPS markers are applicable to marker-assisted selection of leaf margin phenotypes in pineapple breeding.

Genetic Transformation

Agrobacterium mediated pineapple transformation has not been widely used for pineapple improvement due to the technical difficulties and prolonged duration of the transformation process. Despite the time and labour requirements of the process, a few studies have successfully introduced high quality agronomic traits into the pineapple such as delayed flowering through silencing of the ACC synthase gene, herbicide tolerance and blackheart disease control. Pinkglow™ transgenic cultivar was developed, in which the pink flesh accumulates lycopene due to the modification of the carotenoid pathway (Firoozbady & Young, 2015). In addition to this, genetically engineered pineapple has modified traits viz., delayed ripening/senescence, fruit color and shell morphology.

PAPAYA

Origin: Tropical America (Central America)

Taxonomy status:

Order: Brassicales

Family: Caricaceae

Botanical name: *Carica papaya* L.

Chromosome number: n = 9; 2n = 2x = 18

Genome size: 375 Mb

The *Vasconcellea* species can also be used as a source of new genes for papain and resistance to PRSV. The other edible species are *V. pubescens* A. DC. *V. monoica* Desf., *V. erythrocarpa* Heilborn, *V. goudotiana* Solms-Laubach and *V. quercifolia* Benth. and Hook.

Botany

Papaya is a small, sparsely branched, soft-woody tree, usually with a single stem growing from 5 to 10 m in height, with strong apical dominance. The stem is hollow with soft fleshy tissue, covered with smooth grey bark marked externally by numerous large orbicular leaf scars. Leaves are large up to 75 cm across and deeply divided into about 7 palmate lobes again being pinnately lobed. They usually form a crown over the central trunk. It is a trioecious and autogamous species and characterized by various flower types and three primary, polygamous sexual types, viz. pistillate (female; mm), staminate (male; M1m) and hermaphrodite (M2m). The five petalled flowers of papaya are fleshy, waxy, cream to yellow in colour, and slightly fragrant. Flowers are borne singly or on cymose inflorescences in leaf axils. The type of inflorescence depends upon the sex of the tree. In staminate trees, flowers are sessile and are produced in clusters on long, pendulant racemes 60–90 cm long. The individual flower is tubular with ten stamens in two series of five, attached to the throat of the corolla tube, and lacks an ovary. The pistillate trees bear one or two flowers on short, 4–6 cm peduncles at each leaf axil. The five petals are separate but are inconspicuously fused together at the very base of the ovary with the absence of stamens. The hermaphroditic form is between the two unisexual flower types and exhibits numerous deviations. The basic flower on a short peduncle is characterized by an elongated pistil with five stigmatic rays and five petals that are fused to about two-thirds of their length, forming a corolla tube. There are ten stamens in two series of five attached to the throat of the corolla tube. The pistil is normally five-carpellate and elongated, oval or pyriform. Storey (1941) classified papaya flowers into five basic types.

Type I: pistillate or female flower devoid of stamens, with a distinct ovoid ovary terminating in a five lobed stigma.

Type II: hermaphrodite (pentandria) flower having five functional stamens and a globose five-furrowed ovary.

Type III: hermaphrodite (carpelloid) flower having six to nine functional stamens and an irregularly ridged ovary.

Type IV: hermaphrodite (elongata) flower having ten functional stamens and an elongate, smooth ovary.

Type IV+: hermaphrodite (barren) flower having ten functional stamens but the pistil aborts, becomes vestigial and lacks a stigma.

Type V: staminate flower having ten functional stamens only. The ovary is completely absent and flowers are bunched in an inflorescence.

Although five basic floral types are listed, certain male and hermaphrodite trees undergo sex reversal and morphological changes in various degrees under the influence of environmental changes Type V staminate flowers may form functional ovaries and bear fruit during cool months. Type III flowers can form carpelloid fruit when exposed to low temperatures about 40 days before anthesis.

The appearance of a large number of modified flower forms occurs in progenies from appropriate hybridization when grown under a temperature regime that may range from around 13 to 32°C. Therefore, recovery of extreme sex modifications may not be seen in a breeding programme unless conducted in an area with wide fluctuations in seasonal temperatures in highly heterozygous parental types. Stamen carpellody is expressed under cool temperatures, with increasing severity as temperatures lower in the *ca* 40 days before anthesis. Instead of ten stamens in a double whorl, there are only five stamens, with the other five fixed to the normal carpels. The fruits that develop from this carpellody are severely misshaped and unmarketable. Female sterility occurs under warm temperatures, again with increasing severity at higher temperatures in the *ca.* 40 days before anthesis. Excessive nitrogen and moisture also favour stamen carpellody, while plant stress such as N deficiency and moisture stress influences female sterility.

The papaya fruit is a fleshy berry, from 50 g to well over 10 kg, and superficially resembles a melon, being spherical, pyriform, oval or elongated in shape. Fruit shape is a sex-linked character and ranges from spherical to ovoid in female flowers to long, cylindrical or pyriform (pear-shaped) in hermaphrodite flowers. The skin of the fruit is thin and usually green when immature, turning to yellow or orange when ripe. The fruit is normally composed of five carpels united to form a central ovarian cavity, which is lined with the placenta, carrying numerous black seeds. Placentation is parietal, with the seeds attached by 0.5–1 mm stalks. The seeds are dark grey to black when mature and enclosed in a sacrotesta. The ovarian cavity is larger in female fruit than in hermaphrodite. The shape of the cavity at the transverse cut ranges from star shape with 5–7 furrows to smooth and circular in shape.

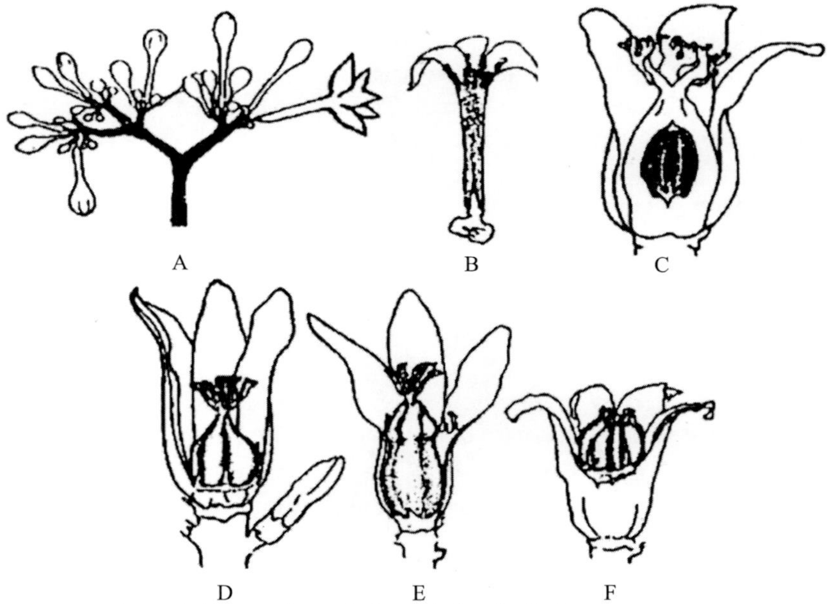

A—male flowers; B—male flower in longitudinal section; C—female flower in longitudinal section; D—hermaphrodite flower of petandria type; E—hermaphrodite flower of elongata type; F—harmaphrodite flower with carpelloid stamens

Fig. 6 Papaya flower types

(Source: Purseglove, 1968)

Floral Biology

The papaya plant flowers throughout the year in India. It takes 21 days from bud emergence to anthesis. The peak anthesis was observed beteween 5.00 AM and 6.00 AM in all the species except in pisitillate flowers of *Vasconcellea cauliflora* and staminate flowers *Vasconcellea goudotiana*. The stigma receptivity is maximum on the day of anthesis. Anther dehiscence is completed 18-36 hrs before flowers open and stigma becomes receptive a day before the flowers open, remaining receptive for 6 days. Wind pollination is minimal as the stamens are packed inside the corolla tube and seldom protrude prominently out of the flower in hermaphrodite population. The hermaphrodite flowers are cleistogamous, i.e. the anthers dehisce and release the pollen prior to anthesis, leading to self-pollination. Varieties such as Sunrise, Kapoho and Eksotika are enforced self-pollinators, and seeds gathered from hermaphrodite fruit will usually breed true to type. Self-incompatibility in cultivars is relatively rare, though there are isolated cases when controlled self-pollinations are done. Self-pollination does not appear to result in any loss of plant vigour.

In mixed planting of pistillate and hermaphroditic trees or in purely hermaphroditic stands, no problem in pollination is experienced. Problem occurs when dioecious cultivars are planted with an inadequate number of male trees. In Australia, the recommended ratio of female to male is 8:1. However, one male tree per 15–20 female trees provides adequate wind pollination if male trees are located considering the prevailing winds. Parthenocarpy in papaya is rare. Seedless fruit or fruit with very low seeds can be produced on female trees. These fruits are generally smaller in size. Auxins have been reported to induce parthenocarpy.

Breeding Objectives

1. To develop dwarf stature, early bearing and low bearing varieties free from branching habits.
2. To evolve gynodioecious varieties with minimum expression of stamen carpellody and female sterility.
3. To develop varieties having the desired fruit characteristics (smooth skin free from blemishes, firm fruit with thick flesh, small and round seed cavity, absence of internal lumps and long shelf life).
4. To breed varieties having good yielding ability, long shelf life of fruits suitable for export.
5. To breed varieties giving high latex yield for papain production.
6. To develop varieties resistant to biotic stress from aphid, fruit fly, red spider mites, thrips, root knot nematode (*Meloidogyne incognita*), reniform nematode (*Rotylenchulus reniformis*), *Pythium,* root, stem and fruit rot (*Phytophthora palmivora*), anthracnose, papaya ring spot virus (PRSV), papaya mosaic virus (PMV) and from abiotic stresses (typhoons, frost and waterlogging).

Limitation in Breeding

1. Lack of diversity.
2. Long juvenile phase.
3. Unstable sex in male and hermphrodite plants
4. Sterility.
5. Self incompatibility.
6. Parthenocarpy.
7. Unpredictable sex expression and identification.
8. Embryo abortion.
9. Lack of established asexual multiplication protocol and maintenance of purity.
10. High heterozygosity and thus need more generations of selfing for attaining homozygosity of traits.

Germplasm Resources

Germplasm collections to support varietal improvement programmes are being maintained. In Malaysia, MARDI currently has 52 accessions of papaya. The USDA/ARS National Germplasm Repository at Hilo, Hawaii holds 37 accessions of *C. papaya* and 13 species of *Vasconcellea*. Other large holdings are in India (90 accessions), Brazil at EMBRAPAmandioca e fruticola (96 accessions), Bahia State at EBDA (82 accessions), Campinas, SP at IAC (169 accessions) and Columbia at UN Mendellin and CORPOICA (83 accessions). About 150 accessions of papaya are being conserved in different field genebank in India at TNAU, Coimbatore, IIHR, Bengaluru (35 accessions), CISH, Lucknow (17), IARI Regional Station, Pusa, Bihar, CHES, Ranchi and CHES, Bhubaneshwar.

Sources for desirable growth and fruiting traits:

Dwarfness: Pusa Nanha, Pusa Dwarf, Red Lady and Waimanalo.

Gynodioecious: Pusa Majesty, Pusa Delicious, Arka Surya, Arka Prabhath, Coorg Honey Dew and CO-7.

Bigger size fruits: Pusa Giant, Pusa Dwarf, Washington and Coorg Honey Dew.

Red fruit flesh: Red Fleshed, Pink Flesh Sweet, Surya, Barwani Red and Arka Prabhath.

High total soluble solids: Solo, Honey Dew, Pusa Delicious, CO-3, CO-4, Pink Flesh Sweet and Washington.

High fruit yield: Pusa Delicious, Pusa Majesty, Pusa Dwarf and Washington.

Good keeping quality: Pusa Majesty, CO-1 and CO-2.

Processing use: Sitiawan.

High papain content: Pusa Majesty, Homstead, CO-2, CO-5 and CO-6, *Vasconcellea cundinamarcensis and Vasconcellea stipulata.*

Export: Solo and Eksotika.

Sources for desirable resistance traits:

Bunchy top: *Vasconcellea monoica.*

Root rot: Wiaminalo-23, Line-8 of Solo, Wiaminalo-24, Ranchi.

Phytophthora rot: *Vasconcellea goudotiana.*

PRSV-P: *Vasconcellea pubescens, Vasconcellea cauliflora*, Cariflora, Rainbow, Tainung No. 5 and Sinta.

Viral diseases: *Vasconcellea cauliflora, Vasconcellea monoica*, Pusa Majesty, *Vasconcellea stipulata, Vasconcellea quercifolia, Vasconcellea pubescens* syn. *V. cundinamarcensis* and Pusa Majesty.

Nematode: Pusa Majesty.

Cold and frost: *Vasconcellea pubescens* syn. *Vasconcellea cundinamarcensis, Vasconcellea quercifolia, Vasconcellea goudotiana, Vasconcellea pentagona,* Fairchild, Punjab Sweet, Petersen, Sunnybank and Arline/57.

High rainfall and waterlogging: Pant Papaya, *Vasconcellea monoica.*

Strong winds: Pusa Giant, Pusa Nanha.

Inheritance Pattern

Precocity and plant stature: Precocity or earliness to fruiting is a factor of the number of nodes produced before the first flowering node. Number of nodes to flowering and internode length determine the height to first fruit. These characters are governed by additive gene effects. In case of hybrids the geometric mean between the two parents determines this. Subhadrabandhu and Nontaswatsri (1997) reported that height of first flower and number of nodes to fruiting are controlled both by additive and non additive genes.

Sex type: Sex in papaya is determined by monogenic inheritance involving three alleles (Hofmeyr, 1938). Monogenic inheritance involves single locus with three alleles — M_1 (male) is dominant for maleness, M_2 (hermaphrodite) and m (female) recessive for femaleness. Male/staminate (M_1m) and hermaphrodite plants (M_2m) are heterozygous whereas female/pistillate plants (mm) are homozygous recessive. Males and hermaphrodites are heteromorphic with sex chromosomes denoted as Y for males and Yh for hermaphrodites, and females are homomorphic with two X chromosomes. The genes that determine the male and hermaphroditic conditions are believed to lie in a male specific region (MSY) and a hermaphrodite specific region (HSY) on the Y and Yh chromosomes, respectively. These sequence differences have been used to develop DNA markers for selection of the desired sex.

Fruit weight and shape: Fruit weight is a quantitative character determined by multiple alleles. The shape of fruit is a sex-linked character. The female flower has a globose ovary that develops into round or ovoid fruit. The elongate or hermaphrodite flower has a slender tapering ovary and this subsequently develops into a fruit that is elongated and cylindrical or pyriform in shape depending on the variety of papaya.

Fruit skin colour: It is governed by a single gene; green (G) being dominant and yellow expressed as double recessive (gg).

Fruit yield: It was significantly correlated with number of fruit, fruit weight, fruiting length and leaf length. Higher estimate of heritability and genetic advance were observed for fruit weight, fruiting length and fruit yield.

Carpelloidy and female sterility: Carpelloid fruit development is significantly influnenced by plant age, genotype and their interaction. Stamen carpelloidy and female sterility do not occur in dioecious cultivars. Carpelloidy and female sterility may involve a number of gene pairs in three loci, two for carpelloidy (c/c, c/c) and one for sterility (s/s). Their normal alleles (+) are partially or completely dominant. The s/+ combination is normal. In c/+, the + is partially dominant over c, so there will be some carpelloidy. The s factor, whether homozygous or not (s/-), is epistatic over the c allele when the carpelloidy factors are heterozygous. The phenotype would be normal but unstable, depending upon environmental conditions. If either of the cofactors is homozygous (c/c), the combined strength of the two alleles at one locus can overcome epistasis, thus exhibiting carpelloidy. High heritability (h^2 = 82%) is obtained for carpelloidy, but effective phenotypic selection may be interfered with by the change-in-rate type of interaction between genotype and plant age. A strong correlation was found between incidence of fruit carpelloidy and internode (r=0.68 to 0.86).

Flavour and total soluble solids (TSS): Muskiness has been reported to be due to the homozygous recessive allele of a single gene. Total soluble solids content is usually associated with sweetness and may range from 8^0 to 16^0 Brix or more in Solo and Eksotika varieties. This trait is governed by quantitative genes with additive effects and hybrids are expected to have intermediate values between the parents.

Pulp colour: Flesh colour is governed by a single gene with yellow (R) being dominant over red (r). Varying intermediate shades of pink may be attributed to modifier gene effects. All red fleshed (rr) varieties like Sunrise Solo and Eksotika will breed true for flesh colour but progenies of heterozygous (Rr) yellow F_1 hybrids like Rainbow will segregate in this trait.

Fruit freckle: Freckle is disorder where superficial water soaked spots of variable size appear on exposed surface of mature fruit. It is governed by genetically controlled partial dominance.

Bacterial dieback-bacterial decline: Co-dominant way involving a few genes.

Papaya Ringspot virus (PRSV): Heritability of tolerance to PRSV in population showed h=61% on disease index. Yield and total soluble solids are positively correlated to the degree of tolerance (Chang and Guo, 2002). The resistance to PRSV-p was controlled by single dominant gene (prsv-1) in *Vasconcellea pubescence*.

Breeding Methods

1. Introduction: Commercial cultivars such as P1 41436, Soniyimma, Malinchly, Maru Ank LG (Nigeria), Sunrise, Wilder, Sunset, T88/211, Kapoho Solo, HCAR 27, 7836, HCAR 16, Sunrise HCAR-29, Maradol HCAR-19 and Cariflora (USA) have been introduced.

There is a need to introduce dwarf accessions Baixinho Sta Amalina and Puertico Rico 217 (Chan, 2009) to use in breeding programmes to develop dwarf hybrids of papaya. There is urgent need to introduce the species such as *Vasconcellea goudotiana, V. cauliflora, V. sphaerocarpa, V. querciflora, V. monoica, V. pubescens and V. stipulata* which have resistance to diseases, pests, frost, etc.(Mitra and Sharma, 2020).

2. Inbreeding and selection: The selection and inbreeding methods have been adopted for varietal development in papaya. Hamilton (1954) studied the inbreeding and cross breeding of two Solo strains and showed that inbreeding has no effect on plant vigour. In the case of dioecious lines, the desirable female was crossed with at least three hermaphrodite plants as the male parent. Sib mating in female and selfing in hermaphrodites may go simultaneously to speed up the breeding programme. This process should be continued for seven to eight generations to achieve satisfactory homozygosity.

The cultivars CO-1, CO-2, CO-5 and CO-6 were developed through inbreeding followed by selection at TNAU. The selections were CO-1 from Ranchi, CO-2 from Local material, CO-5 from Washington and CO-6 from Pusa Majesty. CO-8 was developed in the dioecious group through sib mating of CO-2 which is unique for the red pulp colour. Similar work started in 1966 at IARI Regional Station, Pusa to breed uniform varieties with high yield and good quality for wider adaptability resulted in uniform lines of Pusa Delicious, Pusa Majesty, Pusa Giant and Pusa Dwarf with desirable attributes. The gynodioecious varieties Kapoho, Sunrise Solo, Waimanalo, Kamiya and Eksotika were established by inbreeding by means of the pedigree or backcross breeding method (Manshardt, 1992; Chan, 2009). The other selections are Coorg Honey Dew (from Honey Dew), Punjab Sweet, BR-34Y, Pant-4, Pant-2 and Pant-3. Kamiya is a selection from Waimanalo.

Mass Selection

Mass selection is used for cleaning and stabilizing local varieties selected by farmers. In Malaysia, simple mass selection method was used for improving local varieties, e.g. Subang-6 selected for fresh fruit and Sitiawan for processing.

Recurrent Selection

It involves identification of individuals of superior genotypes and their subsequent intermating to produce a new population, which further avoids progeny test. Parental genotypes are generally maintained by controlled selfing followed by selection and evaluation of progeny for the best one and then intercrossing of superior genotypes followed by selection from the new population for development of dioecious and gynodioecious lines.

The most important commercial papaya cultivars in the world are inbred pure lines such as Kapoho, Sunrise Solo and Eksotika that are propagated from generation to generation from seed obtained by setting hermaphrodite flowers of gynodioecious varieties. No inbreeding depression was recorded during selfing of gynodioecious populations (Hamilton, 1954) or dioecious populations (Aquilizan, 1983). Higher number of fruits and low-to-moderate fruit length is more effective in increasing fruit yield than direct selection. The selection of fruits for higher peel weight, fruit weight and diameter is expected to improve pulp weight in papaya.

Oliveira *et al.* (2012) reported the mixed model methodology for selection in F_2 populations from Tainung and Calimosa hybrids. Estimates of individual heritability were high for fruit length and weight, moderate for fruit diameter, and low for TSS and fruit firmness. The segregating populations from the hybrid Calimosa are more promising for the selection of papaya lines. Ramos *et al.* (2014) stated that a safe way to identify genetic material with characters of interest is the evaluation based on genotypic values, which has an additive nature and greater accuracy. These estimates can be obtained using mixed linear models of the restricted maximum likelihood/ best linear unbiased prediction (REML/BLUP). The REML/BLUP methodology was also reported efficient in identifying the papaya genotypes of the Formosa group to be selected in order to obtain greater mass and number of fruits, pulp thickness and soluble solids content. Due to the greater additive heritability within the progenies and the individual coefficient of variation, the selection practised between and within the half-sib families could provide greater genetic gains when compared with the selection based on the average of the progenies.

3. Hybridization The main methods of breeding papaya are: (i) the Hawaii method used for improving gynodioecious cultivars; and (ii) the Yarwan method which is followed in Australia for improving dioecious cultivars (Aquilizan, 1987; Manshardt, 1992). Each breeding method requires seven or eight generations (about 15 years) to produce a new strain or cultivar.

Some hybrid varieties have been developed by intervarietal or intergeneric crosses. At TNAU, Coimbatore three varieties have been developed, viz. CO-3 (CO-2 x Sunrise Solo), CO-4 (CO-l x Washington), CO-7 (Coorg Honey Dew x CPB-5) and CO-8: It is a red pulp dioecious variety developed by initial selective hybridization of CO-2 (yellow pulped) with red anthered male followed by intermating and repeated selection in segregating population for red pulp colour. However at IIHR, Bengaluru two hybrids named as Arka Surya (Sunrise Solo x Pink Fleshed Sweet), and IIHR-54 (Waimanlo x Pink Fleshed Sweet) were developed. Hybrid HPSC-3 (Tripura Local x Honey Dew) was developed at the ICAR Research Complex, Tripura.

Arka Prabhath: Developed from the crosses (Surya x Tainung-1) x Local Dwarf.

Eksotika (cross of Subang 6 with Sunrise Solo): This is a self pollinated hermaphrodite cultivar.

Eksotika II (Eksotika line 19 × line 20): This is a hermaphrodite cultivar. The fruit is large (600–800g) with a smooth skin and high tolerance to freckles.

Frangi (hybrid of inbred lines LSGC 1 and LSG 2)**:** This was developed by Malaysian Agrifood Corporation Berhad (MAFC). It is a high yielding hybrid.

MSXJ (cross between J hybrid and MST): This is a hermaphrodite and carpellody-free cultivar.

P10 (cross of Eksotika (P4) with Ng Toh Seng (P5)**:** The fruits are bigger (600–1000g) than those of P15 (400–700 g).

Tainung 1 (cross between Costa Rica Red and Sunrise Solo)**:** Its female fruits are oblong and hermaphrodite fruits are long and pear shaped.

Tainung 2 (F_1 hybrid of Thailand and Sunrise Solo)**:** This is a red fleshed gynodioecious variety.

Tainung 5 (cross between Florida (FL-77-5) and Costa Rica Red)**:** Developed in Taiwan it has resistance to PRSV disease.

UH Rainbow (Sunup × Kopho Solo): This is a gynodioecious hybrid developed at the University of Hawaii at Manoa.

Cariflora (cross of K2 x K3 lines): Developed at the Tropical Research and Education Center of the University of Florida at Homestead it is tolerant to PRSV.

Ishigaki Sango and **Ishigaki Wondrous** hybrid have been bred at JIRCAS, Japan.

Intergeneric Hybridization

Attempts have been made since long to develop hybrids between *C. papaya* and *V. cauliflora.* The hybrids lack vigour and rarely survive until flowering. If they do, these are infertile. Molecular analysis has shown that *cauliflora* is the most distantly related *Vasconcella* species from *C. papaya*. Hence the crosses between these two species have met with limited success due to infertility of the hybrid progeny. At Griffith University in Australia numerous hybrids between *Carica* and *Vasconcellea* have been produced, including the species *V. goudotiana*, *V. parviflora* and *V. quercifolia.* Papaya has been hybridized with PRSV-P resistant *V. quercifolia* and the virus resistance backcrossed into papaya (Drew, 2014). Backcross 4 and 5 generations exhibited resistance to both Philippine and Australian strains of PRSV-P. A more durable resistance to PRSV-P has been demonstrated by *V. pubescens* which appears to be immune to PRSV-P. As F_1 populations between papaya and *V. pubescens* are infertile, *V. parviflora* is being used as a bridging species to transfer PRSV-P resistance from *V. pubescens* to papaya.

Iyer *et al.* (1987) reported that hybridization between *Carica papaya* and *Vasconcellea cauliflora* easily produced F_1 plants. After backcrossing with *C. papaya,* they evolved a line, 21–19 showing resistance to mosaic virus with normal fruit quality. Magdalita *et al.* (1998) have also described an efficient protocol for hybridization between *C. papaya* × *V. cauliflora* that involved the use of highly viable pollen of *V. cauliflora* produced during summer, autumn and spring; the use of an isolation time ranging from 90–120 days post pollination of hybrid embryos and the use of the most compatible papaya varieties for crossing with *V. cauliflora*. Subramanyam and Iyer (1981) and Iyer and Subramanyam (1984) stated that F_1 hybrid of *V. cauliflora* × *V. monoica* when crossed with *C. papaya* resulted in fertile hybrids. Crosses of *V. cauliflora* and *V. monoica* were incompatible with *C. papaya*, but that of *V. cauliflora* × *V. monoica* were compatible. Crossing of Arka Surya x *V. cauliflora* was carried out for PRSV tolerance. The progenies were found resistant when field planted after artificial inoculation.

Pujar *et al.* (2019) carried out intergeneric hybridization to know the cross compatibility between papaya varieties (Arka Prabhath, Arka Surya and Red Lady) with *Vasconcellea* species (*V. cauliflora*, *V. cundinamarcensis* and *V. parviflora*). Arka Prabhath was found to be a good combiner with all the three *Vasconcellea* species. Among the male parents, *V. cauliflora* was a good combiner with all three female parents (Arka Prabhath, Arka Surya and Red Lady). The cross combination Arka Prabhath × *V. cauliflora* resulted in a higher mean number of fertile seeds (heterotic combinations) which helps in generating a greater number of intergeneric populations for screening of PRSV.

Interspecific Hybridization

Interspecific hybridization between cultivated and wild relatives of papaya has been reported by many researchers. The interspecific hybridization was found successful among *Vasconcellea* species. However, no success was achieved with papaya. The first systematic hybridization program by using *Vasconcellea* germplasm was reported in Venezuela by Horovitz and Jiménez (1967). Crosses between *V. pubescens* and *V. parviflora* produced vigorous F_1 hybrids which varied in sex expression and comprised female, males and hermaphrodites. Zerpa (1959) made several interspecific crosses, viz. *V. cauliflora* × *V. microcarpa*, *V. monoica* × *V. cauliflora*, *V. monoica* × *V. cundinamarcensis*. Only the cross *V. cauliflora* × *V. cundinamarcensis* produced multivalents at metaphase I, while others produced bivalents indicating high interspecific genetic affinity. Mekako and Nakasone (1975) while making interspecific hybrids using six *Vasconcellea* species obtained viable seeds from the combinations *V. monoica* × *V. goudotiana*, *V. parviflora* × *V. goudotiana*, *V. goudotiana* × *V. monoica* and *V. cauliflora* × *V. peltata*. All F_1 and F_2 plants were resistant to PRSV-P.

Overcoming barriers to Hybridization

Conventional intergeneric hybridization of *C. papaya* with *Vasconcellea* species has been difficult owing to reproductive barriers (Manshardt and Wenslaff, 1989). The interspecific incompatibility is less common within *Vasconcellea* and cases of spontaneous hybridization have been reported. Meiosis is often regular in hybrids (Zerpa, 1959). Poor success can be attributed to sexual incompatibility followed by post-zygotic incongruities including embryo abortion, poor seed viability and sterility in progeny obtained following hybridization between two genera (Horovitz and Jiménez, 1967; Manshardt and Wenslaff, 1989). It was possible to overcome the incompatibility by using techniques such as embryo rescue, use of nutrient solution and bridge species *V. parviflora* (Manshardt and Drew, 1998; O Brien and Drew, 2009; Jayavalli *et al.*, 2011).

Embryo rescue technique: Papaya embryo culture has been applied to produce intergeneric hybrids through *in vitro* embryo rescue. The embryo rescue protocol was developed for 90 day old *C. papaya* embryos (Magdalita *et al.*, 1996). A highly efficient protocol was used for rescue and germination of immature embryos of *C. papaya* × *V. cauliflora*. The *C. papaya* × *V. cauliflora* hybrids lacked vigour and were generally infertile. Subsequently, the protocol was adopted to produce hybrids between *C. papaya* and other PRSV-P resistant species, *V. quercifolia* and *V. pubescens*. Hybrid plants grew vigorously in the field and a few *C. papaya* × *V. quercifolia* plants produced some viable pollen. Hybrid plants between *C. papaya*, *V. goudotiana* and *V. parviflora* were also produced. Plants of all these crosses could be grown in glasshouse or in field. The success of intergeneric crosses depends on the genotype of the papaya parent and the time of year when pollen was collected and the post-pollination time when the hybrid embryos were rescued (Magdalita *et al.*, 1998). Azad *et al.* (2012) achieved plant regeneration and somatic embryogenesis through intergeneric hybridization of the papaya varieties Shahi and Ranchi with *V. goudotiana* and *V. cauliflora*, respectively. The immature 90 day old hybrid embryo of *C. papaya* cv. Shahi × *V. cauliflora* showed the highest percentage of germination as well as plant regeneration.

Use of bridge species: The compatibility between *Carica* and *V. parviflora* can be exploited to bridge hybridization between *C. papaya* and PRSV-P resistant species (Drew *et al.*, 1998). The PRSV-P resistance gene from *V. pubescens* was successfully backcrossed into *V. parviflora* from F_2 hybrids. In the F_1 hybrids between *C. papaya* and *V. parviflora*, 45% pollen fertility was obtained thereby indicating *V. parviflora* as a potential bridging species between *C. papaya* and *V. pubescens.*

Use of nutrient solution: The intergeneric barrier could be broken by applying 10% sucrose on the surface of stigma at the time of crossing before dusting pollen to overcome the pollination barrier between *C. papaya* cv. Pusa Nanha and *V. cauliflora* (Jayavalli *et al.*, 2011).

Shortening the breeding cycle: A short juvenile period is important to enable evaluation of flowering and fruiting traits of hybrid seedlings. The breeding cycle of papaya can be shortened by 10 or 20 days by using embryo culture and by treatment with 100 μl ethrel of 55-day-old fruit. Embryo development was enhanced by ethrel treatment of 55-day-old fruit. The seedlings from seeds of ethrel treated fruit grew more vigorously. The period from pollination to seedling establishment which normally takes 6–9 months could be shortened approximately by 3 months (Tamaki *et al.*, 2011).

4. Backcross breeding: Backcross breeding method has been frequently used in papaya improvement programmes. In dioecious lines, development of inbred lines through repeated back crossing is practised for 6-7 generations followed by sib-mating after keeping male and female in equal proportion. The dioecious papaya varieties such as Pusa Dwarf and Hortus Gold rapidly degenerate due to outcrossing (Saran *et al.*, 2016). Backcrossing is often used for the recovery of desirable traits in such cultivars.

Ram (1993) suggested use of backcrossing in the improvement of inbred lines for use in developing papaya hybrids. This method of breeding was also used successfully for increasing adaptability and fruit size. The variety Eksotika was developed by backcrossing Sunrise Solo and Subang 6 (large fruit) as the non-recurrent parent (Chan, 1987). Poamoho Gold was a backcross F_1 hybrid between Kapoho and Rainbow F_2 (Fitch *et al.*, 2002). Backcrossing was also carried out for transferring PRSV resistance from the donor parent *V. quercifolia* to the cultivated varieties (Drew *et al.*, 2005).

5. Heterosis breeding: Improvement in the F_1 progeny over the best parent involved in a cross has been observed in papaya. Chang and Wu (1974) and Subramanyam and Iyer (1984) demonstrated marked heterosis in crosses between varieties. Ram (1982) reported heterosis up to 11.4% for yield in the cross Solo Yellow × Washington. In crosses Pusa Delicious × Halflong and Pusa Delicious × Homestead, high heterotic responses (47.34% and 39.77%, respectively) were observed with respect to fruit yield over the mid-parent.

Chan (1992) reported heterosis in yield and fruit quality in F_1 hybrids developed from closely related sib crosses. Hybrid vigour is directly related to the degree of genetic divergence of the parents involved. However, high genetic divergence does not necessarily increase the expression of heterosis (Duarte *et al.*, 2003).

The first transgenic papaya from Hawaii Rainbow with resistance to PRSV is a F_1 hybrid between the conventional Kapoho and transformed SunUp (Gonsalves, 1998). Chan (2001) evaluated two hybrids derived from crossing two Eksotika lines (L 19 and L 20) with Sekaki which resulted in the development of Eksotika II. Heterosis (over the better parent) in yield was evident, ranging from 47.5% for L 20 × Sekaki to 72.5% for L 19 × Sekaki. Heterosis in yield increase was mainly due to gain in fruit weight of hybrids rather than increase in the number of fruits.

Heterotic effects for earliness and plant height, as well as fruit yield, weight and number of fruits in different hybrid combinations was reported by Vasconcelos *et al.* (1982). Ram *et al.* (1999) have reported positive heterosis for fruit yield and number of fruits per plant. However, the heterosis was negative for characters such as single fruit weight (average), fruiting length, first fruiting node, plant girth, plant height and days to first flowering. Dwivedi (1998) also indicated a negative influence of fruit length on fruit yield and pulp weight.

6. Combining ability: Ram (1982) observed that Washington × Ramnagar Local had the highest specific combining ability (SCA) for fruit yield per plant followed by Pusa Delicious × Homestead, Waimanalo × Halflong and Pusa Delicious × Halflong. It was the combination of average combiners from both the inbred lines on the one hand and testers on the other which showed the maximum SCA (Ram, 1993). Pusa Dwarf and Sunrise Solo varieties were found to have greater general combining ability (gca) for various characters than cv. Coorg Honey Dew indicating the predominance of additive gene action and the possibility for further selection.

Dinesh *et al.* (1991) observed that the variety Thailand was a good combiner for fruit length, breadth and volume, while Sunrise Solo and Waimanalo were good combiners for quality characters. Pusa Majesty variety showed the highest general combining ability effects for fruit weight and yield. Bhubaneswar Local recorded the highest specific combining ability effect for earliness and quality and Pusa Giant for dwarf growth habit. The cross Pusa Giant × Bhubaneswar Local recorded the highest specific combining ability effects followed by Pusa Majesty × CO.6, and Pusa Delicious × Bhubaneswar Local (Dash *et al.*, 1988).

7. Polyploidy: Hofmeyer (1945) reported polyploidy in papaya. They found that the quality of tetraploid fruit was better than that of the diploid and the fruit was also compact with small seed cavity. But tetraploids were less fertile than diploids as indicated by comparative seed count. However, according to Singh (1955) there was complete sterility in both female and male tetraploids and expressed doubt about their commercial utilization. Zerpa (1957) reported that colchicine induced tetraploid hermaphrodite plants when used as male parent in crosses with a female diploid produced few seeds without endosperm. Two triploid plants were obtained by embryo culture which turned out to be hermaphrodite.

Haploid papaya plant has now been generated by anther culture. The anther derived triploid plants were found to be of short stature and gave high yield of parthenocarpic fruits. The triploids and tetraploids showed lot of potential for exploitation in commercial production of seedless fruits. The aneuploids and polyploids were detected in somatic embryogenesis cultures of papaya and *in vitro* regeneration of polploids was successful.

8. Mutation: Irradiation induced mutation is used as a complementary tool in plant breeding. It is the most appropriate for improving one or two easily identifiable traits in an otherwise well accepted breeding line or commercial variety. Ram and Majumder (1981) developed a dwarf mutant line (160 cm in height) by treating papaya seeds with 15 krad gamma rays. This was later named as Pusa Nanha. This dioecious variety has short petioles and produces about 10 kg fruits per plant. It is suitable for high density planting.

Irradiation of seed of Eksotika was done by 525 Gy for dry seed and by 42.5 Gy for presoaked seed. The M2 population developed by self-pollinating M1 hermaphrodite plants was dwarf and had good vigour as expressed by the longer leaves (Chan and Lam, 2003; Chan *et al.*, 2007). Several M2 mutants showed very good resistance to malformation caused by *Cladosporium*. However, no resistance to PRSV disease was found in the 192 M2 seedlings when inoculated and screened. Seeds of six cultivars (Bh65, Maradol, Shew Mee, Sunrise Solo, Tainung and V3) were irradiated at 0 Gy (control) to 120 Gy using Cobalt-60 with strength of 237.89 Rad min^{-1} by Husselman *et al.* (2016). They observed that irradiation negatively affected seed germination at 60 Gy onwards in most cultivars. Although Sunrise Solo had poor germination, the seed that germinated was of high quality as it produced strong plants with positive characteristics (stunted growth). Irradiation of Bh65 seeds also resulted in stunted plants. Fruit production was negatively influenced by radiation as all treatments produced a lower number of fruit compared with that in untreated control. Radiation did not influence fruit quality.

Kumar *et al.* (2017a) reported that when seeds of Pusa Dwarf were treated with 10 krad gamma rays before sowing produced heavier (1463–1669g) fruits with minimum central cavity and greater pulp thickness than in the untreated control. The treatment also improved fruit quality (total soluble solids, sugars, protein and carotene). In the gynodioecious cultivar Arka Prabhath, treating the seeds with 100 Gy gamma radiation before sowing resulted in first flowering at 73 cm plant height compared with 83 cm in the untreated control. Treatment with 100–500 Gy yielded 31–36 kg fruit per plant compared with 26 kg per plant in the control (Ramesh *et al.*, 2019).

Chan-Tai *et al.* (2003) reported that a hermaphroditic Sunrise papaya mutant (SR*) can be used to develop hermaphroditic progenies by self-pollination and this trait can be transferred to other papaya cultivars.

Hang and Chau (2010) observed that gamma irradiation with 10 to 60 Gy induced precocity and reduced plant height in the local papaya cv. Dai Loan Tim. Four mutants were reported to be superior in plant height at first flowering, plant vigour, fruit yield and quality compared with non-mutants. Pusa Dwarf seeds when treated with 0.50% EMS produced fruits with maximum weight, length and diameter and higher pulp thickness and minimum central cavity. The fruits obtained in plants raised from the seeds treated with 0.50% EMS had significantly higher TSS, sugar, fat, ash, carbohydrate, protein and carotene content and minimum moisture (Kumar *et al.*, 2017b).

9. Biotechnology:

Micropropagation: Micropropagation can stabilize selected dioecious genotypes. Papaya regeneration via organogenesis and somatic embryogenesis has been refined during the past three decades. Embryogenic regeneration from several kinds of immature tissues in the presence of auxins is often difficult to achieve on selective media employed in transformation procedures. However, improved methods of preparing embryogenic cultures are likely to overcome this bottleneck. More attention is needed for production of diploid inbred lines via anther culture.

Marker assisted breeding: The use of marker assisted breeding has great potential to reduce time, cost and effort involved in papaya breeding. Marker assisted selection (MAS) is important for crop improvement. RAPD was exploited to understand the genetic relationship among different varieties of papaya having less defined genetic background. Saraswathi *et al.* (2007) used RAPD markers to determine the pulp colour of gynodioecious varieties Sunrise Solo and Kopho Solo and dioecious selection 9-1D. The extent of polymorphism (47.74%) indicated the existence of a narrow genetic variation among the test types. The arbitrary primer OPC-05 discriminated between the red and yellow types of dioecious papaya. Similarly, primer OPK-13 distinguished the indigenous dioecious form from exotic gynodioecious forms which confirms the hypothesis that the male parent governs the pulp colour in papaya.

MAS is considered to be a simpler strategy for breeding selection than selection based mainly on phenotype. Oliveira *et al.* (2010) demonstrated that the use of SSRs in MAS is an effective procedure for developing inbred lines in that genotyping and selection for desired traits can be carried out in early generations on these homozygous individuals. Priyanka *et al.* (2017) developed expressed sequence tags (ESTs) derived from SSR markers which are considered more valuable as they are derived from the conserved genetic portion. These developed EST-SSR primers are being used for the genetic improvement of papaya, for example to study cross-transferability across genera/species, evaluation of genetic diversity and identification of sex-specific markers.

Genetic transformation: Genetic transformation of papaya using Agrobacterium or microprojectile bombardment has been reported. Type of explants and genotype are important factors for successful genetic transformation using A. tumefaciens of papaya. Fitch *et al.* (1990) developed transgenic papaya plants from somatic embryos after cocultivation with A. tumefaciens strain C58-Z707. Fitch and Manshardt (1990) transformed and regenerated transgenic plants using microprojectile bombarded from immature zygotic embryos and hypocotyls of Sunrise Solo and Kapoho Solo.

Transgenic papaya with pathogen derived resistance carrying the coat protein gene provides effective protection against PRSV strains over a significant period of the cycle of this perennial fruit crop. Additionally, genetically modified papayas with traits for disease resistance and extended shelf life have been extensively screened in field tests. Research to extend postharvest shelf life with sense and antisense versions of the ACC synthase gene is in progress in Hawaii and Australia, and transgenic plants are under observation in the greenhouse.

The papaya genome sequence was published in 2008 and has opened new avenues for papaya improvement by precision breeding, which involves the use of regulatory and functional gene sequences from related genera of the Caricaceae family, and is a logical extension of conventional breeding and genetic transformation. RAPD (randomly amplified polymorphic DNA) map of the papaya genome has been generated and used to identify markers for sex-determination, flowering height, and fruit carpellody.

LITCHI

Origin: South China

Taxonomy status:

Order: Sapindales

Family: Sapindaceae

Subfamily: Nepheleae

Genus: *Litchi*

Botanical name: *Litchi chinensis* Sonn.

Chromosome number: n = 15; 2n = 2x = 28, 30 or 32

Genome size: 550 to 620 Mb

The *Litchi* contains two sub-species that have not been commercialized. Sub-species *philippinensis* is found in the Philippines (Luzon, Sibuyan, Samar and Mindanao) and Papua New Guinea at high elevations, while sub-species *javanensis* has been recorded in the Malay Peninsula and Indonesia. The Philippines lychee has long, oval-shaped fruit with long, thorn-like protuberances. The fruit splits in the middle when ripe displaying an inedible aril that only partly covers the seed. Sub-species *javanensis* is a rare specimen found in Chinese gardens in West Java and Indo-China and has fruit similar to cultivated lychee, but with a thinner aril. It is reported to flower and fruit regularly in the tropics. Many of the Malayan specimens belong to sub-species *chinensis.*

Botany

Tree is medium sized, evergreen, polygamous, reaching 15-20m in height. Branches are brittle and are easily broken by winds. Leaves are arranged alternately, pinnately compound with two to five pairs of leaflets arranged in opposite positions or slightly obliquely along the rachis. The deep green leaflets are elliptical to lanceolate, 2.5–6.4 cm wide and 7.6–17 cm long. The flowers are small (3-6 mm in length), greenish white or yellowish white produced in terminal panicles upto 30cm long, borne in profusion on leafless or leafy, well branched terminal panicles. Flowers are yellow green or brownish yellow and apetatlous with small valvate sepals, a fleshy disc and upto eight stamens. Among the numerous apetalous flowers on the panicle, three sex types are described based on the chronological order of development.

Type I: Functionally male, possesses 6-8 stamens producing an abundance of pollen and lacking ovule differentiation.

Type II: Flower is morphologically hermphrodite functionally, mostly as female with a well developed two carpel pistil and a two lobed stigma. The five to eight stamens do not dehiscence normally and have little viable pollen.

Type III: Flower functions as male, although it is morphologically hernaphrodite with a rudimentry pistil lacking styles and stigma. The six to eight stamens provide viable pollen grains. Approximately 2/3 of the flowers on the panicles are type III and only about 80% of the flowers are potentially fruit providing type (Type II). Their concentration varies with cultivars and temperature.

The fruit is a drupe, 3-4cm long and 3cm in diameter. The fruit skin is covered by pink-red roughly textured rind. The sweet transluscent white flesh (sarcotesta) envelops the seeds. The seed is 2 cm long and 1-1.5 in diameter.

Fig. 7. Panicle, flowers and fruit cluster.

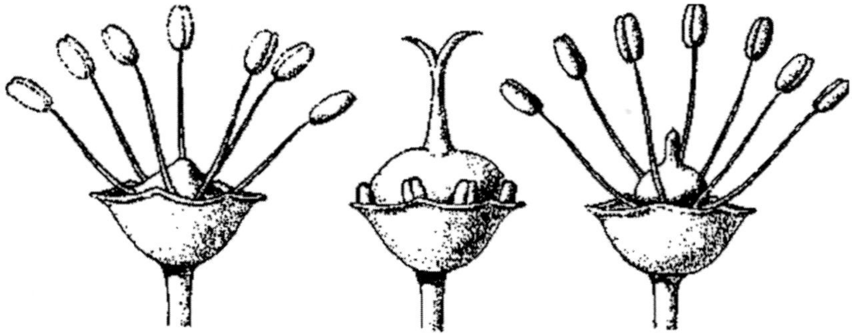

Fig. 8 Flower types (♂, ♀ and ♂)

(Source: Menzel, 2002).

Floral biology

Floral anthesis occurs in cycles of sex types in a period of about 10 days during which only functional male flowers open. Transitional period is of 2-3 days when both staminate and female (type II) flowers open. Then develop only female flowers. Finally male flowers open after 2-3 days of transitional period in 7-10 days cycle depending on inflorescence environment. The anthesis occurs mostly between 04.00 and 08.00 hrs, with a peak at 06.00 hrs. Anther dehiscence was maximum at 10.00 hr and around 90% dehiscence was observed between 07.00 and 12.00 hrs. Anthesis and dehiscence was observed negligible during night. The stigma is receptive 12 hrs from the time they divide into two lobes and anthers remain functional upto 1-3 days after anthesis. Cross pollination is done by honeybees.

Breeding Objectives

1. To breed varieties with wider adaptability to ecological conditions.
2. To breed dwarf and precocious varieties.
3. High and regular fruit yield.
4. To breed varieties bearing high number of fruits of large size.
5. To breed varieties with large fruits, bright red peel and skin colour, firm flesh having acceptable sugar/acid, long shelf life and ability to retain skin colour during storage.
6. To breed varieties with small soft seed, small aborted seeds (chicken tongues) and high aril content.
7. To breed varieties with good fruit quality and high keeping quality.
8. To breed varieties resistant to physiological disorders (fruit cracking) and free from fruit borer.
9. To breed varieties having resistance to water stress, winds, soil salinity, diseases and pests.

Problems in Breeding

1. Long juvenile period.
2. Narrow genetic base.
3. Highly genetic heterozygosity.
4. Lack of information about inheritance pattern of specific traits.
5. Polygenic nature of several important traits.
6. Low female/male flower ratio and premature fruit fruit drop.

Germplasm Resources

More than 500 accessions are preserved in the National Litchi Germplasm Gene Bank located at the Institute of Fruit Tree Research, Guangdong Academy of Agricultural Science, Guangzhou, China, which is the largest litchi germplasm gene bank in the world. National Research Centre on Litchi, Muzaffarpur, India is national repository for litchi in India. It has the largest collection (52 accessions) of litchi. CISH, Lucknow (35), IIHR, Bengaluru, TNAU, Coimbatore, Rajender Prasad Central Agriculture University, Pusa also maintain the germplasm of litchi.

Sources of desirable traits:

Giant fruit: Edanil.

Large fruit size: Bombai and Edanil.

Red skin colour: Early Seedless, Muzaffarpur, Bombai, Gulabi and Dehradun.

Good Quality: Muzzaffarpur, Calcutttia, Late Seedless and Gulabi.

Soft and sweet juice: Early Seedless, Rose Scented, Bombai and Calcuttia.

Pleasant flavour: Early Seedless, Bombai, Calcuttia and Gulabi.

Seedlessness: Late Seedless and Early Seedless.

Small seed: Bedana, Nuomict, Lingshan and Xiangli.

Early maturity: Dehra Rose, Shahi, Shanyuedong, Red Shahi, Early Seedless and Rose scented.

Medium maturity: Do Zoo, Fei, Zi Xiao, Hei Ye, China and Manadaraji.

Late Maturity: Late Seedless, Dehradun, Gulabi, Chi Ye, Huai Zhi, China Late Bedana, Xuehauzi, Late Seedless, Dehradun and Gulabi.

Hot and dessicating winds: Late Seedless, Calcuttia, Swarna Roopa and Late Seedless.

Inheritance Pattern

- High heritability and high genetic advance for fresh seed weight, fruit weight and volume and pulp weight.
- Positive strong correlation between fruit and seed weight and total sugar, ascorbic acid, protein and tryptophan contents.
- Path analysis indicates the direct positive effect of the peel and seed coat and aril size and negative effect of aril on embryo.

- Positive correlation between small fruit and bright aril colour.
- Positive and strong correlation between bright fruit colour and TSS and dark colour and TSS.

Breeding Methods

1. Introduction: There is need to introduce promising cultivars Fay Zee Siu, Kwai May, No May Chee, Haak Lip from China.

2. Selection: Most of the commercial cultivars of litchi are selections from open pollinated seedlings. Rajendra Sabour variety was developed by clonal selection. Kaimana is a open pollinated seedling selection from Hei Ye. The other selections are Groff (from Haklap), Brewster (from Chenzi), CHES-2 (clonal selection from Bombai at CHES, Ranchi), Ambika Litchi-1, Indira Litchi 2 (selections made at College of Agriculture and Research Station, Ambikapur), Saharanpur Selection (local selection), Rajendra Sabour Bedana (Seedling selection from Bedana) and Swarna Roopa (local selection at HARP, Ranchi).

3. Hybridization: Intervarietal crosses at BAU, Sabour between Purbi and Bedana gave hybrids namely Rajendra Hybrid-235 (Purbi x Late Bedana), Sabour Madhu (Purbi x Bedana) and Sabour Priya (Purbi x Bedana). Among the hybrid progeny, some of the plants produced seedless fruits, which could have interesting implications for selection of superior seedless, interspecific fruit.

Longan (*Dimocarpus longan*) and litchi are partially sexually compatible, and intergeneric hybrids can be recovered, but only if litchi is the maternal parent. Intergeneric crosses between *Dimocarpus longan* and *Litchi chinensis* proved to be successful with similar type of phenotypes. A seedless litchi variety was produced by conventional breeding in Hainan, China.

4. Biotechnology:

Micropropagation: Depending on the species and cultural conditions, *in vitro* propagation can be achieved by enhanced axillary shoot proliferation (shoot culture), nodal culture and de novo formation of adventitious shoots through shoot organogenesis. *Litchi chinensis* is recalcitrant *in vitro*. Shoot or bud culture is very difficult to achieve. Embryogenesis and plantlets have been obtained by culture of anthers and immature embryos and these protocols would be the best basis for further *in vitro* studies. Somatic embryogenesis and plant regeneration has been claimed in genotypes of litchi.

Molecular markers: Use of molecular markers for identification and detection of cultivars and hybrids can help in reducing the breeding cycle and making it more efficient. Constructing a high dense molecular linkage map for litchi based on AFLP and RAPD markers can provide a strong foundation for QTL location or molecular assisted breeding.

A limited number of SSR markers were developed for lychee and used to evaluate germplasm collections. In contrast to RAPD and AFLP marker studies, Viruel and Hormaza (2004) analyzed 21 lychee cultivars using SSR markers and found a very narrow genetic base. Li *et al.* (2006) also found limited genetic diversity by using SSR markers in evaluating 58 lychee cultivars from Hainan Island.

Genetic tranformation: *Agrobacterium tumefaciens* mediated genetic transformation has also been reported. Cloning of ACC oxidase gene to enhance the storage life of fruit has been attempted.

Conservation: Production of synthetic seeds in litchi by encapsulating somatic embryos, shoot buds, etc. can be helpful in propagation and short term storage and germplasm exchange and needs intensive research. *In vitro* conservation of embryogenic callus has been identified as one of the methods to conserve the germplasm of litchi. It reduces space requirement and labour cost.

POMEGRANATE

Origin: Regions between Transcaucasian-Caspian and northern Turkey.

Taxonomy status:

Order: Myrtales

Family: Lythraceae

Genus: *Punica*

Botanical name: *Punica granatum* L.

Chromosome number: 2n = 2x = 16, 18 (double flower type)

Genome size: 274 Mb

The genus *Punica* contains two species, *P. granatum* L. and *P. protopunica* Balf. f. *P. granatum* is the only species cultivated while *P. protopunica* is a wild species endemic to Socotra Island, Yemen. Some taxonomists classify the dwarf *P. granatum* var. *nana* as a distinct species, namely *P. nana* L. *Punica granatum* has two subspecies viz. *Chlorocarpa* and *Prophyrocarpa.*

Botany

Pomegranate is shrub or small deciduous or evergreen tree growing 5 to 10 m tall with multiple spiny branches. The colour of the bark of young branches depends on the variety. The bark colour varies from pink to purple, while in others it is light green with pink-purple spots or stripes. Leaves are opposite or sub-opposite, glossy, narrow oblong or oblanceolate, entire, 3–7 cm long and 2 cm broad. The flowers are short peduncled or subsessile. Flowering occurs about 1 month after bud break on newly developed branches of the same year, mostly on spurs or short branches. Flowers can appear solitary, in pairs or clusters. Flowers are of three types viz., staminate, intermediate and hermaphrodite. The calyx of the hermaphrodite flower is urceolate with a broad well developed ovary. Male flowers are smaller with a companulate (bell shaped) calyx and rudimentary ovary. The intermediate flowers are tubular. The intermediate forms exhibit various degrees of ovary degeneration. The sepals are 5 to 8 fused in their base and form a red fleshy vase. The sepals will not drop with fruit set but will stay as an integral part of the fruit as it matures, generating a fruit crowned with a prominent calyx. The flower has 5 to 8 petals. Their number usually equals the number of sepals. The petals, which alternate with the sepals, are separated and have a pink-orange to orange-red colour depending on the variety. The petals are obovate, very delicate and slightly wrinkled. The multiple long stamens are inserted into the calyx walls in a circle and frequently number more than 300 per flower. They have an orange red filament and yellow bilocular anthers that remain attached to the prominent calyx. Heterostyly is common. Hermaphrodite flowers are usually homostylous or pin eyed (i.e. the stigmas are on the same level or higher than the anthers) and male flowers are thrum eyed (i.e. the stigmas are beneath the level of the anthers). Nectaries are located between the stamens and the ovary base. Ovary is inferior with many locules (6) usually superimposed in two whorls, the lower with axile, the upper with parietal placentation. The fruit is botanically known as balusta. The pomegranate fruit is berry-like with a leathery rind (or husk) enclosing many seeds surrounded by the juicy arils, which comprise the edible portion of the fruit. The aril juice sack is composed of many epidermal cells. According to cultivar, arils range from deep red to virtually colourless, whereas the enclosed seed varies in content of sclerenchyma tissue, which affects seed softness. The number of locules and arils (and enclosed seeds) varies. The fruit has a prominent calyx, which is maintained until maturity and is a distinctive feature of the pomegranate fruit. The husk is comprised of two parts, the pericarp, which provides a cuticle layer and fibrous mat, and the mesocarp, which is the spongy tissue and inner fruit wall where the arils attach. Septal membranes are the papery tissue that further compartmentalizes groups of arils, but arils do not attach to this tissue. The number of seeds in a pomegranate can vary from 200 to about 1,400.

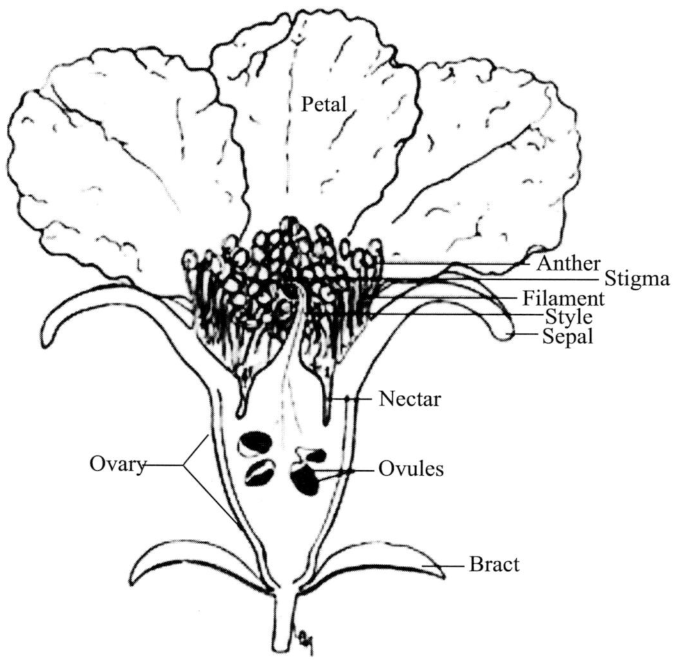

Fig.9. L.S. of pomegranate flower

Floral Biology

Flowering in pomegranate occurs almost throughout the year in tropical climate whereas in the subtropics and temperate region, it flowers once a year. Time span between the start of flower bud elongation (growth) and anthesis varies between 14 and 28 days depending on the variety and climatic conditions. Thus, in subtropical climate of the northern hemisphere, flowering occurs from the last week of March till the second week of May. Several distinct flushes of flowering on the same tree occur quite frequently. In subtropical central and western India, there are three distinct seasons for flowering, i.e. Ambe Bahar (January-February flowering), Mrig Bahar (June-July flowering) and Hasth Bahar (September-October flowering). The flower size varies in different types of flowers in same cultivar and among different cultivars. The maximum anthesis occurs between 6 AM and 10 AM. Dehiscence may start a few hours prior to anthesis or somewhat later depending on the cultivar. It is completed 7 to 15 hrs later or may take 24 hrs for completion. The exact time span depends upon cultivars and climatic conditions. High atmospheric humidity and temperature may hasten dehiscence to a certain degree. Stigmas are found to be receptive a day before anthesis. The peak receptivity coincides the time of anthesis. Later the receptivity, as determined by pollination tests, declines gradually and ceases altogether at the end of the 3rd day. Self and cross pollination occur but cross pollination results in higher fruit set. Wind pollination seems to be insignificant because the concentration of pollen in the atmosphere was found to be quite low for wind pollination. Large numbers of insects visit the flowers which include black ants, honey bees and lemon butterfly.

Breeding Objectives

(i) To develop vigorous growing plants with thornless twigs

(ii) To develop high yielding, fruit cracking resistant cultivar with reddish or blood red bold arils, soft seeds and sweet juice (high sugar and less acid).

(iii) To develop short duration varieties.

(iv) To develop high yielding varieties having big size dark red fruits and bold high acid arils for making anardana.

(v) To develop resistant cultivars to fruit borer (*Virachola isocrates*), fruit rot (*Phomopsis* sp.) and soil borne fungi, especially *Fusarium*.

(vi) To develop variety with attractive rind color.

(vii) To develop a cultivar having very smooth but tough outer rind to protect from thrips and mites infestation and withstand long distance transport.

(viii) Development of oily spot resistant variety.

Limitation in Breeding

1. High heterozygosity.
2. Long juvenility.

3. Inter and intra-species incompatibility.
4. Sterility.
5. Information on inheritance pattern is lacking.

Germplasm Resources

More than 1,000 accessions were assembled in the Turkmenistan Experimental Station of Plant Genetic Resources. A collection of 200-300 accessions is maintained in Azerbaijan, The Ukraine, Uzbekistan and Tajikistan. Cultivars of diverse genotypes (238 reported cultivars) are reported within China. The USDA National Clonal Germplasm Repository (NCGR), in Davis, California has nearly 200 accessions. The collection includes accessions from all over the world including Turkmenistan, Russia, Iran, and Japan.

The ICAR-National Research Centre on Pomegranate, Solapur, is the national repository for pomegranate in India. It has the largest collection of 362 accessions. Pomegranate accessions are maintained in field genebanks at CIAH, Bikaner (195 accessions), IIHR, Bengaluru (265), PAU, Ludhiana, RRS, Bawal, CCSHAU, Hisar and MPKV, Rahuri.

Sources for useful traits:

Ornamental: Double Flower, Nana.

Early ripening: Yushiliu 4.

Soft seeded: Hongmanaozi.

Good quality fruit: Zaoxuan 018 and Zaoxuan 027 and Duo Hong 1, Duo Qing 11 and Duo Bai 2.

Red colour: Kamel, Akko.

Yellow skin: Lefan, Kabul Yellow.

High yield: ME15, ME16, Agaria de Blanca, Alber de Bianca and Borde de Albatera.

Vitamin C: Siyah.

Less fruit cracking: Jalore Seedless, Karki, Guleshah, Shirvan, Burachni, Asperonskii Krasnyi and Bedana.

Tolerance to sandy alkaline soils: Yushiliu 1 and Yushiliu 2.

Tolerance to moisture stress: Taishan, Dahongshiliou and Goma Khatta.

Tolerance to bacterial blight: Daru and Nana.

Inheritance Pattern

1. Recessive markers for yellow foliage colour and rosette-form growth, originated from a mutant of Kabul Yellow, was used for studying the mode of pollination.
2. High acidity was always dominant to low acidity, pink aril colour was dominant to white colour, and hard-seededness was dominant to soft seededness.

3. The hard-seeded cultivars as a group had significantly higher fruit weight and volume than semisoft and soft-seeded pomegranates.
4. The rind weight, acidity percentage, fruit weight, arils per fruit, yield per tree and number of fruits per tree exhibit high heritability and high genetic advance.
5. No specific gene has been identified for yield and it is considered to be a polygenic trait involving most genes.
6. Dominant gene (R) confer pink or red aril colour and the recessive gene (W) imparts yellow pigmentation of the petiole base, leaf margin, flower bud and fruit rind.
7. Soft seededness is controlled by multigenes.
8. Bacterial nodal blight is controlled by recessive genes.

Breeding Methods

1. Introduction: The introduction of deciduous varieties namely Sellimi, Roman, Chokab, Suffami, Wellingi, Ras-el-Baghi from Baghdad, Palestine and other Mediterranean countries failed to establish in the warm Indian climate. Similarly, introductions like Gulsha Rose Pink, Shirin Anar and Gulsha Red from the erstwhile USSR gave poor performance owing to their deciduous habit. However, some introductions like Gulsha Red Pink, Appuli, Gulsha Red, Lupania, Bedana Sedana, Kandhari and Khog are being used in breeding programmes to induce dark red aril pigment in the popular cultivars. Some commercial cultivars having soft seeds and dark red arils, viz. Wonderful from USA, Males, Be Hastah, Alah, Agha Mohammad Ali, Post Sephid, Sirin from Iran and Rannyiz G-1-8-23, Rannyiz G-1-3- 34, Cherenyj G-1-8-7 and few cultivars from Tunisia have been introduced.

2. Selection: Genetic improvement in pomegranate has for centuries depended on selection from seedling variability followed by their clonal propagation. Hence, seedling selection from the existing variability assumes great significance.

Jalore Seedless: Clonal selection from Ahore area of Jalore.

Ganesh: This soft seeded selection was made at Ganeshkhind Gardens, Pune from open pollinated seedlings of hard seeded Alandi by Dr. Cheema in 1936. It is also known as GBG-1.

RCRI: It is a selection from Alandi cultivar.

P-13: It is a selection from Muscat variety.

P-16: This is a selection from Muscat variety. The tree is evergreen and spreading type. Fruit is round and smooth and pink in colour with reddish tinge.

P-23: It is a selection from Muscat variety.

G–137: It is a clonal selection in 1989 from Ganesh at MPKV, Rahuri.

Jyoti: The variety was released from GKVK campus of UAS, Bengaluru. This is a selection from the seedlings of Bassein Seedless.

CO-1: It is clonal selection from Tamil Nadu Agriculture University, Coimbatore.

Yarcaud-1: It is a clonal selection made at Horticulture Research Station, Yercaud, Tamil Nadu.

Phule Bhagawa Super (Selection–4): It is a selection from Bhagwa cultivar at MPKV. It performed better in respect of colour of fruit, juice percentage, bold and soft arils.

Granada: Originated in Lindsay, California as bud mutant of Wonderful. Ripens one month earlier than Wonderful.

***Cultivars developed in Israel*:**

Emek: A red and very early ripening cultivar.

Shani-Yonay: It is a seedling selection from an unknown source.

Amel: It is a selection akin to Wonderful.

***China*:**

Zaoxuan 018: It is a chance seedling from Dahongpao

Zaoxuan 027: It was selected from Daqinpi.

3. Hybridization

The technique: Select appropriate young bud, which will open the next morning. Insert forceps through petals or remove one petal and take out the stamens. Take care that no anther is left inside the flower, and the ovary is not injured. This can be done easily as stamens emerge from within upper half or more of hypanthium (calyx tube) and are distant from ovary. Bagging and tagging is done. Collect the desired anthers in yellow stage and incise them with needle and transfer the pollen the next morning with the help of a brush on to the receptive stigma of the emasculated bud. Bagging and labelling is done. Crossing is done in the morning hours before 11 AM. The steps involved are emasculation (removing of anthers) by using forcep, pollination with the pollen of desired male parent, covering with butter paper bag and labelling.

Hybrids developed at several research organizations are described below:

Bhagwa (Kesar): It is a selection from F_2 progeny of a cross between Ganesh and the Russian cultivar Gul-e-Shah Red. This is the most popular variety of pomegranate in Maharashtra and covers maximum area under cultivation. The variety is known by various names viz., Shendri, Asthagandh, Kesar, Jai Maharashtra, Red Diana and Mastani.

Mridula: It is a seedling selection from an open pollinated F_2 population of a cross between Ganesh and Gul-e-Shah Red.

Arakta: Seedling selection at MPKV., Rahuri in 2003. The arils are bold with deep blood red colour.

Ruby: This variety is a complex hybrid between Ganesh, Kabul, Yercaud and Gulsha Rose Pink. It was released after selection from the progeny of {[(Ganesh x Kabul) X Yercaud]-F_2} X {(Ganesh X Gule Shah Red)-F_2}.

Phule Arakta: It is a selection from F_2 progeny of the cross between Ganesh and the Russian cultivar Gul-e-Shah Red.

Phule Bhagwa Super: Selection from Bhagwa. It was developed by MPKV, Rahuri in 2013.

Solapur Lal: Bhagawa x [(Ganesh x Nana) x Daru]

Anardana type:

Amlidana: It is a hybrid (F_1) between Ganesh and Nana (dwarf ornamental miniature variety) released from IIHR Bengaluru.

Goma Khatta: It is a hybrid between Ganesh x Daru.

Solapur Lal: Bhagawa x [(Ganesh x Nana) x Daru]

4. Heterosis: Karale and Desai (2000) measured heterosis manifested in fruit characters by the individual hybrids over mid parental value. They found that heterosis values were maximal in percentage juice and aril weight.

5. Polyploidy: Pomegranate tetraploids have been produced by colchicine treatment of shoots. The tetraploid plants that were generated had shorter roots, wider and shorter leaves and flowers with larger diameter as compared to diploids.

Shao *et al.* (2003) obtained tetraploid plants in pomegranate in Nana cultivar by colchicine treatment of *in vitro* propagated shoots.

6. Mutation: The varieties like Hongma-naozi, Taihanghong and Hongyushizi were isolated as natural mutants (bud sports) from different varieties in China.

Promising mutants were bred by treatment with N, N-dimethyl N-nitrosourea. Among the most promising forms produced was the soft seeded seedling "Sverkhrannil" (meaning super early) with fruit ripening in August.

7. Biotechnology:

Micropropagtion: Attempts have been made to regenerate plants by using leaf and shoot tip explants. Somatic embryogenesis was done from petal explants. Zeatin riboside (ZR) was reported to induce bud break and axillary shoot proliferation in mature nodal segments of cultivar Ganesh. The immature or mature zygotic embryos from unripened or ripe fruits, respectively, have been utilized for the induction of somatic embryogenesis.

Molecular markers: Enzyme based marker was used to identify the genetic variability among the existing genotypes. Molecular markers, such as AFLP, RAPD, and ISSR, were reported by several groups. Pomegranates are highly polymorphic. But others reported that the degree of polymorphism in pomegranate was surprisingly low. In some of these studies, the apparent phenotypical differences observed among pomegranate cultivars were not reflected in the polymorphism of the molecular markers. Obviously many more markers should be isolated from pomegranates to make them useful for breeding and for evolutionary studies. Work in this direction involves the construction of pomegranate genomic libraries potentially containing microsatellites. Till now about 26 SSR primer pairs have been used for screening pomegranate genotypes.

Genetic transformation:

Pomegranate is a difficult-to-transform plant. *Agrobacterium* mediated transformation method developed for molecular analysis confirmed the integration of the nptII transgene in transformed shoots (ValizadehKaji and Ahmadreza, 2017). The transgenic shoots were rooted and successfully acclimatized. So far there is only one report of development of transgene and that too in a dwarf ornamental pomegranate.

DATE PALM

Origin: Mesopotamia (southern Iraq along with adjacent portions of northern Syria, southeast Turkey and southwest Iran).

Taxonomy status:

Order: Palmae

Family: Arecaceae

Subfamily: Coryphyoideae

Tribe: Phoeniceae

Genus: *Phoenix*

Botanical name: *Phoenix dactylifera* L.

Chromosomes number: n = 18; 2n = 36

Genome size: 605.4 Mb

Related species are *P. acaulis, P. canariensis, P. paludosa, P. reclinata, P. rupicola, P. roebelenii* and *P. sylvestris.*

Botany

Date palm trees can reach about 21–23 m in height and 40–50 cm trunk diameter. It has a non-branching trunk. The leaves are 4–6 m long, with spines on the petiole, and pinnate with about 150 leaflets. The leaflets are 30 cm long and 2 cm wide. *Phoenix dactylifera* including other *Phoenix* spp. are dioecious and produces clusters of male and female flowers on separate palms. These emerge from the axils of leaves of previous year's growth and are enclosed in a hard, fibrous cover (the spathe) during the early stage of flower development. The Inflorescences of female and male trees differ in morphology. The unisexual flowers are pistillate (female) and staminate (male) in character. They are borne in a big cluster (inflorescence) called spadix or spike, which consists of a central stem called rachis and several strands or spikelets (usually 50-150 lateral branches). Each spikelet carries tiny flowers as many as 8,000 to 10,000 in female and more in male inflorescence. The male inflorescence is crowded at the end of the rachis, while branches of the inflorescence of the female cluster are less densely crowded at the end of the rachis. These characteristics allow the recognition of the sex of inflorescences before opening.

The male flower is sweet-scented, yellow in colour and normally has six stamens, surrounded by waxy scale-like petals and sepals (3 each). Each stamen is composed of two little yellowish pollen sacs. The female flower has a diameter of about 3 to 4 mm and has rudimentary stamens and three carpels closely pressed together and the ovary is superior (hypogynous). The three sepals and three petals are united together so that only tips diverge. On opening, the female flowers show more yellow colour while the male ones show white colour dust, produced on shaking. The pollen sacs usually open within an hour or two after the bursting of the spathe. Only one ovule per flower is fertilised, leading to the development of one carpel which in turn gives a fruit; the other ovules abort. The aborted carpels persist as two brown spots in the calyx of ripe fruits. The fruit is drupe, oval-cylindrical, 3–7 cm long, and 2–3 cm diameter, range from bright red to bright yellow in colour, fleshy pericarp with a single grooved seed about 2–2.5 cm long and 6–8 mm thick.

1. Female flower 2. Male flower

Plate 3 Types of Date flower

(Source: Chao and Krueger, 2007)

Floral Biology

The inflorescence arises pushing through the sheaths in the axils of the leaves. The sheaths crack longitudinally just before flowering. Only the portion of the rachilla that bears flowers is exposed. The fruit stalk elongates 50 to 60 days after anthesis and pushes out the portion of the inflorescence that does not bear flowers to a length of 60 to 120 cm. Date palm is wind pollinated but insect pollination is also possible. Natural pollination occurs when about equal number of male and female plantsare there in the orchard. However, pollen produced from one male palm can pollinate up to 100 females. Therefore, commercially, few male trees are grown in date garden and pollen collected from these is used for artificial pollination. Pollination of 60% to 80% of the female flowers results in adequate fruit set. Cultivars differ greatly in fruit set percentage. Incompatibility or partial incompatibility between different female and male cultivars is common but is not well understood.

The fruit normally develops after fertilization from one of the three carpels within each pistillate. Natural fruit drop occurs 25 to 35 days after spathe cracking. Some cultivars have a second fruit drop about 100 days after spathe cracking.

Pollen from the palms of other *Phoenix* species is compatible for pollination. Pollen from different sources can have direct effects on fruit size, fruit shape, yield and ripening time. This pollen effect on tissue outside the embryo and endosperm is called metaxenia. The metaxenia can reduce fruit ripening time from 10 to 60 days depending on the growing season, resulting in more returns because of higher price gains for early fruit. Also it may be beneficial in regions where heat units available is low for satisfactory fruit ripening because the period is cut short by monsoon rains.

Breeding Objectives

1. To develop precocious lines.
2. To develop cultivars having dwarf growth habit.
3. To develop cultivars having reduced number and size of spines in the leaves.
4. Identification or development of hermaphroditic or monoecious lines.
5. To increase the length and flexibility of fruit stalks.
6. To develop high yielding cultivars having early fruit maturity and good fruit quality.
7. To develop cultivars having small seed size.
8. To develop cultivars adapted to mechanical harvesting.
9. To develop cultivars resistant to black nose disorder, red palm weevil (*Rhynchophorus ferrugineus*) and diseases (bayoud disease caused by *Fusarium. oxysporum* f. sp. *albedinis* and khamedj disease (inflorescence rot) caused by *Mauginiella scaettae* Cav.
10. To develop the cultivar tolerant to salinity, poor drainage and cold, extreme heat, high humidity and rain damage

Limitation in Breeding

1. Long life cycle.
2. Long juvenile phase.
3. Heterozygosity.
4. Tall tree height.
5. Dioecism.
6. Pollination is cumbersome.
7. Incompatibilty.
8. Require large number of hybrid offshoots for assessment.
9. Inheritance pattern is not known.

Germplasm Resources

Bettencourt *et al.* (1992) listed 15 gene banks, mainly in Algeria, India, Iraq, Nigeria and the USDA ARS National Clonal Germplasm Repository for Citrus & Dates, Riverside, California, USA. In India, ICAR Central Insititute for Arid Horticulture, Bikaner is the national repository for date palm. It has a collection of 64 accessions. Futher PAU, RRS Abohar, GAU RRS, Mundra and RAU, Bikaner also maintain germplasm of date palm.

Sources for useful traits:

Dwarfness: *Phoenix roebelenii* and Khadrawy.

Sparse spines: Khadrawy.

Precocious: Khadrawy.

Long fruit stalks: Kush Zebda and Horra.

Black fruit colour: Abada.

Large fruit: Medjool, Tadala and Badrayah.

Superior quality: Deglet Noor, Amir Hajj, Kush Zebda and Barhee.

Early ripening: Barhee, Khadrawy, Halawy, Deglet Noor and Deglet Beida.

Late ripening: Thoory, Medjool and Tadala.

High yield: Barhee and Medjool.

Tolerance to humidity and moisture: Khadrawy, Thoory, Dayri, Halawy.

Tolerance to bayoud disease: Bou Sthammi, Iklane, Tadement, Sair-Layalet and Bou Feggous Ou Moussa.

Inheritance Pattern

1. The large fruit size of Medjool inherited in progeny produced by Medjool BC pollen.
2. The character of moisture tolerance was with the mother variety.
3. Characteristics of long fruit stalks of Deglet Noor was often transmitted to its offspring.
4. Low pollen production increased shattering, albino leaf and depression in vigour and offshoot production by males.
5. Sex ratio is believed to be controlled by a single gene. A sex chromosome with nucleolar herterochromatin has been identified which may be used for sex determination.

Breeding Methods

1. Introduction: The cultivars like Halawy, Khadrawi, Hayany, Khalas, Khuneizi, Chip Chop, Braim, Zahidi, Barhee and Medjool have been introduced in India from USA, Egypt, Algeria, Iraq, Iran and Tunisia.

2. Selection: The systematic efforts have been made to select promising types from seedling populations in North Africa, Middle East, Spain, Peru, USA and Mexico, etc. These cultivars have superior fruit characteristics, hardiness and tolerance to saline soil, drought, diseases and pests. In India, the date palm groves of Kutch are of 200 years old. From the variability in these groves more than 200 selections have been made. Based on observations, superior seedling male palms, Barhee A19, Boyer No. 11, Crane, Fard 4 and Jarvis 1, have been identified in USA.

3. Hybridization: In USA, date palm breeding strategy was taken in two phases. In first phase, production of "varietal" males was taken to BC5 by sequential backcrosses to develop male trees genetically similar to the female parent. The pollen from such palms in intervarietal cross would impart characteristics of female ancestor. This was done from 1948 to 1971. During the second phase, intervarietal crosses were done to combine the characteristics of original fruit bearing varieties. This was done from 1971 to 1982.

Intraspecific/interspecific hybridization

Hybrids were obtained from crosses *P. dactylifera* x *P. sylvestris* (India), *P. dactylifera* × *P. canariensis* (Morocco, Algeria and Israel and *P. dactylifera* × *P. reclinata* (Senegal).

Research is needed to understand the basis of metaxenia effects of different male genotypes. This could lead to the selection of specific male cultivars to pollinate the given female cultivars to improve fruit size, yield and even to produce seedless fruits. Pollen from *Phoenix* species *P. canariensis*, *P. roebelenii*, *P. sylvestris*, *P. loureiroi* and *P. pusilla* have been tested. The response to different pollens depended on the female cultivar. The shape of fruits differed from oval to dumb bell-shaped when pollinated by *P. pusilla* pollen. Nixon (1928) showed that fruits and seeds of Deglet Noor were small when pollinated by *P. canariensis.* Similar was the result when pollen of *P. loureiroi* was used. The pollen of *P. pusilla* induced small fruit size in Barhee but these were of equal size in Medjool. However, it was possible to have larger fruits by pollinating with *P. sylvestris* pollen. There was no effect on colour change in fruits pollinated by *P. pusilla*. Ripening time was also influenced by pollen. The *Phoenix dactylifera* after pollination with *P. pusilla*, appear infertile. The seed stops growth in fruits developed from *P. pusilla* pollen and was observed in the the cultivars Barhee, Medjool and Sultana. According to Nixon (1928), the seed colour changes when *P. canariensis* is used as a source of pollen, compared to date palm pollen; whereas there was no effect on colour change in fruits pollinated by *P. pusilla*. Fruit ripening was delayed when pollinated by *P. roebelenii, P. canariensis* or *P. pusilla* whereas the fruits ripened earlier when pollinated with *P. loureiroi* compared to that in palms pollinated by *P. dactylifera* pollen.

Intervarietal hybrids from the crosses Abada x Medjool (BC2), Khalasa x (Khalasa BC2), Thoory x Deglet Noor BC4, Thoory x Khadraway BC3, Thoory x Halawy BC3, Thoory x (Dayri x Deglet Noor BC3, Medjool x (Dayri x Deglet Noor BC3) and Horra x (Dayri B2) have been developed in USA.

4. Backcrossing: The backcross breeding programme was initiated at USDA Date and Citrus station, Indio, Calfornia in 1948 to develop male lines for producing hybrids. A number of male backcrosses were produced that had morphological similarities to the named female cultivars. But no useful progenies developed because of sterility and low vigour owing to inbreeding depression. The female cultivars derived from these backcrosses with named females were also produced.

5. Breeding for resistance to Bayoud disease: Two approaches have been adopted: i) the selection of elite cultivars from wild genotypes in date palm groves and ii) controlled hybridization of superior quality females with *Bayoud* resistant males or females known for their good results in terms of fruit set.

6. Biotechnology:

Micropropagation: *In vitro* culture of date palm is well established. For micropropagation, shoot tips are primarily used as explants. NAA is the most effective to induce roots in *in vitro* shoots. BA or 2iP are good shoot inducers. The best shoot multiplication occurs when $AgNO_3$ or STS is incorporated into a BA-containing MS medium. GA_3 is an effective shoot elongation agent. The protocol can be successfully applied for the acclimatization of *in vitro* plantlets. The embryogenic callus obtained from axillary or terminal shoot tips is useful for genetic transformation.

Embryo culture: Conventional breeding incorporated with plant tissue culture techniques was successfully implemented to develop tall (T) x dwarf (D) date palm hybrids in Kuwait. Female palms of Barhee, Madjool, and Sultana were pollinated with pollen of dwarf date palm (*Phoenix pusilla*). Fertilization followed by fruit set was achieved in all three female palm cultivars. But seed development was arrested and embryo abortion followed. Consequently, seedless fruits were obtained in all the three cultivars. Immature embryos of the hybrids from the three cultivars were successfully isolated from the sterilized immature fruits and cultures were established *in vitro*. Acclimatized T x D hybrids was planted in the field until flowering. This is believed to be the first successful T x D hybrid production in date palm.

Molecular markers: *Phoenix* species readily hybridize naturally in the wild or when grown near each other in cultivation. This created uncertainty about the validity of certain taxonomic names. Achievement in understanding the taxonomy of the genus *Phoenix* was realized through a study of microsatellite markers which affirmed the validity of 14 species within the genus, and that *P. dactylifera* was initially domesticated from wild populations of the same species and not from the other *Phoenix* species.

DNA fingerprinting is being used to confirm the identity of date palm cultivars. Genetic analysis on cv. Medjool confirmed its status as a landrace variety in Morocco. AFLP and MSAP markers have also been used to study intracultivar genetic variation or variation within offshoots and plants derived from tissue culture. The existence of a genetic map makes it possible to choose useful genes, clone them and transform them into date palms to improve yield or agronomic qualities.

Genetic transformation: An efficient *Agrobacterium* mediated genetic transformation was successfully carried out for cultivar "Khalasah" using mature somatic embryos. Morphologically advanced mature somatic embryos developed on MS medium fortified with TDZ (1.0 mg/l) were co-cultured with A. tumefaciens strain LBA 4404 harboring binary vector pBI 121, containing uidA (GUS) and npt II genes and incubated for 4 days. Later the germinating somatic embryos were inoculated and put on plantlet conversion MS medium supplemented with BAP (0.75 mg/l) + kanamycin (100 mg/l). Prolific shoots developed from putatively transformed mature embryos showed 47.5 % transformation efficiency (Aslam *et al.,* 2015).

7. Germplasm conservation: In view of the diminishing numbers of traditional cultivars and wild germplasm of date palm, conservation efforts are needed. An appropriate conservation strategy for date palm germplasm requires a holistic approach, combining the different *ex situ* and *in situ* techniques including nature reserves, gene banks, etc. in a complementary manner. While field gene banks provide easy access to conserved material for use, they run the risk of destruction by natural calamities, pests and diseases. Therefore, safety duplicates of the living collections should be established using alternate strategies such as *in vitro* tissue culture techniques. This also enables conservation in a pathogen tested state thereby facilitating safer distribution.

The conventional method used to conserve genetic resources of date palm has now been supplemented by biotechnology techniques. The *in vitro* techniques for storage of date palm germplasm may be either by keeping the sterile plant tissues or plantlets on nutrient gels (growth procedure) or by cryopreservation in liquid nitrogen. Thus date palm germplasms can be preserved by storing at low temperatures or by cryopreservation.

BER

(INDIAN JUJUBE)

Origin: India

Taxonomy status:

Order: Rosales

Family: Rhamnaceae

Tribe: Paliureae

Genus: *Ziziphus*

Section: Perdurans

Series: Cymosiflorae

Botanical name: *Ziziphus mauritiana* Lamk.

Chromosome number: n=12; 2n= 48 (tetraploid) cv. Umran; 2n=8x=96 (octaploid) cv. Illaichi

Chinese jujube: 2n=2x=24

Genome size: 300.77 to 640.94 Mb (Chinese jujube)

423.55 Mb (sour jujube)

Ziziphus jujuba (Chinese jujube) is the related species with its putative centre of origin located in the Yellow River basin in northern China, together with other 13

species in the genus *Ziziphus* including *Z. nummularia, Z. oenoplia, Z. rugosa, Z. sativa, Z. vulgaris and Z. xylopyrus.*

Botany

It is thorny shrub or small tree grows upto 12 m in height. Leaves are alternate, simple, 3-nerved, intimately denticulate, obtuse, broadly oval to rounded elliptical, densely, tomentose, underside and has stipuler thorns, Infloresence is axillary cyme and sub-capitate in the leaf axils. Flowers are small, greenish-cream, bisexual actinomorphic, pentamerous, sepals 5, petals 5, stamens 5, opposite petals, stigma-bilobed, ovary superior, 2 locular. Fruit is drupe, ellipsoid to sub globose, greenish yellow to golden yellow, 2.2-5 x 2-3.5 cm in size.

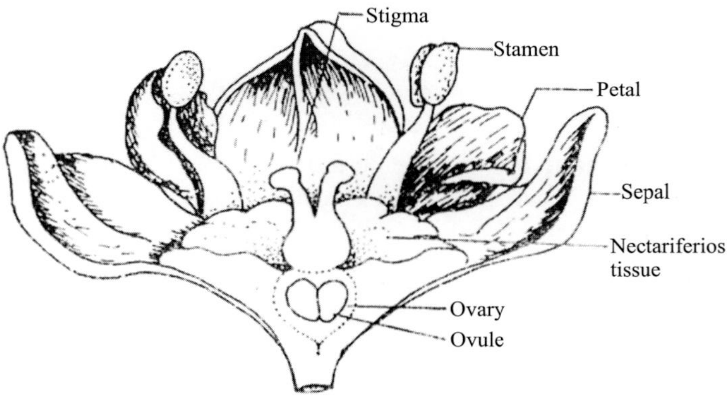

Fig. 10. L.S. Flower of chinese ber.

(Source: MacGregor, 1976)

Floral Biology

The time of anthesis varied in different cvs. It occurred between 7.30 and 8.30 AM in cultivars Seb, Jogia, Ponda, Chhuhara, Jhajjar Selection, Aligagnj and Illaichi and between 12.00 and 13.00 PM in Gola, Mundia, Banarsi Karka, Umran and Ajmeri. The anthesis was followed by dehiscence within one or two hrs and continued for 4 to 6 hrs. The stigmatic receptability occurred on the day of anthesis. It is cross pollinated by insects.

Breeding Objectives

(i) Development of dwarf stature and less thorny plant suitable for high density plantation.

(ii) To develop high yielding cultivars with earliness and good fruit quality attributes.

(iii) To develop varieties resistant to biotic (fruit fly and witches broom (associated with phytoplasma), leaf rust (*Phakopsora zizyphi-vulgaris*), *Alternaria alternata*, *Phoma destructiva*, *Fusicoccum* sp., *Erwinia jujubovora* and various fruit rot diseases, powdery mildew) and abiotic stress (cold).

Limitations in Breeding

(i) Long juvenile phase.
(ii) Heterozygosity.
(iii) Polyploidy.
(iv) Small size of flower.
(v) Insufficient understanding of the reproductive biology and mating system.
(vi) Self and cross incompatibility.
(vii) Male sterility.
(viii) High level of cultivar duplicates in the germplasm.

Germplasm Resources

The ber collection is maintained in national repository at the ICAR Central Institute for Arid Horticulture, Bikaner. It has the largest collection of 373 accessions. The field genebanks of ber are maintained at PAU, Ludhiana, RRS, Bawal, CCSHAU, Hisar and MPKV, Rahuri.

The largest collection of Chinese jujube is in the National Chinese Jujube Germplasm Repository located in Taigu County, Shanxi Province and the National Foundation for Improved Cultivars of Chinese Jujube located in Cang County, Hebei Province, besides several other smaller local collections.

Sources of useful traits:

Attractive colour and soft fruit: Gola, Thar Sevika, Mundia Murhara and Kathaphal.

Early maturity and fruit quality: Gola and Thar Sevika.

Juicyness: Reshmi.

Resistance to fruitfly: Katha, Tikidi, Dodhia, Mehrun, Chinese, Sanaur-1, Ilaichi, Mirchia, ZG-3, Umran, Thornless, Narikela, Thar Malti and Rohtaki Gola.

Resistance to powdery mildew: BS-1, Chhuhara, Safeda Selected, Nazuk, Glory, Sanaur-2, ZG-2, ZG-3, ZG-4, Safed Rohtak, Sanaur-5, Khatha Phal, Darakhi-1, Darakhi-2, Thar Malti and Guli Villati.

Resistance to leaf spot: Safed Rohtak, Jhajjar Selection, ZG-3, Thar Malti and Sanaur-1.

Frost tolerance: Tikadi.

Cold tolerance: Chinese ber (*Z. jujuba*)

Drought tolerance: *Z. jujuba*, BS 7S-3, Gola, Sindhura Narnaul Thar Sevika and Thar Bhubhraj.

Salinity tolerance: *Z. jujuba* and Banarsi Karaka.

High Yielding: Umran, Banarsi Karaka and Ponda.

Dehydration quality: Raja, Babu, Jeevan and Sukhawni.

Long transport: Umran, Katha, Maharwali, Banarasi Karaka.

Leaf fodder: *Z. nummularia* and *Z. mauritiana.*

Dwarfing: *Z. nummularia.*

Inheritance Pattern

(i) The highest heritability values were seen for fruit set and days from fruit set to ripening and pruning to sprouting, fruit drop and shoot length.

(ii) High heritability coupled with genetic advance for yield (kg/plant), pulp/stone ratio indicated that high heritability obtained for these characters was probably due to additive gene effect.

(iii) High heritability was expressed by yield (kg/plant) and pulp/stone ratio in addition to the TSS. Heritability was high for fruit weight and size, pulp/stone ratio and seed weight.

Breeding Methods

1. Introduction: Being home of ber, lot of variabilty occurs in India with regard to seed and fruit size, quality and production (number and weight). There is need to introduce Chinese jujube cultivars. Cultivar Thai Apple has been introduced from Thailand that bears fruits of about 100-150g and looks similar to green apples.

2. Selection: From the wide range of variability that exists in India for all important characters suggests substantial scope for improvement by selection. Most of the commercial cultivars are the result of selection made by local people. The promising ones are Gola, Seb, Banarsi Karka, Barnarsi Peondi, Sanauri-1, 2, 3, 4, Katha, Umran and Mundia. Thar Bhubhraj selected from the collection made from from Bhusawar area of Rajasthan was released by CIAH, Bikaner. It has juicier and sweet fruits and is also an early bearer besides being resistant to fruit rot. Recently Thar Malti has been selected from CIAH and produce 60 kg fruit /plant. A clonal selection Goma Kirti from Umran was identified at Central Horticultural Experiment Station, Godhara, Gujarat. It bears fruits earlier than Umran, flowering three weeks earlier, maturing along with the early cultivar Gola. It has been released as Goma Kirti. Mukta is selection from local ber seedlings made at MAU, Pharbani.

3. Hybridization: Breeding in ber for resistance to fruit fly (*Carpomyia*) was attempted at CAZRI, Jodhpur. The cultivar Tikadi which shows resistance to fruitfly was crossed with Seb and Gola at Jodhpur. CIAH hybrid 1 (Seb x Katha) was found to be a promising precocious and prolific bearer. It has good fruit set at high temperature under arid conditions at Bikaner and is early mature. For fruit fly resistance, BCF1 (Seb x Tikadi F_1 x Seb) is being evaluated. Thar Sevika (Seb x Katha), released by CIAH, Bikaner, is resistant to fruit fly and has superior fruit quality and early maturity. At CCSHAU, Bawal 13 hybrids (F1 Gola × BS-1, Kaithali × BS-1 and their reciprocals) are being evaluated for fruit characteristics (weight, size, TSS, yield), resistance to fruit fly and powdery mildew.

Interspecific Hybridization: Viable embryos were obtained from *Z. mauritiana* × *Z. jujuba* crosses suggesting that there is interspecific cross compatibility.

4. Mutation: Mutants were selected following N-methyl-N-nitrosourea treatment (0.02-0.04 % for 12 h) of pregerminated seeds of cultivar Ma Hong, which maintain the early maturity of its parental cultivar, Gia Loc, allowing two harvests per year. Fruits, however, are round rather than oval, pink rather than yellow and sweet rather than sour. The Dao Tien cultivar was developed in 1986 at Vietnam that produces round and larger (25 instead of 20g) fruits with better flavour than those of its parent cultivar Thien Phien.

5. Polyploidy: Attempts have been made in ber to transfer powdery mildew resistance trait from the resistant diploid genotypes by inducing polyploidy. Significant differences were observed in stomatal dimension, frequency of their occurrence and chloroplast number per two stomatal guard cells among the genotypes of different ploidy levels. However, no significant differences were observed between pentaploid and octoploid genotypes in these attributes, indicating deleterious effect beyond pentaploid level. In a triploid *Z. jujuba* (*Z. sativa*) cultivar the water saturation deficit, transpiration rate and daily maximum transpiration rate were lower than in the two diploid cultivars.

6. Biotechnology:

Micropropgation: *In vitro* multiplication protocols have been developed for clonal propagation of *Z. nummularia, Z. mauritiana* and *Z. jujuba* Methods using stem explants of mature trees. Benzyl adenine promoted bud differentiation and IBA promoted root growth in stem segments of *Z. jujuba* lines A17, A27 and A80 cultured in MS medium. Gola and Seb cultivars of *Z. mauritiana* took 150 days from initial culturing to transplanting. Standaridization of callus culture and organogenesis from mature as well as juvenile plant parts of *Ziziphus maurtiana* has been successful.

Molecular Markers

To study genetic diversity in ber (*Ziziphus* spp.) by using 11 primer pairs in AFLP analysis detected 952 fragments, of which 789 (83.8%) were polymorphic. The coefficients ranging from 0.14 to 0.86 suggested that the 41 *Ziziphus* accessions used in this study were divergent. The wild relative commonly used as a rootstock showed greater diversity than the cultivated species. The range (0.37 – 0.86) and average (0.62) similarity among the 33 cultivated accessions indicated broad genetic base of the commercial cultivars. Cluster analysis revealed the complete separation of the 33 cultivated species accessions and the eight wild species accessions into two distinct groups. Morphologically similar cultivars clustered together with a high degree of similarity. A high correlation of the similarity matrix using a single primer combination with that based on all eleven primer combinations suggested that one or a few primer combinations could be used to estimate diversity in ber.

AONLA (INDIAN GOOSEBERRY)

Origin: Tropical Southeast Asia, particularly central or southern India, Pakistan, Bangladesh, Sri Lanka, Malaya, Southern China and to Mascarene Islands.

Taxonomy status:

Order: Malpighiales

Family: Phyllanthaceae

Botanical name: *Phyllanthus emblica* L. syn. *Emblica officinalis* Gaertn.

Chromosome number: 2n=26; 28 (Perry, 1943); 23-104 (Ammal and Raghvan, 1957).

Genome size: ?

The related edible species is *Phyllanthus indofischer* and *P. acidus.*

Botany

Aonla is evergreen or deciduous, small to medium size tree having glossy greenish grey to brown exfoliating bark which peels off in thin flakes. The tree is characterized by phyllanthoid branching habit with two types of shoots. On the basis of growth characteristics, these have been characterized as long (indeterminate) and short (determinate) shoots. The determinate shoots bear small (10-13 mm long, 2-3 mm wide) leaves, arranged so closely that apparently appears to be a pinnately compound leaf. Flowers are unisexual, pale green, 4-6 mm in size, borne in leaf axils in clusters of 6-10. Male flowers appear first; perianth 6, yellowish green or deep pink in colour with valvate aestivation. Androecium consists of 3 stamens, each profusely branched, and filament attachment is basi-fixed or innate type, short, and cohesion of anther is syngenesious. Female flowers have tiny green perianth with two branched stigma. Ovary is hypogynous, carpels 3-4, three chambered axile placentation and two ovules per locule. The fruit is botanically capsular (drupaceous) berry, pediceless, globose, depressed, round globose or oblate, indented at the base with fleshy exocarp, smooth to obscurely 6 lobed capsule and 5.7g in weight. The stone of fruit is six ribbed, splitting into three segments each usually containing two seeds. Seed is 4-5 mm long, 2-3 mm wide each weighing 572 mg.

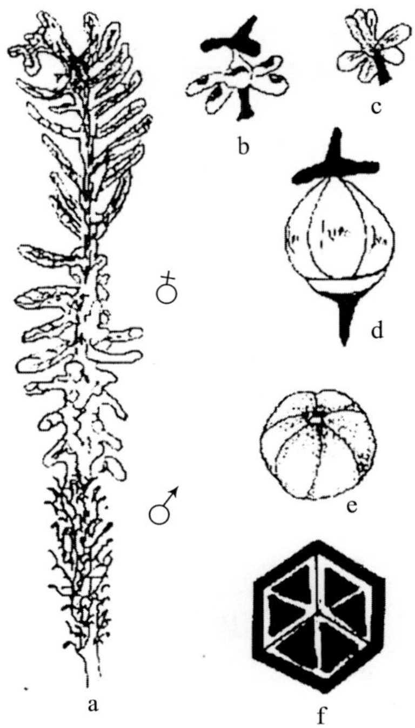

(a) determinate shoot of aonla showing male and female flowers (b) female flower; (c) fruit let; (d) mature fruit; (e) transection of ovary

Fig. 11 Determinate shoot of aonla showing male and female flowers.

(Source: Pathak, 2003)

Floral Biology

The flowering period varies in different varieties from 17 to 26 days. Flowering occurs during April-May in North India whereas it occurs twice a year, February-march and June-July in South India,. Male flowers appear first in the form of clusters at the basal part of the determinate shoot followed by female flowers in the axil of leaves at the distal end of the same shoot. The ratio of female to male flowers varies from 1: 109 to 1:501 depending on the cultivars. The time of anthesis is between 5.30 PM to 7.30 PM with the peak from 6.10 PM to 7.30 PM. The anther dehiscence takes place 10-15 minutes after the anthesis. Stigma becomes receptive on 3rd day of anthesis and remains receptive for 48 hrs. Aonla is highly cross pollinated being monoecious. The pollination takes place by wind and honeybees.

Breeding Objectives:

1. To breed varieties having wider geographic adaptability.
2. To develop dwarf tree stature for high density planting.
3. To develop precocious and prolific bearing cultivars.
4. To breed varieties with high sex ratio, i.e. more number of female flowers.
5. To develop excellent fruit quality for processing (for value added products).
6. To develop varieties resistant to pests and diseases.
7. To develop cultivars resistant to frost, heat, drought and humid soils.

Limitation in Breeding

1. Highly heterozygous nature.
2. Long juvenile phase.
3. Complex taxonomy.
4. Cross and self-incompatibility.
5. Chromosomal aberrations.
6. Lack of knowledge on inheritance pattern.
7. Small flower size.

Germplasm Resources

ICAR- Central Insititute for Arid Horticulture, Bikaner is national repository for aonla. It has the largest collection of 50 accessions. NDUAT, Faizabad, CISH, Lucknow (35) also maintain the germplasm of aonla.

Sources for useful traits

Spreading growth: Banarasi, NA-7.

Early bearing: Banarasi, Krishna, Goma Aishwariya.

Mid season bearing: Francis.

Late bearing: Chakaiya, NA-6, NA-7.

High sex ratio: Kanchan, Chakiya.

Fibreless fruit: Goma Aishwarya

Processing quality: Banarasi.

Necrosis free fruit: Kanchan, Chakaiya, NA-7.

Drought: Goma Aishwarya.

Inheritance Pattern of useful traits:

Incompatibility: Dominant gene.

Yield: Polygenes.

Breeding Methods

1.Introduction: Being the home of aonla, India has lot of variabilty with regard to seed size, fruit size, number of fruits and quality.

2. Selection: Varietal improvement has so far been confined to selection of promising genotypes from the existing seedling variability.

NDUAT, Faizabad has made seedling selection from the widespread variability. These are NA-4, NA-5, NA-6, NA-7 and NA-10. Selections made by AAU, Gujarat are Anand-1, Anand-2 and Anand-3. The other selections are BSR 1 from TNAU, Coimbatore, Goma Aishwarya from CHES, Godhra. Besides, Neelum from NA-7, Amrit from NA-6, Anand-2, Laxmi-52, Goma Aiswarya, BSR1, ATPS-1, NA-20, NA-25, NA-26, CHES-1, Gujarat Aonla-1 and CIAH/EO-24 are other promising selections.

Banarasi: It is a seedling selection from Varanasi district of Uttar Pradesh.

Francis: It is a seedling selection, also known as Hathi Jhool.

Chakaiya: Seedling selection from Banarasi.

Kanchan (NA-4): Itis chance seedling of Chakaiya.

Kanchan: A seedling selction from Chakaiya in Pratapgarh.

Krishna: A seedling selection from Banarasi in Pratapgarh.

Amrit (NA-6): Chance seedling selection from Chakaiya.

Neelam (NA-7): Selection from Francis.

Laksmi-52: A seedling selection from Francis.

Krishna (NA-9): A seedling selection from Open seedling selection from Francis.

Balwant or Agra Bold: Chance seedling from cultivar Banarasi.

Bhawani Sagar (BRS-1): A selection from Thimban, a local cultivar of Tamil Nadu.

3. Hybridization: Occurrence of xenia effects were reported from the crosses Chakaiya x Banarasi, Banarasi x NA-9, Francis x NA-7, Kanchan x NA-6 and NA-6 x NA-9 for size and weight of fruit.

4. Biotechnology: Micropropagation of aonla from mature tree has been demonstrated. But the protocol needs to be upscaled. Since the protocols are highly genotype dependent, cultivar specific protocols needs to be developed.

GRAPES

Origin: Region between Caspian Sea and Black Sea.

Taxonomy status:

Order: Vitales

Family: Vitaceae

Genus: *Vitis*

Subgenus: *Euvitis* Planch.

Species: *vinifera*

Botanical name: *Vitis vinifera* L.

Chromosome number: 2n=2x=38; 2n=2x= 40 (*Muscadinia*)

Genome size: 500 Mb.

Planchon (1887) divided the *Vitis* species between two sub-genera, namely *Euvitis* (bunch grapes) and *Muscadinia* (muscadine grapes). *Muscadinia* has a genome of 2n = 40. *Muscadinia* contains three species; *M. munsoniana* and *M. rotundifolia* native to the south-eastern USA and *M. popenoeii* native to Central America. *V. vinifera* has 2 wild subspecies, *sylvestris* Gmel. and *caucasis* Vav., and one cultivated subspecies *sativa* DC. Majority of the table and wine grapes grown commercially are of *V. vinifera* origin. In the USA, other local *Vitis* spp. are also grown commercially. A number of American *Vitis* spp. of economic importance in the past are still used in breeding programmes, especially to develop phylloxera resistant rootstocks and disease resistant wine and table grape cultivars. Of these, *V. labrusca, V. berlandieri, V. aestivalis*, *V. riparia* and *V. rupestris* are important. *Vitis* spp. native to tropical America, like *V. smalliana, V. caribaea* and *V. shuttleworthii* are included in breeding programmes to develop cultivars adapted to tropical climate. In the East European and Asian countries, *V. amurensis* is often included in breeding programmes to induce cold hardiness and disease resistance.

Botany

Grape is deciduous, woody vine, has a tight, shedding bark and un-branched tendils. Leaves are lobed, simple, rounded and broadly ovate with coarsely serrate edges. Three flower types have been recognized: functionally hermaphroditic or perfect (both the stamens and pistils are functional), pistillate (stamens are non-functional, reflexed) and functionally male (staminate, pistil is incompletely developed and non-functional). Flowers are small, greenish, hermaphrodite, pentamerous, 5 sepals, 5 petals and are borne opposite to leaf. Petals fused or joined at the top. At flowering it gets loosened at the base and falls off like a cap. There are five stamens opposite the petals. The ovary consists of carpels having the locules. Each locule has two ovules. The fruit is borne in small, loose clusters of 3-40 berries. The fruits are round-ovute and have a thick and rough skin containing upto 1-5 hard and oblong seeds.

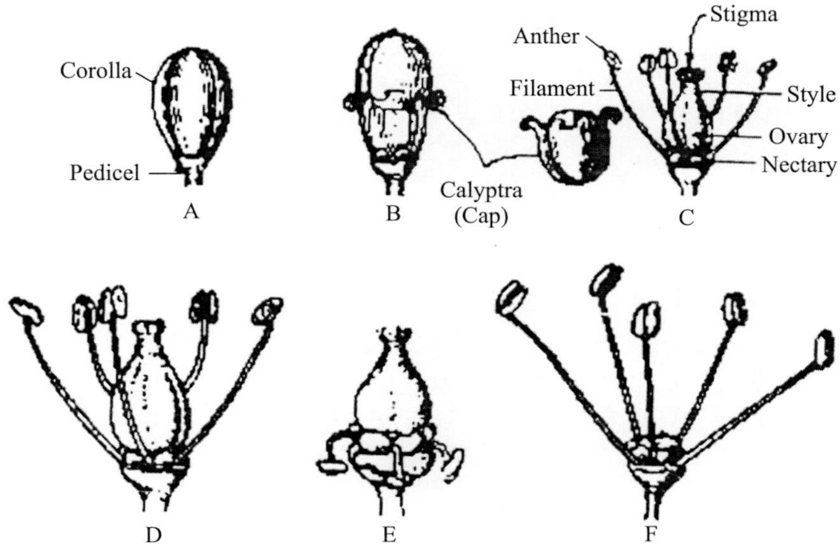

A—calyptra attached; B—calyptra separating; C—open flower; D—hermaphrodite flower; E—female flower, and F—male flower.

Fig. 12 Bloom sequence of flower and flower type.

(Source: Babo and Mach, 1909)

Floral Biology

Anthesis starts early in the morning and continues beyond 5 PM, the peak anthesis being between 8 and 10 AM under South Indian conditions. The peak anthesis is between 7 and 9 AM in north Indian conditions. Anthesis is comleted in maximum number of flowers before noon and later declines sharply. Anther dehiscence is completed within 5 to 15 minutes depending upon weather conditions. The stigma receptivity is characterized by the presence of sugary secretion on stigma giving it a bright appearance. Production of pollen grains per anther is comparatively higher in Muscat, Pachadraksha and Kishmish than in Anab-e-shahi and Bangalore Blue. *V. vinifera* have perfect flowers. Some native North American species have imperfect flowers. *Muscadinia* and American species are dioecious and bear male and female flowers on separate vines. Flowers are self pollinated or pollinated by wind, but honeybees and some other insects may also be active pollinators.

Breeding objectives in tropical environment:

1. To develop varieties with less degree of apical dominance.
2. To develop varieties with high fruitfulness of basal buds.
3. To develop high yielding and high quality varieties.

4. Varieties suitable for table, raisin, wine and juice making.
5. To develop varieties resistant to diseases.

Limitation in breeding in tropics

1. High heterozygosity.
2. Long juvenile phase.
3. Low genetic base.

Breeding objectives for scion

(i) To evolve suitable varieties for table, raisin and wine purpose.

(ii) For table grapes, seedlessness, large berry size, good eating quality and keeping quality characteristics (tentuier) are required.

(iii) Develop suitable varieties for tropics that have medium vigour, productive basal buds and less degree of apical dominance.

(iv) To develop high yielding and superior quality varieties for subtropics having short growth cycle and early fruit ripening.

(v) To evolve varieties resistant to physiological disorder like berry cracking and pink berry disorder.

(vi) To develop tolerant varieties to cold, salinity and drought.

(vii) To develop varieties resistant to diseases like anthrcnose, downy mildew (*Plasmopara viticola*) and powdery mildew (*Uncinula necator*) and to phylloxera (*Daktulosphaira vitifoliae).*

Breeding objectives for rootstock

1. To impart dwarfness.
2. To efficiently utilise nutrients from soil and water.
3. Good propagation abilty.
4. To develop rootstocks resistant to abiotic (cold, salinity and drought) and biotic stresses.

Limitation in Breeding

(i) Long juvenile phase.

(ii) Heterozygous nature.

(iii) Narrow genetic base.

(iv) Dioecious vines.

(v) Pronounced inbreeding depression.

(vi) Parthenocarpy.

(vii) Self and cross incompatibility.

(viii) Lack of information on inheritance pattern.

Germplasm Resources

The National Clonal Germplasm Repository (NCGR) at Davis, California houses germplasm collection of >2800 different grape accessions, including one of the worlds finest *Vitis* species collections, ~600 *V. vinifera* wine grape varieties, and ~600 table grape varieties. The sister repository in Geneva, NY houses the cold hardy grape germplasm collection.

In India, field gene banks of grapes are maintained in at the National Research Centre on Grapes, Pune. The germplasm contains 465 accessions. Germplasm are maintained in the Division of Fruits and Horticultural Technology at IARI, New Delhi, Indian Institute of Horticulture Research, Bengaluru (72 genotypes) and Ganeshkhind Botanical Garden, Pune etc.

Sources for useful traits for resistance to biotic stress:

Phylloxera: Rupestris du Lot (*V. rupestris*), Harmony, Dog Ridge, *V. riparia,* St. George, 99-R, *V. riparia*, *V. argentifolia, V. champinii, V. berlandieri, V. monticola, V. aestivalis, V. cinerea, V. cordifolia* and *Muscadinia rotundifolia.*

Pierce's disease: *Vitis arizonica, V. aestivalis V. berlandieri, Muscadinia* spp., Eudora, Delicious and Southern Jewel.

Powdery mildew: Nistru, Bangalore Blue, Bangalore Purple, Catawba, Concord, Isabella, Champion, James, Champanel, *V. lanata, V. parviflora, V. aestivalis, V. arizonica, V. berlandieri, V. californica, V. candicans, V. lincecumi, V. longi, V. palmata, V. riparia, V. rupestris, V. tiliefolia*, Muscat of Hamburg, Rangspay, *V. bryoniifolia, V. davidii* and *V. piasezkii.*

Downy mildew: *V. amurensis*, Krivlyanskii, Muscat, Onitskanskii, Survchunskii, Bangalore Blue, Catawba, Khalili, Concord, Champion, James, *V. aestivalis, V. arizonica, V. berlandieri, V. californica, V. candicans, V. cinerea, V. cordifolia, V. lincecumi, V. longi, V. palmate, V. riparia, V. rupestris, V. tiliefolia* and Marroo Seedless.

Anthracnose: Arka Majestic, Arka Hans, Arka Neelamani, Arka, Shyam, Arka Trishna, ARI-27, Pusa Navrang, *V. parviflora*, Temple, Bangalore Blue, Muscat, Hamburg, Gulabi, Phakadi, Bangalore Purple, Catawba, Concord, Champion, James, *V. lanata, V. aestivalis, V. arizonica, V. longi, V. palmata, V. riparia, V. rupestris* and *V. tiliefolia.*

Black rot: *V. berlandieri*, *V. candicans*, *V. cinerea* and *V. riparia.*

Crown gall: V54-62, V 64-127, V 65-1, V 63-30, V 52-28, Chardony, Gordan, Krey, Pinct, Rhine and Riesling.

Cercospora leaf spot: Champion, Champach and Australia No. 2.

Aphids: *V. cinerea.*

Thrips: Karachi and Fakdi.

Nematode: Dog Ridge, Salt Creek, 1613 (Solonis x *V. riparia*), Teleki-5-A Jaos Beli, *V. champinii, V. candicans, V. solonis, V. rotundifolia,* SO 4 (*V. berlandieri* x *V. riparia*), 1103 Paulsen (*V. berlandieri x V. rupestris*), Richter (*V. berlandieri* x *V. rupestris*) and 5 BB (*V. berlandieri* x *V. riparia*).

Sources for useful traits for resistance to abiotic stress:

Hot tropics: Isabella, Kyoho, Campbell Early, Villard blanc, *V. smalliana, V. caribaea, V. shuttleworthii,* Ribier, Perlette, Muscat Hamburg, Beauty Seedless, Bhokri, Cheema Sahebi, Kali Sahebi and the *V. labrusca,* Cardinal, Waltham Cross, Muscat Hamburg and Criolla hybrids *(V. caribaea).*

Cold: *V. amurensis,* Regale, Sterling, *V. labrusca, V. riparia* and *V.argentifolia.*

Frost: Liana-12 and *V. amurensis.*

Heat: *Vitis lincecumii, V. aestivalis* and *Muscadinia rotundifolia.*

Drought: Dog Ridge, Ramsey, 101-14 Mgt, *V. arizonica, V. monticola, V. vinifera* and *V. aestivalis.*

Soil salinity: Digraset, St. George, Salt Creek, Dog Ridge Solonis-1616, 1616-C, 216-C and 1103 P.

Soil acidity: *V. riparia, V. rupestris* and *V. berlandieri.*

Chloride toxicity: Dogridge, *V. berlandieri, V. champini.*

Calcarious soil: 420 A, 161-49, Hybrid 99, 110, Hybrid 150-15 and *V. berlandieri.*

Wet soil: 101-14 Mgt, 3306C, Ramsey and Dog Ridge.

Sources of useful traits for growth and fruiting

High yield: Rubired, Beauty Seedless, E-26-8, E-30-14, E-29-5, Dilkush, Angur Kalan and Anab-e-Shahi.

Earliness: *V. parviflora, V. riparia, V. labrusca, V. amurensis*, Pearl of Casaba, Beauty Seedless, Flame Seedless, Perlette, New Perlette, Beauty Seedless, Pusa Navrang and Pusa Urvashi.

Basal fruitful buds: Perlette, Beauty Seedless, Muscat Hamburg

High TSS: Beauty Seedless, Thompson Seedless, E-29-7, Golden Queen, Athens, Bangalore Blue and Concord.

Seedlessness: Beauty Seedless, Perlette, Thompson Seedless, Arkavati, Arka Shyam and Arka Neelamani.

Berry splitting: Black Muscat and Pusa Navrang.

Sources of useful traits for processing purpose

Table: Perlette, Beauty Seedless, Arka Shyam, Arka Kanchan, Shweta Seedless, Arka Neelamani, Manjiri Naveen and Red Globe.

Juice: Concord, Pearl of Casaba, Pusa Urvashi, Pusa Navrang, Arka Krishna and Arka Shyam.

Raisin: Thompson Seedless, Pusa Seedless, Sultana, Tas-A-Ganesh and Sonaka.

Wine: Arkavati, Arka Kanchan, Arka Neelamani, Arka Hans, Arka Soma, Arka Trishna, Beauty Seedless, Rubired and Black Champa.

Inheritance Pattern

Sex: Multiallelic and epistatic phenomena. Single pair of genes has dominant gene for hermaphrodite flowers; **mf** recessive gene for male flower and **sfa** recessive gene for pistillate flowers.

Thick and large cane, internode, vigour, drooping habit, pentagonal leaf shape, early maturity, round berry, black berry, low bunch weight, oval shape of seed: Dominant genes.

Loose bunch: Highly variable, quantitative gene.

Berry skin colour: Black fruit colour is dominant and epistatic to that for red and white. Red fruit colour is dominant to white which is recessive for both genes, controlled by two pair of genes with epistatic action.

Berry size: Highly variable, quantitative genes.

Aroma: Mostly for *V. vinifera* muscat flavour. Reynolds *et al*. (1982) proposed a three gene, dominant and complementary system for methyl anthranilate (MA) and a two gene system for total volatile esters (TVE).

TSS: Total sugar content has additive effect with broad sense of heritability 0.61-0.84.

Malic acid: Additive with broad sense heritability 0.73-0.89.

Tartaric acid: Non-additive, broad sense heritability 0.59-0.84.

Total acid: Strongly additive, broad sense heritability 0.53-0.90.

Seedlessness: Three independently inherited recessive genes controlled by a dominant regulator gene.

Berry maturity: Highly variable, quantitative genes.

Phylloxera, nematodes, PD and fungal diseases: High degree of dominance.

Downy mildew, anthracnose and gray mold: Seems to be controlled by several oligogenes or by polygenes with considerable dominance effect.

Chloride: Single dominant gene controlling chloride exclusion ability in *V. berlandieri.*

Pre-selection indices and early evaluation:

One year old vines were grown under a restricted water regime (1/9 maximum evapotranspiration) in small containers inside a greenhouse. The parameter used as indicator for water absorbing activity of the rootlets of each plant is F/rs (F = total active leaf area; 1/rs = stomatal conductance of active leaves) for pre-selection test for drought resistance (Carbonneau, 1985).

Shortening the breeding cycle: Shotening of juvenile period in hybrid progenies is one of the strategies to advance the selection process. This can be done by the treatments described below.

(i) In France, hydroponic culture has been used for induction of early fruiting in glasshouse grown seedlings.

(ii) Budding and grafting can help to speed up maturity of hybrid seedlings and thus shortening the breeding cycle.

(iii) Fruiting is induced in single node cuttings: Canes of uniform diameter were made into cuttings keeping varying number of nodes (1-6). The uppermost bud on each cutting was retained and all others were pinched off. Plants were grown in a glasshouse under natural illumination during spring and summer. The inflorescences of woody vine cuttings usually stop growing soon after bud burst and they shrivel and die. Atrophy of the inflorescences is prevented by removing expanding leaves from the base of the new shoot and partial defoliation allows production of small bunches of grapes by single node cuttings.

(iv) Srinivasan and Mullins (1978) showed that isolated tendrils (modified inflorescences) cultured *in vitro* developed into recognizable flowers when various cytokinins were applied directly. Later, they demonstrated this conversion *ex vitro* and obtained fertile flowers and fruit set. Flowering and fruit set could be obtained from 3 month old seedlings, demonstrating the potential utility of this technique in reducing the time for recurrent hybridization and selection. Cytokinin application to staminate flowers causes developing pistils that would otherwise be non-functional to become fertile. Such flowers can be pollinated to set fruit. This technique allows normal male vines to be utilized in breeding since they can then be crossed with other male or hermaphroditic vines.

(v) Seed dormancy can be overcome by soaking the seeds for 70 days in a solution containing 1500 ppm GA_3, 2500 ppm kinetin for one day. Bearing in the seedlings commences one year in advance.

Grape improvement work started in India in 1960 at IARI, New Delhi. Subsequently it started at IIHR, Bengaluru, PAU, Ludhiana, HAU, Hisar, ARI, Pune and APAU, Rajendra nagar, AP.

Breeding Methods

1. Introduction: The cultivars Red Globe, Flame Seedless, Fantasy Seedless (B 36–27 x P 64–18), Autumn seedless (Muscat of Alexandria x Thompson Seedless), Perlette, Thompson Seedless, Delight, Cardinal, Pearl of Casaba, Early Muscat were introduced from USA, former Yugoslavia, former USSR and Australia. Anab-e-Shahi has been introduced from the Middle East. The red wine cultivars viz., Cabernet Sauvignon, Merlot, Shiraj/Syrah, Zinfandel, Pinot Noir, Cabernet Franc and Cinsaut have been introduced from France and Spain. Similarly, white wine grapes viz., Chardonnay, Chenin Blanc, Sauvignon Blanc and Clairette have been introduced from different countries of the world. Rootstock cultivars like Dogridge, B 2/56, Salt Creek, St. George and Teleki 5A were also introduced.

2. Selection: Open pollinated seedlings segregate for a large number of characters. Obviously, the seedling population from open pollinated seeds is a potential source for desirable types.

Thompson Seedless (Sultanina, Sultanine, Kishmish, Sultana, Sultanina Marble, Sultanine Blanche): It is an ancient seedless white grape variety selected and originated from Asia minor.

Cheema Sahebi: Seedling selection from Pandhari Sahebi. Berry is pale yellow, oval, cylindrical, small and has 2-3 seeds, 20-21°B TSS. It is highly susceptible to downy mildew and anthracnose.

Rao Sahebi: Selected from clones of Cheema Sahebi, it produces long berries. Some promising seedlings were also selected from open pollinated population, e.g. cultivar Selection-49, Pandhari Sahebi and Kabul Monukka.

Clonal selection: The concept of clonal selection was applied to grapes by Bioletti and important clonal selections were made some of which are listed below:

Dilkhus: A selection from Anab-e-Shahi. It produces loose bunch which are golden yellow in colour and the berries are elongated.

ARI-302: A selection from Anab-e-Shahi made at Agharkar Research Institute, Pune. Berry seedless, golden yellow, ellipsoidal, TSS 22-24° B, but is susceptible to downy mildew, powdery mildew and anthracnose.

Pusa Seedless: A selection from Thompson Seedles made at Indian Agricultural Research Institute, New Delhi (Singh and Murthy, 1993). Elongated berry, TSS 20-22°B, suitable for table as well as raisin making but is highly susceptible to anthracnose.

Tas-A-Ganesh: Clonal selection fromThompson Seedless. Berry size and clusters are bigger than in Thompson Seedless and it gives better response to GA_3 dip treatment.

Manik Chaman: Selection from Thompson Seedless. Berry is elongated, golden yellow having medium thin skin.

Sonaka: Selection from Thompson Seedless. Berries are seedless, sweet, TSS 20 - 22° B, more elongated than in Tas-A-Ganesh and has good keeping quality.

Mauriti Seedless and Ambe Seedless, Sonaka: Clonal selection from Thompson Seedless.

Sharad Seedless: A clonal selection (bud sport) from Russian Kishmish Cherynyi (Black Sultana). High yielder, berries are bold, crispy and brilliantly black in colour, TSS 22-24° B, good keeping quality and moderately resistant to powdery mildew.

Manjari Kishmish: Clonal selection from Kishmish Rozavis.

Kishmish Rozawis White (K.R.White): A clonal white mutant from the Russian red variety Kishmish Rozavis made at National Research Centre for Grapes, Pune. Attractive bunches and berries, suitable for close planting, responsive to hormone sprays. It is an export quality grape.

Manjiri Naveen: A mutant selection from Centennial Seedless made at National Research Centre for Grapes, Pune. Early ripening (25 days before Thompson Seedless), bold and seedless berries, mild muscat flavoured, TSS 17-18° B, good keeping quality, moderately tolerant to downy mildew and anthracnose. It is also an export quality grape.

Sultanine-11: Clonal selection from Perlette. Bunches are moderately compact, responds well to GA treatment, early ripening and a promising type for export.

Ruby Roman: It is an open pollinated progeny of Fujiminori.

Degrasset: A selection from *Vitis champinii* Planch. for use as rootstock.

Dogridge: A selection from *V. champinii* Planchon for use as rootstock.

St George: A selection from *V. rupestris* Scheele for use as rootstock.

3. Hybridization:

The technique: Using a funtionally female varieties as seed parent to eliminate the tedious emasculation task and selection of proper parent combinations have helped in developing a large number of hybrids in grape.

***IIHR, Bengaluru*:**

Arkavati: Black Champa x Thompson Seedless.

Arka Hans: Bangalore Blue x Anab- e– Shahi.

Arka Shweta: Anab-e –Shahi x Thompson Seedless.

Arka Shyam: Bangalore Blue x Black Champa.

Arka Neelamani: Black Champa x Thompson Seedless.

Arka Chitra: Angur Kalan x Anab-e-Shahi.

Arka Kanchan: Anab- e– Shahi x Queen of Vineyards.

Arka Krishna: Black Champa x Thompson Seedless.

Arka Majestic: Angur Kalan x Black Champa.

Arka Trishna: Bangalore Blue x Convent Large Black.

***IARI, New Delhi*:**

Pusa Navrang: Mandeleine Angevine x Rubired.

Pusa Urvashi: Hur x Beauty Seedless.

Pusa Aditi: Banqui Abyad x Perlette.

Pusa Trishar: (Hur x Bharat Early) x Beauty Seedless.

Pusa Swarnika: Hur x Cardinal.

***PAU, Ludiana*:**

Punjab Purple: Catawba x Beauty Seedless.

***NRC on grapes, Pune*:**

A-18/3: Black Champa x Thompson Seedless

Manjri Medika: Pusa Navrang x Flame Seedless.

Autumn Seedless: Calmeria x Muscat Alexandria x Thompson Seedless.

Blush Seedless: Emperor x Davis Z-4-87.

Autumn Royal: Autumn Black x C 74-1.

Fantasy Seedless: Complex hybrid derived from a cross between B36-27 X P64-18.

***Agharkar Research Institute, Pune*:**

ARI-27: Hybrid of Diamond Jubilee × Rubi red. It is tolerant to the mildews and anthracnose and can be used for making red port wine and juice.

ARI-144: ARI-144 is an inter-specific hybrid of *V. vinifera* var. Cheema Sahebi × *V. labrusca* var. Catawba. It has good fruit quality and disease resistance.

ARI- 516: Catawba x Beauty Seedless. It is early ripening hybrid tolerant to fungal diseases.

***Bordeaux, Rhône valley, Burgundy, France*:**

Cabernet Sauvignon: Derived from a crossing between Cabernet Franc and Sauvignon Blanc from Bordeaux.

Chardonnay: Result of crossing Gouais Blanc with Pinot.

Syrah (Shiraz): Result of crossing the Mondeuse Blanche and Dureza grape varieties.

Pinot Noir (Blauer Burgunder): Cross between Pinot Noir and Gouais Blanc.

***USDA Agriculture Research Station, California*:**

Crimson Seedless: Resulted from a cross between 'Emperor and C33-199. It was the only late-ripening red seedless grape and released by D. Ramming and R. Tarailo.

Flame Seedless: It was bred by J. Weinberger and F. Harmon. It resulted from a cross made in 1961 that involved selections with Cardinal, Sultanina, Red Malaga, Tifafihi Ahmer, and Muscat of Alexandria in their pedigree.

Table varieties bred by H. P. Olmo, University of California, Davis

Early Niabell: Campbell Early 4x x Niagra.

Niabell: Campbells Early 4x x Niagra 4x.

Royalty: Alicante Ganzin x Trousseau.

Rubired: Alicante Ganzin x Tinta Cão.

Scarlet: Golden Muscat x Teinturier.

Beauty Seedless: Scolokertek kiralynoje x Black Kishmish.

Blush Seedless: Emperor x Z4-87 (Emperor, Queen of the Vineyards, Ribier, IP 75).

Centennial Seedless: Gold x Q25-6 (Complex parcentage of Gold, Emperor, IP 75).

Christmas Rose: S44-35c x 9-117D (Emperor, Hunisa, and IP 75).

Delight: Scolokertek kiralynoje 26 x Sultanina marble.

Early Muscat: Muscat Hamburg x Scolokertek kiralynoje

Gold: A3-94 x K3-78 (Muscat Hamburg x Sultanina) x (Muscat Hamburg x Scolokertek kiralynoje.

July Muscat: 126-11 x K4-41 (Muscat of Alexandria x Flame Tokay) x (Muscat Hamburg x Scolokertek kiralynoje.

Perlette: Scolokertek kiralynoje 26 x Sultanina marble.

Queen: Muscat Hamburg x Sultanina.

Ruby Seedless: Emperor x Pirovano.

Red Globe: L12-80 x S45-48 (Emperor, Hunisa and Nocera).

Canner: Hunisa x Sultanina.

Dawn Seedless: Gold x Perlette.

Emerald Seedless: Emperor x Pirovano 75.

Calzin: Zinfandel x Refosco.

Carmine: F2-7 x Merlot.

Emerald Riesling: Muscadelle (CA) x Rieslsing.

Flora: Semillon x Gewürztraminer.

Helena: Zinfandel x Refosco.

Ruby Cabernet: Carignan x Cabernet Sauvignon.

Symphony (tetraploid): Grenache gris x Muscat of Alexandria.

Sunny Rouge: A cross between Pione and Red Pearl.

***Cornell University, Ithaca, New York*:**

Everest Seedless: This variety includes genes from two Japanese varieties Kyoho and Pione as well as from the American Concord and Niagara. It descends from several other Cornell-developed grapes such as Himrod and Ontario. This grape variety is cold-tolerant, Concord-type blue colored grape having berries weighing up to 7g, roughly twice the size of the traditional Concord.

***National Agriculture and Food Research Organization, Institute of Fruit Tree Science, Akitsu (NIFTS), Japan*:**

Kaiji: Registered as a new cultivar in 1977, it was developed by S. Uehara, a private breeder, by crossing Flame Tokay with Neo Muscat. It has attractive, large (10g) red berries with crispy flesh, a neutral flavour and no cracking.

Rosario Bianco: A cross between Rosaki and Muscat of Alexandria.

Seto Giants: A cross between Guzal Kara and Neo Muscat.

International Fruit Genetics, Bakersfield, California:

The varieties viz., *Funny Fingers* (long berries and a very dark skin color), *Sweet Sapphire*, *Jack's Salute*, *Sweet Celebration* and *Cotton Cand* are created by Cain and IFG. These new varieties bring new attributes to table grapes.

Interspecific hybrids:

Southern Home: *V. vinifera* and *V. rotundifolia.*

French-American hybrids: Developed in from crosses between the American grapes species viz., *V. rupestris, V. lincecumii, and V. riparia* and *V. vinifera* in France and used primarily for wine production e.g. Baco Noir, Chambourcin, Chancellor and Seyval Blanc.

Florida hybrid bunch grapes: *V. vinifera* crossed with *V. aestivalis* ssp. *smalliana a*nd other native species of south-eastern USA. These are resistant to Pierce's disease.

Cultivars like Shan, Oz, Nava, Elul Odem have been evolved by hybridization followed by selection.

Interspecific diploid hybrids:

Shine Muscat: It is a cross between Akitsu 21 X Hakunan.

Oriental Star: It is a diploid table grape cultivar derived from *V. labruscana* and *V. vinifera.*

Neo Muscat: A cross between Muscat of Alexandria and Koshu-sanjaku, developed by a private breeder M. Hirota.

Muscat Bailey: A cross between *V. labruscana* and *V. vinifera* cultivars made in 1921.

Interspecific tetraploid hybrids:

Kyoho: Cross between Ishiharawase (tetraploid vinifera) and Centennial (labruscana type). It was first produced by a private breeder named Y. Ohinoue in 1945.

Pione: An offspring from a cross between Kyoho and Muscat of Alexandria 4*x*.

Fujiminori: An offspring from Ikawa Selection 682 × Pione.

King Dela: Cross between Red Pearl and King Dela Muscat of Alexandria.

Intergeneric hybrids:

Muscadine grape (*M. rotundifolia*) adapted to the humid warm climate is a rich source of resistance to many diseases and pests and also possesses a unique and delightful flavour and aroma. Hybridization to combine the pest and warm climate resistance traits of *M. rotundifolia* with the large cluster and desirable berry quality traits of *V. vinifera* has been carried on for more than a century. The crosses between *Vitis* and *Muscadinia* which differ in chromosome number are made with difficulty, but most of the resulting hybrids remain sterile.

4. Polyploidy: Polyploidy may occur spontaneously from a latent bud near a pruning wound or can be induced by colchicine treatment. For development of tetraploid grape variety, the germinated seeds of diploid hybrid grape were treated with 0.6% colchicine at 25°C. Chromosome counting in root tips indicated that these descedants are $2n=4x=76$ while their parents were $2n=2x=38$. These were considered to be tetraploids based on the size and shape of pollen and stomata and the number of chloroplasts in guard cells. Three tetraploid cultivars named Zaoheibao, Qiuheibao and Wanheibao were generated using this strategy, all having black purple colour, Muscat flavour, good adaptability to environment and high disease resistance and maturing in early season (August) and late season (September). The tetraploid cultivars Jing Xiu and Red Globe were developed by colchicine treatment of shoot segments.

Polyploidy is effective to increase berry size in table grapes. The tetraploid table grape cultivar Fujiminori was used in Japan to promote seedlessness and enhance berry size. Allotetraploids are more promising than autotetraploids. Tetraploids are used in breeding to obtain triploids. The triploid grapes are highly sterile and have potential for use as rootstocks. Some tetraploid varieties are Marvel Seedless, Campbell Early, Early Giant, Black King, Baumann, Muscat and Cannon Hall.

5. Mutation: Natural mutation has played a vital role in economic characters of grapes like hermophroditism, seedlessness, larger berry size, fruit colour and flavour. ARI-302 is a seedless mutant of Anab-e-Shahi. New Perlette is an induced mutant of Perlette. Other induced mutants are Red Nigara from Nigara and Robin Cardinal from Cardinal. Anab-e-Shahi is a bud sport having a satellite chromosome.

6. Biotechnology:

Micropropagation: The technology for micro-propagation using shoot tip culture is well established and the resulting plants appear to be genetically identical to the mother plant. But the role of tissue culture in commercial propagation is likely to be limited because standard propagation systems based on grafting are well established.

Molecular marker: Confusion in cultivar names is a major problem in viticulture. Often the same cultivar is grown under different names or sometimes several cultivars have same name. Therefore molecular markers such as RFLP, DNA RAPD and micro-satellites (SSR) have been widely used for cultivar identification.

Protoplast culture: Several attempts have beenmade for protoplast culture, but difficulties were encountered to regenerate plants. The first report of success was from Reustle *et al.* (1994) who regenerated plants from protoplasts derived from embryogenic material of the hybrid cultivar Seyval blanc.

Somaclonal variation: Many clones of established cultivars of wine grapes, those were selected and are of ancient origin and are still grown and have been vegetatively propagated over a long period. These have provided somatic heterogeneity to increase variability in the existing cultivars thus thwarting genetic erosion and to serve as a source for selection particularly for resistance to biotic and abiotic stress. Schneider *et al.* (1996) presented the first molecular evidence for somaclonal variation in grapevine. Three plants with altered phenotype derived from protoclones of Seyval Blanc showed modified RAPD profiles.

Somatic embryogenesis:

Mullins and Srinivasan (1976) obtained plants via somatic embryogenesis from unfertilised ovules of grape cultivar Cabernet Sauvignon. This was the first report on successful development of plantlets via somatic embryogenesis in *Vitis* spp. and inter-specific hybrids. The most spectacular application of somatic embryogenesis in grapevine was the separation of the chimaeric periclinal L1 and L2 layers in *V. vinifera* Pinot Meunier and obtaining plants having phenotypes different from the original plant (Franks *et al.*, 2002). Pinot Meunier has a L1 cellular layer responsible for its characteristic hairiness and also carries a mutation in a gene homologous of the *Arabidopsis* GAI (gibberellin inhibitor) gene. The mutation of this gene gives the plant a dwarf phenotype in which all the tendrils are converted into inflorescences capable of flowering and bearing fruit (Boss and Thomas, 2002). These characteristics confer considerable interest in future grapevine genetics and breeding on this genotype.

Embryo rescue:

Development of improved seedless varieties by conventional breeding methods is complicated by the impossibility of using seedless vines as female parent. Seedless vines can be used only as pollen parent. This restricts the number and type of desirable germplasm combinations that can be achieved, since the possibility of using seedless x seedless crosses or seedless vines as female parent is eliminated. On using seeded vines as female parent, the degree of seedlessness is low. Also, the frequency of resulting progeny having acceptably small and soft ovules at maturity (that can be regarded as seedless) is low.

The development of embryo rescue technique enables the breeders to develop offsprings from two stenospermocarpic seedless parents. Cain *et al.* (1983) reported the first successful embryo rescue of grapevine, followed soon after by Emershad and Ramming (1984) and Spiegel-Roy *et al.* (1985). Seed from early ripening seeded cultivars usually have low germination rates and embryo rescue techniques were successfully applied to develop seedlings from such cultivars. Since the seedless trait is only found in *Euvitis*, embryo rescue technique was also employed to develop plants from crosses between *Euvitis* and *Muscadinia*. Ramming *et al*. (2000) reported the first *V. vinifera* × *V. rotundifolia* seedless seedling, although it is difficult to regenerate progeny from such crosses.

Genetic transformation:

Researchers need to be able to select transformants from nontransformants and to confirm whether stable introduction of a foreign gene into *Vitis* has taken place. Most commonly the *npt*II and *hpt* genes conferring antibiotic (kanamycin or hygromycin) resistance have been used as selectable marker genes, while the *uid*A gene conferring glucuronidase (GUS) expression and the *gfp* gene conferring green fluorescence were used to confirm transformation. However, the use of antibiotics in the selection process is not widely accepted.

Though some researchers use efficient biolistic (micro-projectile bombardment) transformation systems, the majority make use of the *Agrobacterium* mediated transformation system. This technology is based on the ability of *Agrobacterium* to insert genes into plant cells during the infection process. Although *A. tumefaciens* causes crown gall in grapevine, disarmed strains that do not induce crown gall are used for genetic transformation. The most suitable plant material for co-cultivation with *Agrobacterium* is embryogenic cell lines, of which the quality and developmental stage have a strong effect on transformation efficiency. The first report on transformed plants with *V. rupestris* St. George was from Mullins *et al.* (1990). A protocol was standardized to regenerate six grape cultivars through meristematic bulk (MB) induction and was used for genetic transformation. Transformation efficiency was the greatest in Chardonnay (51.7 %), followed by Thompson Seedless (42.3 %), St. George (41.6 %), Redglobe (40 %), Cabernet Sauvignon (35.6 %) and 101-14 Mgt (29.9 %). It was found that MB induction was a fast and simple alternative for genetic transformation of grape cultivars (Xie *et al*., 2016). Trans-graft protection against Pierce's disease mediated has been achieved by using transgenic grapevine rootstocks. The non-transgenic scions were grafted onto rootstocks engineered to produce an antimicrobial peptide and a protein that inhibits cell wall degradation. These proteins were transported to the scion through the xylem, resulting in the enhanced mobilization of water and nutrients and a 30–95% reduction in pathogen-induced mortality (Dandekar *et al*. 2019).

VitisGen project has been launched in 2011, with a grant from the National Institute of Food and Agriculture Specialty Crop Research Initiative (NIFA-SCRI), at Cornell University with collaboration of 25 scientists from 11 different institutions of USA to accelerate the development of the next generation of grapes. The project will expand in developing novel economic, phenotyping and genetics knowledge and tools related to new grape cultivars, and translate these innovations into new applications for improving grape breeding programs. (Fellman and Walter-Peterson, 2015).

SAPOTA

(SAPODILLA)

Origin: Central America

Taxonomy status:

Order: Ericales

Family: Sapotaceae

Botanical name: *Manilkara zapota* (L.) P. Royen syn. *Achras sapota* L.

Chromosome number: n=13; 2n=2x=26

Genome size: The complete chloroplast genome is 159,853 bp in length.

Related species of sapota are *M. hexandra, M. kauki* and *M. elangi.*

Botany

Sapota is evergreen, upright to spreading, low branching and 15-20 m tall tree having latex. Leaves are alternate, elliptic lanceolate to ovate lanceolate, glossy above and pubescent below and trend to cluster at the end of the shoot. Hermphroditic flowers are yellow or white with three outer and three inner brown hairy sepals enclosing a pale green to white tubular, six lobed petals. Androecium consists of 12 stamens arranged in two whorls of 6 each (6 petaloid stamenoids and 9 stamens) on the rim of the corolla tube (epipetalous). The filaments of stamens are short and dorsifixed. Pistil is syncarpous and composed of 4-5 fused carpels. The style is long green and stigma is capitate. Fruit is round to ovate berry, 4–8 cm in diameter with a pale yellow to an earthy brown flesh of grainy texture. Fruit contains one to six seeds. The seeds are hard, glossy and black.

Fig. 13. Sapota flower shoot.

(Source: Hutchinson, 1969)

Floral Biology

Flowers appear more or less throughout the year in three flowering seasons; June-August, October-December and March. Flower bud takes 36 days from its emergence to opening. Anthesis takes place in the morning (5-6 AM) which is followed by anther dehiscence (7-9 AM). Stigma was receptive a day before anthesis and remained receptive a day after anthesis. However, under North Indian conditions anthesis occurred from 5 AM to 7 AM with maximum at 6 AM. The highest per cent fruit set on pollination between 7-10AM on the day of anthesis indicated the peak stigma receptivity. It was also observed that stigmatic exudates acted as a trap for flower visitors. Sapota is protogynous. Style and stigma grow out of the bud about 2 days before flower opening. Sapota trees produce many flowers, but the number of fruits reaching maturity is very small. The major fruit drop occurs during the first five weeks following fruit set and as little as 1.6% flowers produced on a tree may develop into fruit. Pollinatlon is carried by wind, thrips and *Silvanopsis* sp. Shape of fruit is markedly affected by cross pollination. The self pollinated fruits are oblong to oval whereas the cross pollinated ones are round.

Breeding objectives for scion:

(i) Wide adaptability.
(ii) To develop dwarf tree and precocious.
(iii) Synchronized flowering.
(iv) To develop cultivars having high yield of large seedless fruits or with few seeds, melting sweet and deliciously flavoured pulp and good keeping quality.
(v) To develop early maturing cultivars.
(vi) To develop salinity tolerance.
(vii) Resistannce to pests and diseases.

Breeding objectives for rootstock:

(i) Compatible with scion.
(ii) Induce dwarfness, earliness, good fruit yield and fruit quality in the scion cultivar.
(iii) High propensity in propagation.
(iv) Wide geoclimatic adaptability.
(v) Tolerance to biotic and abiotic stresses.

Constraints in breeding:

(i) Long juvenile phase.
(ii) High heterozygosity.
(iii) Narrow genetic base.
(iv) Self and cross incompatibility.
(v) Sterility.
(vi) Heterostyly and protogyny.
(vii) Lack of information on inheritance pattern.

Germplasm Resources

The germplasm of sapota is maintained at New South Wales & Darwin in Australia (6 &15 cultivars respectively), Costa Rica (118 accessions), Cuba (20 primitive cultivars), Philippines (73 accessions), Mexico (31 accessions), Puerto Rica (32 cultivars) and Venezuela (27 cultivars).

In India, germplasms of sapota are maintained *in situ* in germplasm collection at IIHR, Bengaluru (52), TNAU, Coimbatore (14), UAS, Dharwad (12), GAU, Gandevi (9), TNAU, Periyakulam (26), MPKV, Rahuri (7), SKLTSHU, Rajendra Nagar (12), SKLTSHU, Fruit Research Station, Sangareddy (17), Dr YSR Horticultural University HRS, Venkataramannagudem (30) and UHS, Bagalkot (40).

Sources of useful traits:

Resistance to leaf spot: CO-1, Cricket Ball, Kalipatti and CO-2.

High TSS: Kirtibarthi, Thagarampudi, Dwarapudi.

Good quality: Dhola Diwani, Kalipatti, Bhuripatti, Pala, Pot Sapota, Thagarampudi.

Good transportability: Kirtabharthi and Ayyangar.

Oval fruits: Pot Sapota, Oval, Thagarampudi, Kalipatti, Dhola Diwani, Pala, Krithabharthi and Vavi Valasa.

Round fruits: Jonnavalasa Round, Cricket Ball, Calcutta Round and Baramasi.

Rose scented: Ayyangar.

Inheritance pattern:

Oval fruit shape: Dominant character.

Large fruit size: Is not linked with low yield.

Breeding Methods

1. Introduction: The cultivars introduced in India are Seedless, Cuba, Ponderosa, Common, Lamad and Gonzalez from the Philippines; Alano from Hawaii; Hasya, Morena, Molix and Oxkutzcab from Mexico; Makok from Thailand and Modello, Prolific, Russell and Seedless from USA. There is need to introduce promising cultivars Sawo Kulan, Sawao Betawi (Round fruit) from Indonesia; Pineras, Sao Manila from the Philippines; large fruited cultivars Krasuey and Kai Hahn from Thailand and Prolific, Russell and Brown Sugar from Florida (USA). Other exotic genotypes that exist in India and possess desirable traits can be utilized in improvement programme or released directly as cultivars if considered suitable.

2. Selection: Attempts have been made to make selections from the commercial cultivars like CO-2 (clonal selection from Baramasi) at TNAU, Coimbatore; PKM 1 (clonal selection from Guthi), PKM (Sa)-4 (four open pollinated clone of PKM 1) and PKM (Sa)-5 (from open pollinated seedlings) at HC&RI TNAU, Periyakulum; DHS-1-2/1, a clonal selection from DHS-1 at UHS, Bagalkot and Rajendra Sapota selection from local material at RAU, Pusa, Samastipur, Bihar.

3. Hybridization: Hybrids have been developed and released for commercial cultivation, e.g. CO-1 (Cricket Ball x Oval), CO-3 (Cricket Ball x Vavilavalasa),

PKM-2 (Guthi x Kirthibarthi) and PKM-3 (Guthi x Cricket Ball) developed at HC & RI, TNAU, Periyakulam DSH-1 (Kalipatti x Cricket Ball) and DSH-2 (Kalipatti x Cricket Ball) developed at UAS, Dharward.

4. Mutation: Ravishankar and Purohit (1991) reported bud sport mutant from the population of 'Cricket Ball' variety. This mutant is large fruited and leaf spot tolerant.

5. Biotechnology: Attempts have been made to devise protocols for micropopagation from mesocarp, endosperm and embryo by culture on modified MS medium supplemented with various levels and combinations of auxins and nutrients.

CUSTARD APPLE

(SUGAR APPLE OR SWEETSOP)

Origin: Tropical America (Central America)

Taxonomy status:

Order: Magnoliales

Family: Annonaceae

Subfamily: Annonoideae

Tribe: Annoneae

Botanical name: *Annona squamosa* L.

Chromosome number: n=12; 2n=2x=14: *A. glabra*, which is a tetraploid species ($2n = 4x = 28$).

Genome size: 656.78 Mb (soursop)

The other annona species producing edible fruits are Cherimoya (*A. cherimola* Mill.), Atemoya (*A. squamosa* × *A. cherimola*), Pond apple (*A. glabra* L.), mountain soursop (*A. montana* Magfady), Bullock's heart or custard apple (*A. reticulata* L.), Ilama (*A. diversifolia* Staff.) and Sour sop (*A. muricata* L.).

Botany

Custard apple tree is deciduous, much smaller than the soursop, reaching a maximum height of 6 m with many lateral branches. Lenticels are present in the stems, young shoots are pubescent and the old ones are smooth. The leaves are oblong elliptical, measuring 5 to 17 cm in length and 2 to 7 cm in width, with an obtuse or acuminate apex. The flowers are hermaphroditic and are produced singly or in small clusters in a compound inflorescence on an enlarged receptacle. Flowers arise on the current season's growth and occasionally on old wood. The flowers measure 2.0 to 2.5 cm in length and are much smaller than in soursop. Flowers are borne solitary or in short lateral clusters of 2–4 about 2.5 cm in length, greenish yellow and are borne on a hairy, slender and long stalk. Flowers are hermaphrodite having 3 green outer petals, purplish at the base, oblong, 1.6 cm to 2.5 cm long, and 0.6 cm to 0.75 cm wide, three inner petals reduced to minute scales or are absent. Stamens are numerous, crowded and white and ovary is light green. Styles are white, crowded on the raised axis. Each pistil forms a separate tubercle (small rounded wartlike protuberance which matures into the aggregate fruit. The fruit is rounded, heart shaped, ovate or conical, 5 to 7.5 cm in diameter, 6 to 10 cm in length weighing 120 to 330g. The white custard like pulp has a pleasant sweet sour flavour. The fruit contains 35 to 45 black seeds, each 1.5 to 2.0 cm in length and 0.6 to 0.8 cm in width.

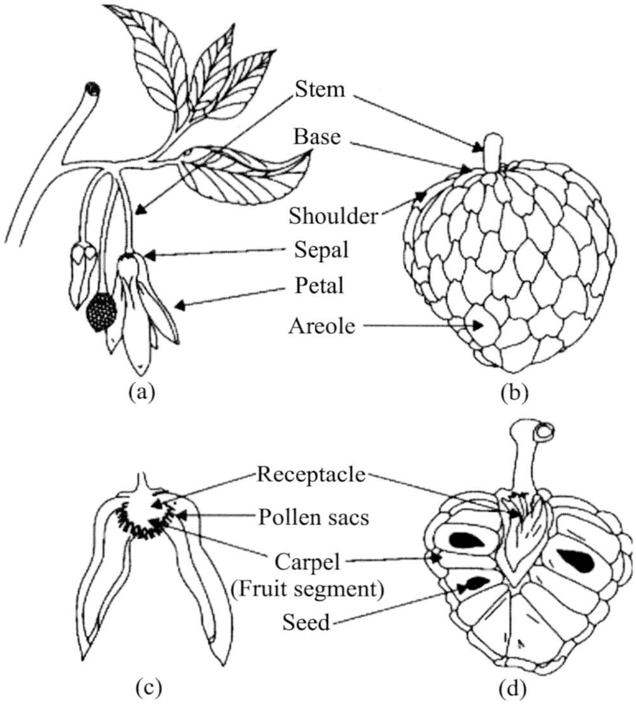

Fig. 14. Leaves, flower and fruit of custard apple.
(Source: Wongs-Aree and Mongkut's, 2011)

Floral Biology

Flowering in most of the annona species appeared in different flushes although April and September were observed to be major bloom periods under tropical humid eastern coastal region of India. The flowers are hermaphrodite and dichogamous in nature. The anthesis takes place in sugar apple flowers at morning 5:30 AM to 6:30 AM and anther dehiscence at 12 PM to 2 PM of next day of the anthesis and stigma receptivity start a day prior to anthesis and continue upto day after anthesis but found maximum receptivity at the day of anthesis. The anthesis takes place in reticulata flowers at morning 6 to 8 AM and anther dehiscence at 12 PM to 2 pm of next day of the anthesis and stigma receptivity start a day prior to anthesis and continue upto day after anthesis but found maximum receptivity at the day of anthesis . In Soursop, the flowers exhibited protandrous dichogamy, the peak anthesis occurs in the afternoon 2 PM to evening 8

PM, anther dehiscence at 4 AM – 8 AM and stigma receptivity start after completion of male phase. The flowers are hermaphrodite and dichogamous in nature, both the male and female flowers are borne in the same flower, but the female part matures before the male part (protogyny) in most of the species except pond apple (protandry), therefore the annonas possess highly cross pollination.. Natural pollination in annona mainly carried by only through insects, such as beetles, nitidulid beetles play a major role in pollination of sugar apple, insects rarely visits the annona flowers because the flowers are not attractive and have low mild fragrance to attracts the insects, which suggests that canthorophylous pollination plays only a secondary role in the annona species. Pollen grains occur in clumps and are somewhat sticky, wind pollination is not common. Pollen germination is low and may influence final fruit set, which varies from 5.4% to 5.6%. Hand pollination improves fruit set and substantially boosts fruit yield per tree by increasing the fruit size and number, the major yield components. Hand pollination is normally carried out before 8 AM using a small brush. Pollen can be collected in the morning between 5 and 8 AM from fully open flowers, when the sacs have turned from white to cream. The collected pollen is used to pollinate half-open flowers whose pistils are already receptive.

Breeding Objectives

1. To develop variety having dwarfness.
2. To develop precocious variety.
3. To develop high yielding variety having fruit with round and symmetrical fruit shape, attractive red colour, good flavour, texture and quality (sweetness, seedlessness and high flesh:seed ratio), average in fruit size, smooth skin, tolerant against bruising and lack russeting and good shelf life.
4. To develop variety having resistance to pests and diseases (*Armillaria melle, Rhizoctonia* spp. anthracnose and *Phytophthora*).
5. To develop variety tolerant to heavy and poor soils.

Problems in Breeding

1. Long juvenile phase.
2. Heterozygosity.
3. Dichogamy and protogyny.
4. Self and cross incompatibility.
5. Narrow genetic base.
6. Unknown Inheritance pattern.

Germplasm Resources

A total of 1,741 germplasm accessions of eleven identified species and one interspecific hybrid have been documented and maintained in germplasm respositories. About 36 accessions of custard apple are being conserved in field genebanks at IIHR, Bengaluru (48), The germplasm also maintained at MPKV, Rahuri and FFRS, Sangaredddy and ICAR-Central Institute for Arid Horticulture, Bikaner.

Sources of useful traits:

Dwarf growth: Lal sitaphal.

Red flesh colour: *A. purpurea.*

Less seed number and good fruit quality: *A. cherimoya.*

Good fruit quality: *A. montana, A. cherimoya.*

Good processing quality: *A. muricata.*

Resistance to fruit cracking: *A. purpurea.*

Suitability in lowlands: *A. diversifolia.*

Tolerance to low temperature: *Annona cherimola* and *A. atemoya.*

Tolerance to waterlogging: *A. glabra.*

Tolerance to drought: *A. squamosa.*

Cold tolerant: *A. cherimola.*

Suitability to sea side: *A. reticulata.*

Tolerance to hard and stony soils: *A. squamosa.*

Inheritance Pattern

Yield: Controlled by quantative character.

Red skin colour: Single or double recessive gene.

Breeding Methods

1. Introduction: The cultivars like British Guinea, Brabados Seedling, Brandy, Mammoth, Washington PI 98797 and African Pride have been introduced.

2. Selection: Selection done in Florida (USA) is red sugar apple. Similarly, Anand Selection at Anand, NKM-1 and 2 in Maharashtra, TP-7 at MAU, Parbhani, Phule Parander and JWC-1 by a farmer Shri Navnath Malhari Kaspate from Solapur district of Maharastra are custard apple selections. Among the four varieties, Annona-2, NMK-1, NMK-2 and NMK-3, NMK-1 commonly called as Golden gained importance in south India because of colour and quality of its fruits. APK(ca)-1 is a selection from *A. reticulata.* The other sugar apple selections are Balanagar, Mammoth and Red Sitaphal. Mossman and Whaley from Australia; Libby, Lisa, Ott and Red sugar apple from USA; Reretai from New Zealand; Molate and Lobo from Philippines; Leone from Italy; Abd El-Razik from Egypt; Madeira and Mateus 1 from Portugal; IPA selection from Brazil and Negrito from Spain are also selections of sugar apple.

Table 14. Characteristics of promising selections.

Species	Desirable characteristic	Selection name	Origin
Annona squamosa	Red skin colour	Red sugar apple	Florida
	Large size, light yellow skin	Noi	Thailand
	Seedless	Acc.no. 6333	Philippines
	Big, sweet, few seeds	Lobo	Philippines
	Seedless	Cuban Seedless	Cuba
Annona diversifolia	Pink skin colour and flesh	Imery	Florida
Annona reticulata	Red skin and flesh, smooth skin	Camino Real	Guatemala
	Purple skin colour	Fairchild purple	Florida
	Orange skin and flesh	Dr Leon	Guatemala
	Sweet type/low grit	Young	West Java
	Red skin and flesh	El Remate	Mexico
	Red skin and flesh	Canul	Belize
Annona cherimola	Orange skin	Citamex	Mexico
	Seedless or low-seeded types	Seedless	Peru and Ecuador

Selected cultivars of cherimoya:

Australia: Andrews, Kempsey and Mossman.

California, USA: Bays, Booth, Libby, Lisa and White.

Chile: Bronceada and Concha Lisa.

New Zealand: Burtons and Wonder.

Spain: Campa, Cristalino.

Ecuador: E-8

Selected cultivars of atemoya

Australia: Maroochy Gold, KJ Pinks and Pinks Mammoth.

Southern Africa: African Pride.

Florida, USA: Bradley, Page, Geffner, Kabri and Malalai.

3. Hybridization:

Intervarietal and interspecific hybridization

Intervarietal crosses have been made between the best selections of atemoya and cultivars of African Pride, Pinks Mammoth and Hillary White.

Interspecific crosses have also been made between the species *A. cherimola* (cherimoya), *A. squamosa* (sugar apple), *A. reticulata* (bullock's heart) and *A.*

diversifolia (Ilama). P. J. Wester produced the first hybrid Atemoya at Florida in 1908 which is a hybrid between *A. squamosa* and *A. cherimola*.

The studies conducted in Serdang, Selangor, Malaysia, involved six selected species of Annonaceae family, namely *A. muricata, A. squamosa, A. glabra, A. montana, A. rollinia and A. reticulata*. Hybridizations were attempted between the species along with their reciprocal crosses. Although fruits formed in crosses between some species, the seeds were not viable except from two hybridizations: *Annona muricata* x *Annona montana* and *Rollinia mucosa x Annona muricata*. The former F_1 hybrids produced plants that flowered but did not set fruit. Atemoya x *A. reticulata* and cherimoya x *A. reticulata* crosses produced late fruit maturing progeny.

More than 2000 inter and intra-species *Annona* hybrids F_2 backcrosses were developed at IIHR, Bengaluru and evaluated which resulted in the release of the hybrid Arka Sahan. It is an outstanding recombinant isolated from the family of Island Gem (atemoya) and Mammoth custard apple. Its light green fruit skin is moderately thick (0.5 cm) with large segments and flat eyes and has a waxy bloom. The creamy white flesh is tender and juicy with mild pleasant aroma and scarce seeds (9/100 g). For improving fruit set, fruit size and shape, pollination with *Annona squamosa* pollen was found to be most effective.

4. Polyploidy: An unexpected high proportion of triploid genotypes were found in the progeny of an interspecific cross between the diploid species *A. cherimola* and *A. squamosa.*

5. Mutation: There is very limited work has been done in annona for development of mutant varieties but the Natural mutants obtained are Brazilian seedless and Thai seedless (Ts). Thai seedless produces fully seedless fruits of normal size. This feature has the potential to improve annona fruit production.

6. Biotechnology: There are reports of micropropagation and plantlet production using nodal explants of *A. atemoya, A. cherimola* and *A. squamosa*. Organogenesis and rooted shoots have been obtained from hypocotyls of *A. atemoya, A. muricata,* and *A. squamosa*. Organogenesis from petioles and zygotic embryos of *A. cherimola* has also been reported. Plantlets have been obtained following anther culture of *A. squamosa*.

Protocols for regeneration via organogenesis could form the basis for the development of a transformation system for *Annona* spp. Molecular markers would be useful both for documenting genetic diversity and elucidating the confusion surrounding the naming and origin of many commercial cultivars. Diversity analysis by DNA markers has now taken the place of the conventional phenotypic based system. The study with RAPD and SSR markers suggested that at DNA level there is considerable genetic diversity among *Annona* genotypes. Investigations based on biochemical parameters also showed significantly high diversity.

A. muricata, a rich source of seed oil, can be exploited in oil industries. The genotypes characterized based on molecular marker and biochemical gives an opportunity to fruit breeders to alter the fruit qualities of genus *Annona*.

AVOCADO

Origin: Central and South America.

Taxonomy status:

Order: Laurales

Family: Lauraceae

Genus: *Persea*

Subgenus: *Persea*

Species: *americana*

Botanical name: *Persea americana* Mill.

Chromosome number: n = 12; 2n = 2x = 24

Genome size: 920 Mb (*P. americana* var. *drymifolia*); 912.6 Mb (*P. americana* cv. Hass).

The subgenus *Persea* consists of three species: *schiedeana, parvifolia* and *americana. Persea americana* has been subdivided into three ecological or horticultural races:

Mexican: *Persea americana* var. *drymifolia.*

West Indian: *Persea americana* var. *americana.*

Guatemalan: *Persia americana* var. *guatemalensis*.

The racial classes also differ phenotypically in fruit size and shape, skin colour and thickness, seed size and fruit ripening.

Botany

The avocado is erect, evergreen tree and grows up to 9 to 18 m high with bole diameter from 30 to 60 cm. The dark green, leaves are of shapes including lanceolate, elliptic, oval or ovate. These may be alternate, glossy on the upper surface and whitish on the underside. Panicles of flowers with deciduous bracts arise on new growth from the axils of leaves. The small, pale green or yellowish green flowers are borne on multibranched axillary panicles terminating in a shoot bud. The flowers are inconspicuous, bisexual, yellow or greenish yellow, 5–10 mm wide. Six parted perianth consists of two whorls with nine stamens, six of which form the outer circle and three in the inner circle. At the base of the inner circle are located a pair of nectaries, alternating with three staminodes, which also secrete nectar. Each stamen has four pollen sacs, which release cohesive pollen. The single pistil contains one carpel and one ovule. Flowers are protogynous and exhibit a unique mechanism of synchronous dichogamy for alteration of sexes. The avocado fruit is a one seeded berry. The skin colour of the ripe fruit has several shades from green to yellow green, reddish to maroon and light to dark purple. The buttery flesh (mesocarp) of ripe fruit is greenish yellow to bright yellow to cream.The single large seed surrounded by a thick fleshy mesocarp has two cotyledons enclosing an embryo.

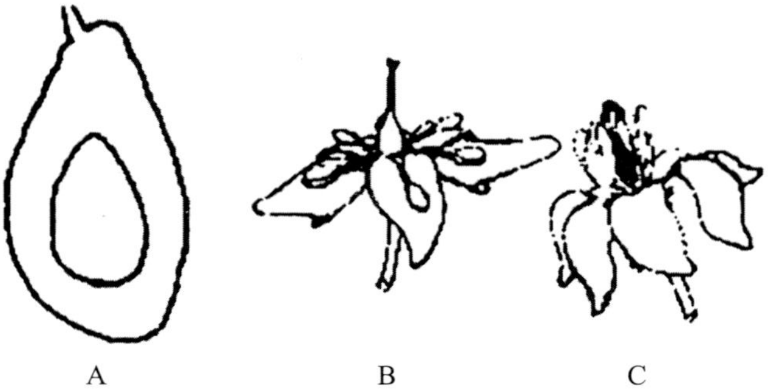

A—fruit in longitudinal section;
B—flower at female stage; C—flower at male stage.

Fig. 15. Avocado flower types with fruit.

Floral Biology

Floral initiation occurs during the autumn. The avocado has a unique flowering behaviour, described as "complementary synchronous dichogamy". It is an outbreeding mechanism which promotes cross pollination thereby maintaining genetic diversity. The flower is perfect having both male and female sexes. The flower opens twice a day, in morning and afternoon, for 2 consecutive days; first time as a functional female (stage I) and second time as functional male (stage II). The pistil ripens earlier than the stamen (protogyny) fall into one of the.

Based on the pattern of opening of flower all avocado cultivars and seedlings, irrespective of race, are classified into two complementary groups designated A and B. In type A, flower opens first as functional female during the morning hours, close near midday and reopens as functional male in the afternoon of the following day. In the type B, cultivars exhibit the first opening of female flowers in the afternoon and the second opening of functional male in the morning of following day. As a result of synchronous dichogamy this offers the best opportunity for fertilization. This type of reproductive behavior promotes outcrossing. However, considerable self pollination is known to occur in commercial plantings. The flowers are entomophilous and depend on honey bees for pollination. The relative humidity may play some role in prolonging the receptivity of pistil. Various pollinating agents such as honey bee, various species of wild bees, wasps, flies and humming birds visit its flowers.

Breeding objectives for scion

1. To develop varieties having wide adaptability, ease in propagation, precocity and spreading growth habit.
2. To develop varieties having good fruit set, high quality of medium size fruit (250-300g) having smooth skin, fibreless pulp and high oil content(>18%) and small seed.
3. To develop varieties having short fruit maturation period and heavy and regular bearing.

4. To develop varieties having resistance to diseases (root rot, sun-blotch) and insect pests and tolerance to cold, heat and salinity.

Breeding objectives for rootstock

1. Dwarfing or semi-dwarfing.
2. Vigorous and easily propagated.
3. Impart high quality fruiting in scion.
4. Tolerant to *Phytophthora* and other organisms and free from sun blotch.
5. Tolerance to salinity.
7. Tolerance to chlorosis.
8. Tolerant to drought.

Limitations in Breeding

1. Long juvenile period.
2. Large tree size.
3. Typical flowering behaviour.
4. High level of heterozygosity.
5. No controlled crosses.
6. Inheritance pattern is not understood.

Germplasm Resources

The USDA-ARS Subtropical Horticulture Research Station (SHRS) Miami, Florida, curates a large avocado germplasm collection with 256 unique cultivars. The Fairchild Tropical Botanic Garden, Williams Farm (Homestead, FL) has a collection of 100 unique cultivars. The University of California, Riverside (UCR) maintains germplasm collections at two locations. The first is at UC South Coast Research and Extension Center (SCREC) in Irvine has a) twelve *Persea* spp. relatives, b) over 200 cultivars consisting of heritage cultivars and material from previous scion breeding programs (GERMCA), and c) over 149 unique individuals from the UCR rootstock germplasm collection including hybrids, advanced rootstock selections and material imported from several countries including Chile, Guatemala, Mexico, South Africa, Australia and Israel. The second UCR collection is at the Agricultural Operations Station (AgOPs) in Riverside which maintains a scion collection and 10 open pollinated rootstocks.

More than 50 collections of avocado are maintained by IIHR at CHES, Chethalli and Bengaluru (47). TNAU Horticultural Research Station, Thadiyankudaisai also maintains the germplasm of avocado.

Sources for useful traits:

Dwarfing: *P. schiedeana,* Fuerte.

Excellent peeling: Pinkerton.

Low oil content: Pollock.

Tolerance to *Phytophthora cinnamomi* (root rot): *P. caerulea, P. cinerascens, P. pachypoda, P. donnell-smithii,* Duke 7, Thomas, D 9, Barr Duke, Toro Canyon, Toro, Canyon, Duke-7, Lulo, G 6, Topu Topa, Latas (Merensky I), Dusa (Merensky II), Martin Grande (*P. americana var. guatemalensis* x *P. schiedeana*) and Evstro (Merensky III).

Cold hardiness: Fuerte, Toro Canyon, Duke 7, Brooksville, Grainesville and Mexican race.

Salinity tolerance: West Indian race, Borchard, Toro Canyon, Latas (Merensky I), Dusa (Merensky II) and Evstro (Merensky III).

Tolerance to calcareous soil: West Indian race, Borchard.

Iron resistance: *Eriodaphne.*

Inheritance Pattern

1. Tree size, flowering intensity, fruit weight, fruit density and harvest duration shows addiative and non-additive genetic variance.
2. The fruit skin colour, flowering group and anise scent was found to be controlled by several loci having several alleles in each locus.

Breeding Methods

1. Introducation: The cultivars Fuerte, Bacon, Zutano, Hass, Booth-7, Booth-8, Sharwil, Ettigner, Pollock, Waldin, Nabal, Linda, Pubela, Gottfried, Collins Purple, Green, Peradeniya Puple hybrid and Trapp have been introduced from USA and Mexico.

2. Selection: Most of the cultivars in avocado growing countries of the world originated as selections from chance seedlings as given below:

Hass: Selection from seedlings of Guatemalan race.

Edranol: Selection from seedlings of Guatemalan race. Originated in California

Nabal: Selection from seedlings of Guatemalan race.

Galil: Selection from Oshir cultivar in Israel.

Gwen: Selection from Hass seedlings at University of Calfornia.

Zutano: Selection from seedlings of Mexican race Originated in California in 1932.

Sir Prize: Selection made at University of Califonia.

Reed: Selection by Mr. James Reed from Natal seedlings in California in 1977.

Jim: Selection from Bacon seedlings.

Alboyee: Selection from open pollinated seedlings

Ettinge: Selection from Fuerte seedlings.

Mcfie: A selection from Fuerte seedlings.

Teteluku and Aitken: Selection from Fuerte seedlings.

Frankish, Clarke, Doornkop: Selections from seedlings of Guatemalan race.

Table 15. Important cultivars of the three races of avocado.

West Indian	Guatemalan	Mexican
Butler	Benik	Duke
Pollock	Dickinson	Ganter
Ruchle	Anaheim	Gottfried
Russell	Edranol	Mexicola
Simmonds	Pinkerton	Puebla
Waldin	Reed	Zutano
Trapp	Taylor	

3. Hybridization: Breeding methods may be by self pollination or by hybridization two selected parents parents cvs., the choice depending upon objectives of breeding programme. Selfing can be achievd by use of cage with pollinator bee enclosed within the tree or by using isolated trees of desired parent cultivar. Outcrossing is reported nil where trees cvs are separated by 100 m or more.

Hybridization is the only way to obtain progeny with two or more desirable traits that are not present in an available breeding line. Hybridization has major disadvantages compared with selfing: These are a) obtaining positive hybrids is far more difficult and expensive and b) the breeding worth of each parent tends to be obscured by the contribution from the other parent.

Hybridization is necessary when two or more desirable traits are to be combined or when traits intermediate between two cultivars are desirable, but this has proved difficult with avocado crop. Selfing has proven to be generally much more efficient, as seedlings can be obtained from isolated or buffered trees at a cost no higher than that of the fruits. Selfing has advantages:-

1. better evaluation of breeding worth,
2. elimination of harmful recessive alleles for future breeding,
3. greater homozygosity and so more predictable breeding behaviour,
4. reduction of excess vegetative vigor, and
5. requiring self-fertilization, to reduce the commercial need for cross-pollination.

Selfing can be achieved by the use of cages with pollinator bees enclosed within the tree or by using isolated trees of a desired parent cultivar. Outcrossing is reported to be nil where trees of a cultivar are separated from other genetic lines by 100 m or more. Variations in daily weather conditions during the flowering period may disrupt the diurnal synchrony of the flowers, allowing self-pollination.

Hybrids

Ardith: Mexican x Guatemalan hybrid from University of California, Riverside.

Gwen: It is a Mexican x Guatemalan hybrid from University of California, Riverside.

Ryan: It is a Mexican x Guatemalan hybrid bred in California.

Fuerte: It is a natural Mexican x Guatemalan hybrid.

Booth 7 and Booth 8: Natural hybrids of Guatemalan and West Indian races.

Lamb Hass: Gwen x Thill hybrid from University of California.

Gil: Cross between Tova and a unknown pollen donor.

Eden: Pinkerton x NISI 2 hybrid.

Ardith: Mexican x Guatemalan hybrid.

Bacon: Mexican x Guatemalan hybrid.

Choquette: Guatemalan x West Indian hybrid.

Ettinger: Mexican x Guatemalan hybrid.

Fuerte: Mexican x Guatemalan hybrid.

Pinkerton: Mexican x Guatemalan hybrid.

Sharwil: Mexican and Guatemalan hybrid.

Interspecific hybrids:

Martin Grande: *Persia americana* var. *guatemalensis* x *P. schiedeana.*

4. Mutation: Spontaneous mutants have been observed to occur in nature over a long period of cultivation. Natural mutants of Fuerte, Hass and Jim have been recognized. Only a few attempts have been made to induce mutations artificially with chemical mutagens or irradiation.

5. Biotechnology:

Micropropgation: Micropropagation of shoots from axillary branching has been used in the improvement programmes by selection to generate a limited number of selected genotypes for fungal resistance. But it has not been used commercially.

Genome research:

Nearly 10,000 EST sequences have been collected from early development of avocado flower. EST avocado fruit and transcriptome project initiated in Mexico has resulted in the identification of 5233 unigenes. Some of the most abundant genes expressed in criollo avocado fruit have been identified. Among these are metallotheoneins involved in metal homeostasis, two putative transcription factors, Ethylene Response Factor and Auxin-repressed protein like gene, as well as pathogen related genes, i.e. an endochitinase and a protease inhibitor. Other genes involved in the interactions of avocado fruit with the fungus *Colletotrichum gloeosporioides* as well as the interaction of avocado root with the pathogen *Phytophthora cinnamomi* have also been identified. These studies would facilitate designing more efficient strategies to manage these diseases.

Genetic transformation: Methods for transforming avocado embryogenic cultures are available and, in some cases, plants have been recovered from transgenic embryos, although the methodology to convert transformed somatic embryos into plants needs to be improved. Embryogenic avocado cultures have been transformed with different gene constructs in order to address different breeding objectives. Avocado is being *Agrobacterium rhizogenes* mediated plant transformed with genes encoding for different antimicrobial proteins, including chitinase, glucanase, the antifungal protein (AFP) and magainin, and the first greenhouse planting of transgenic plants containing the gene for AFP has been established. As part of a strategy for controlling avocado sunblotch disease, embryogenic cultures have been transformed with the pac1 ribonuclease gene from yeast. In order to control fruit ripening and extend the shelf life of avocado fruit, embryogenic cultures have been transformed with the gene for SAM hydrolase, a bacterial gene that mediates the breakdown of S-adenosylmethionine (SAM), a precursor of ethylene.

6. Conservation: Efforts are needed to apply these techniques to avocado shoot tips or embryonic axes, in order to use this material as backup for avocado germplasm collections and to facilitate international movement of genetic resources, particularly from the great genetic repositories in USA, Israel, Mexico and Australia. Use of the available cryopreservation techniques could be useful for preserving selected embryogenic cell lines.

JACKFRUIT

Origin: Western Ghats of India.

Taxonomy status:

Order - Rosales

Family - Moraceae

Botanical name: *Artocarpus heterophyllus* Lam.

Chromosomes number: n=14, 4x=56 (tetraploid).

Genome size: 273 Gb.

Related edible species are *Artocarpus altilis, A. camansi, A. mariannensis. A. integer*, *A. anisophyllus*, *A. lacucha* and *A. odoratissima.*

Botany

Jackfruit is a large evergreen monoecious tree growing about 25 m in height. Leaves are oblong or oval. The inflorescences are formed on the trunk, branches or twigs. The inflorescences are pedunculated, cylindrical to ellipsoidal or pear shaped, about 10-12 cm long and 5-7 cm wide. Inflorescences are initially completely enveloped in egg shaped cover sheets which rapidly slough off. The flowers are very small. There are several thousand flowers in an inflorescence, which sit on a fleshy rachis. The male flowers are greenish, hairy and the perianth ends with two 1 to 1.5 mm membrane. The individual and prominent stamens are straight with yellow, roundish anthers. After the pollen distribution, the stamens become ash-gray and fall off after a few days. Later all the male inflorescences also fall off. Some male flowers are sterile. The greenish female flowers, with hairy and tubular perianth, have a fleshy flower like base. The female flower contains an ovary with a broad, capitate or rarely bilobed scar. The ellipsoidal to roundish fruit is a multiple fruit formed from the fusion of the ovaries of multiple flowers. They possess a hard, gummy shell with small pimples surrounded with hard, hexagonal tubercles. The very large and variously shaped fruit is 30 to 100 cm in length and 15 to 50 cm in diameter and weighs 10-25 kg or more. There may be 100-500 seeds in a fruit. The seed coat consists of a thin, waxy, parchment like and easily removable testa (husk) and a brownish membranous tegmen. The cotyledons are usually unequal in size. The endosperm is minimally present.

Floral Biology

In South India flowering occurs throughout the year but in North India it occurs just after the rainy season. Separate male (staminate) and female (carpellate) inflorescences are found. Thus jackfruit is primarily outcrossing. The time of flower opening is 2-4 AM and 4-6 PM. The majority of buds open at 0.60 hrs, but anthesis occurs throughout the day until 18.00 hrs. Anther emergence occurs between 09.00 and 11.00 hrs and dehiscence is completed by 15.00-17.00 hrs after 3-5 days.

Breeding Objectives

1. Wide adaptation.
2. Precocity with a moderate canopy.
3. Early fruiting.
4. Good fruit size and fruit quality traits (aroma, edible percentage, light pulp-seed ratio, flesh firmness, colour and flavour).
5. High yield.
6. Regular and prolific bearing.
7. Long post harvest life.
8. Resistance to fruit borer.

Limitations in Breeding

1. Long juvenile phase.
2. Seed germination is poor.
3. Protogyny.
4. Monoecious.

Germplasm Resources

The jackfruit collections held in Bangladesh are 130, 947 in India, 155 in Indonesia, 350 in Nepal, 10 in Pakistan, 77 in Sri Lanka, 87 in Thailand, 202 in Vietnam and 30 in Pacific Islands. In India, the Institutes like IIHR, Bengaluru (171), KAU, Thrissur, CCARI, Goa, NBPGR Research Station, Thrissur and Dr. BSKKV, Dapoli also conserve the germplasm of jackfruit.

Sources of useful traits

Excellent fruit quality: *Artocarpus odoratissima.*

Inheritance Pattern

There is hardly any information on heritability of economic characters like texture of edible flesh, sweetness, seed size and number and size of fruits per tree and thickness of the rind.

Breeding Methods

1. Introduction: The cultivars Jack-2, Jack-29, Jack-31, Jack-31, Jack-27 and Jack-30 were introduced from Malaysia. Na-29 and Na-31 were introduced from Indonesia and Singapore or Ceylon Jack from Sri Lanka.

2. Selection: Jackfruit trees are cross pollinated and are mostly propagated by seed. They exhibit a wide range of variation in morpho-agronomic characteristics. Variation exists in fruit weight, size and shape and also in sweetness, flavour and taste. This has been used for selection in various countries.

Myanmar: Talaing, Kala Sri lanka Vela, Varaka (Waraka), Peniwaraka, Kuruwaraka, Singapore Jack (Ceylon Jack).

Bangladesh: Bangladesh Topa, Hazari, Chala, Goal, Koa and Khaja.

India: Swarna Manohar and Swarna Poorti from Jharkhand, Bihar, Orissa and UP; Muttam Varikka Khujja or Karcha from Kerala; Ghila or Ghula, Hazari, Rudrakshi, Gulabi, Hazar, NJT I, NJT 2, NJT 3 NJT 4 from Faizabad.

The other selection are Koozha Navarikka or Pazam Varikka, Safeda, Khaja, Bhusila, Bhadaiyan, Handia, Mammoth, Everbearer, Rose-scented, Kooli, Varika, Gerissal, Barica, Ghila, Karcha, Rudrakshi, Champa, Sindoor, Safeda, Khaja, Bhusila, Bhadaiyan, Handia, T-Nagar, Jack, Velipala, Ghulabi and Champa.

Palur-1 Jack or PLR 1: It is a clonal selection made by Tamil Nadu Agricultural University (TNAU). It produces two crops (March-June and October and October-December).

PPI Jack: It is a clonal selection from Ulagumoodee by TNAU.

Konkan: It is a clonal selection from Mulagummoodu local.

Swana: From UAS, Bengaluru.

3. Hybridization:

Hybrid Jackfruit: Singapore Jack x Velippala.

Surliar-1: Singapore Jack x Velippala.

4. Biotechnology:

Micropropagation: Multiplication of rooted plantlets of *A. heterophyllus* (jackfruit) by micropropagation, using apical and axillary buds of juvenile tissue as explants, has been reported.

Molecular Markers

AFLP makers have proved effective in characterizing the genetic basis of *A. heterophyllus* for assessing genetic diversity and relation between accessions. Microsatellite markers have been used to characterize cultivars and wild relatives of breadfruit, jackfruit and other *Artocarpus* species. Twenty four transcriptomes of breadfruit and its wild relatives revealed positive signals of possible local adaptation or natural selection.

JAMUN

Origin: India, Myanmar and Sri Lanka.

Taxonomic status:

Order: Myrtales

Family: Myrtaceae

Genus: *Syzygium*

Species: *cumini*

Botanical name: *Syzygium cumini* (L.) Skeels.

Chromosome number: x = 11; 2n = 40.

Genome size: 156,2299 (Chloroplast(cp)

158, 954 bp (Malaya apple)

Related species are *S. aqueum, S. malaccense, S. samarangense, S. jambos* and *S.densiflorum*.

Botany

It is a large, evergreen tree attaining height up to 25-30 metre with grayish white trunk. Leaves are simple, opposite, elliptic in shape, pinnately veined.The inflorescences in jamun are generally borne in the axils of leaves on 5 months to one year old branchlets. It bears with a few flowers in a panicle, light yellow in colour, hermaphrodite, calyx lobes 4, petals 4, stamens many, ovary inferior and 2 celled. Fruit is a berry, purplish red, ovoid and edible.

Floral Biology

The maximum anthesis was recorded between 10 A.M. and 12 noon. Peak period of anther dehiscence was recorded between 9.00 AM - 2.00 PM. The stigma became receptive one day prior to anthesis and remained receptive upto 5 days after anthesis. Honeybees, houseflies and wind do the cross pollination.

Breeding Objectives

1. To develop varieties with precocious, early bearing, high quality fruits with high yield and synchronized ripening.

2. To develop varieties having quality fruits (high pulp percentage, colour, large size fruit with small seed, good shape, sweet aromatic taste) and enhanced shelf life.
3. To develop variety having resistance against pests like fruit borer and fruit fly and diseases like anthracnose and leaf spot.

Problem in Breeding

1. Highly heterozygous nature
2. Long juvenile period.
3. Lack of inheritance studies.

Germplasm Resources

The CISH, Lucknow has maintained (38 accessions). In India, the Institutes like CIAH, Bikaner (2), IIHR, Bengaluru (95), NDUAT, Faizabad, and Dr. BSKKV, Dapoli also conserve the germplasm of jamun.

Inheritance Pattern

There is no report in jamun on studies of the genetic inheritance of traits.

Breeding Methods

1. **Selection:** Attempts have been made to select superior strains from the seedling populations:-

   CIAH, Bikaner

   Goma Priyanka (GJ - 2): This was released at CHES, Godhra.

   GJ - 8: Selection made at CHES, Godhra.

   GJ - 19: Selection made at CIAH, Bikaner.

   GJ – 23: Selection made at CIAH, Bikaner.

   GJ – 40: Selection made at CIAH, Bikaner

   Katha Jamun : Selection made at CIAH, Bikaner.

   CISH, Lucknow

   CISH J - 37: Selection made at CISH, Lucknow.

   CISH J - 40: It is promising genotype collected at CISH, Lucknow.

   CISH J – 42: Selection made at CISH, Lucknow

   Selection have also been made at several other stations/institutions:

   CHES, IIHR, Chethalli: *Gokak, Dhoopdal.*

   NDUAT, Faizabad: *Narendra Jamun* 6.

   GAU, Gujarat:

   Paras: It is a seedling selection known for its large, sweet and juicy fruits.

***KKV, Dapoli*: Konkan Bahdoli.**

***Bihar Agricultural College, Bhagalpur, Bihar*:**

Rajendra Jamun-1

***Institute of Agricultural Sciences, Banaras Hindu University, Varanasi*:**

Selection 1

***KRC College of Horticulture, Arabhavi, Karnataka* : AJG- 85**

Clonal selection:

Central Sub-tropical Horticulture Institute, Lucknow has identified two promising clonal selections, *viz.* **Jamwant-37** and **CISH J-42** (seedless).

2. **Hybridization:** Interspecific hybridization was done by Lancaster and Bose (1965). The parents were Alba variety of *S. javanicum* (wax apple) and *S. jambos* (rose apple). In the resulting F_1 progeny, one hybrid had high and precocious fruiting with fruits having fragrance and sweetness of rose apple. But the fruits were seeded and seed size was more than those in either parent.

3. **Biotechnology**

   Micropropagation: Multiple shoots could be induced from nodal and shoot tip segments of 10-15 days old seedlings of *S. cumini* on modified MS medium supplemented with BA alone and when combined with NAA or IBA. Excised shoots were placed for root induction on modified MS medium containing NAA or IBA and then transferred to MS basal medium to form complete plantlets. Somatic embryogenesis was found successful for multiplication of jamun plants.

   Molecular markers: Molecular characterization of available jamun germplasm was done at CISH, Lucknow by using multilocus marker system, as there is dearth of sequence information. The cluster analysis (UPGMA) based on 133 RAPD markers produced a dendogram of genetic relatedness of the selected jamun accessions. The accessions could be discriminated with the help of these primers. Accession Nos. J -33 and J -30 clustered together and were most similar (similarity coefficient 0.898). The dendogram separated the accessions into four main clusters according to their geographical origin. J -34 and J -37, which are selections from seedling populations of Lucknow region, shared close similarity. J-36 and J-22 occupied distant branches sharing the least similarity with other accessions.

BAEL

Origin: Indian subcontinent and Southeast Asia.

Taxonomy status:

Order: Sapindales

Family: Rutaceae

Subfamily: Aurantioideae

Tribe: Aurantieae

Subtribe:Balsamocitrinae

Genus: *Aegle*

Species: *marmelos*

Botanical name: *Aegle marmelos* (L.) Correa

Chromosome number: n= 9; 2n = 18.

Genome size: ?

Botany

It is a deciduous shrub or small to medium sized tree with slender drooping branches. The bark is pale brown or grayish, smooth or finely fissured and flaking, armed with long straight, 1.2–2.5 cm spines, singly or in pairs. The leaf is trifoliate, alternate, each leaflet 5-14 x 2–6 cm, ovate with tapering or pointed tip and rounded base. The flowers are 1.5 to 2 cm, pale green or yellowish sweetly scented, bisexual, in short drooping unbranched clusters at the end of twigs and leaf axils. They usually appear with young leaves. The calyx is with 4 or 5 sepals. The four or five petals of 6–8 mm overlap in the bud. Many stamens have short filaments and pale brown, short anthers. The ovary is bright green with an inconspicuous disc. The fruit has diameter between 5 and 12 cm. It is globose or slightly pear shaped with a thick hard rind. The fruit has 8 to 15 or 20 sections filled with aromatic orange pulp, each section with 6-8 to 10-15 flattened oblong seeds.

Floral Biology

Flower bud emergence takes place by end ofMay and continues till last week of May. Anthesis occurs from 2 to 9 AM. Maximum anthesis takes place from 6 to 8 AM. Peak anther dehiscence ranges from 6.30 to 8.30 AM. The stigma becomes receptive from one day before anthesis and continues till one day after anthesis. Maximum stigma receptivity has been recorded on the day of anthesis.

Breeding Objectives

1. Dwarf plant stature.
2. Medium size fruits of about 750-1250 g.
3. Thin shell of fruit.
4. Less number of seeds.
5. Low mucilage content.
6. High sugar content.
7. Low phenolics.
8. Pleasant flavour.
9. Less prone to fruit drop and cracking.
10. Resistance to biotic and abiotic stresses.

Limitation in Breeding

1. Highly heterozygous nature.
2. Long juvenile phase.
3. Inheritance pattern is not understood.

Germplasm Resources

The CIAH, Bikaner is the national repository for bael. It has the largest collection of 57 accessions, CAZRI, Jodhpur (5 accessions), NDUAT, Faizabad (13 accessions), MPKV, Rahuri and IIHR, Bengaluru are also maintaing the germplasm of bael.

Sources for useful traits

Drought : Thar Neelkanth, Thar Divya, Goma Yashi.

Dwarfness: Pant Aparna, Pant Shivani, Pant Urvashi, NB-9, NB-16.

Early maturity: CISHB-1,Pant Shivani, Goma Yashi.

Late maturity: CISHB-2, NB-5, NB-7.

Globose fruit shape: Goma Yashi, Pant Shivani, Pant Urvashi.

Yellow pulp colour: Pant Shivani, Pant Sujata, Pant Aparna, Goma Yashi, NB-17.

Thin shell thickness: GomaYashi, NB-5, CISHB-1.

Low Mucilage content: Goma Yashi, CISHB-1, NB5, NB-17.

High TSS content (>38°Brix): Pant Shivani, Pant Urvashi, Goma Yashi.

Low crude fibre (%): Goma Yashi, NB-5.

Inheritance pattern: Information is not available.

Breeding Methods

1. Selection: Bael is an open pollinated fruit crop and considerable variability exists in UP, Bihar and West Bengal. Attempts have been made to select superior genotypes with respect to tree stature, fruit yield and quality from the seedling population.

NDUAT, Faizabad:

Narendra Bael –5: Average fruit weight 1.43 kg, fibre 80 g/kg/pulp, TSS 38°B, acidity 0.21 %, ascorbic acid 15.50 g/100 g of edible portion.

Narendra Bael –9: Average fruit weight 2.1 kg, fibre 6.5 g/kg pulp, TSS 25°B, acidity 0.12%, ascorbic acid 14.50 g/100 g of edible portion.

NB- 6, NB -7, NB-16 and **NB-17** are other selections from Faizabad.

CIAH, Bikaner:

Thar Divya: Selection made at CIAH, Bikaner.

Goma Yashi: Selection made at CIAH, Bikaner.

Thar Neelkanth: Selection made at CIAH, Bikaner.

CISH, Lucknow:

CISH Bael -10, CISH Bael-11, **CISH Bael-16, CISH Bael-21** and **CISH J – 42.**

GBPUAT, Pantnagar:

Pant Bael-1, Pant Aparna, Pant Shivani, Pant Sujata and Pant Urvashi: These are selections made at Department of Horticulture, GBPUAT, Pantnagar.

BAU, Sabour:

Sabour Bael-1: It has been released by BAU, Sabour, Bihar.

2. Biotechnology:

Researchers have developed micro propagation protocols for bael. Clonal micro propagation technique of bael from mature tree using nodal explant has been developed for two improved cultivars i.e. CISH B1 and CISH B2 (Pati and Muthukumar, 2013). Early experiments targeting the genetic transformation of medicinal plants were carried out using *Agrobacterium rhizogenes* to obtain hairy root cultures, or *Agrobacterium tumefaciens* to produce transformed cells to be maintained in culture or to regenerate whole plants. Transformation was successful in earlier stages of cell lines but further regeneration was not achieved.

PHALSA

Origin: India and South East Asia.

Taxonomy status:

Order: Malvales

Family: Malvaceae

Sub-family: Grewioideae

Genus: *Grewia*

Botanical name: *Grewia asiatica* L.

Chromosome number: 2n=2x=36

Genome size: ?

Related species are *G. tenax, G. hirsute, G. oppositifolia, G. tiliaefolia, G. rothii and G. orantalis.*

Botany

Plants are small trees or deciduous shrubs growing up to 8m tall. The leaves are alternate, simple and coarsely serrated, 5–18 cm long and broad with a 1–1.5 cm long petiole. The flowers are produced together in several cymes. The flowers are actinomorphic, about 2 cm in diameter, yellow with 5 large (12 mm long) sepals, polysepalous or connate and basally velvate. Petals are pale yellow to greenish yellow and five in number. Stamens are free and numerous. Anthers are two celled and dehiscence occurs longitudinally. Ovary is bi-carpillary, syncarpus and superior. The fruit is an edible drupe, 5–12 mm in diameter and purple to black when ripe.

Floral Biology

Bud development to full bloom takes 24 days. Most of the flowers open between 9 and 11 AM during April. Anthesis is accelerated by high temperatures and low humidity. Dehiscence precedes anthesis and reached peak between 7 and 8 AM. Maximum stigma receptivity occurs on the day of anthesis after which the stigma becomes non-receptive.

Breeding Objectives

1. Development early bearing cultivars.
2. Development of cultivars with synchronous ripening and good keeping quality.
3. Development of high yielding cultivars.
4. Development of cultivars having resistance against biotic and abiotic stress.

Limitation in Breeding

1. Heterozygous.
2. Long juvenile phase.
3. Less gene pool.
4. Lack of information on inheritance pattern.

Germplasm Resources

CIAH, Bikaner is the national repository for phalsa. It has the largest collection of phalsa accessions (57). CISH, Lucknow, NDUAT, Faizabad, MPKV, Rahuri and IIHR, Bengaluru are also maintaing the germplasm of phalsa.

Sources for useful traits: Not known.

Inheritance pattern: Information on Inheritance pattern is not available.

Breeding Methods

1. Selection: Not much attention has been given for the improvement of phalsa. There is no recognized variety of phalsa except a few local favourites in some growing regions. In Hisar, two local types, tall and dwarf, are grown. The dwarf type is more productive than the tall one. The fruits of dwarf type have higher contents of total and non-reducing sugars while the tall type has more reducing sugar. The seed protein content differs in both the types.

KARONDA

Origin: India.

Taxonomy status:

Order: Gentianales

Family: Apocynaceae

Botanical name: *Carissa carandans* L.

Chromosomes number : 2n=2x= 22

Genome size: ?

Related species are *C. bispinosa C. edulis C. grandiflora, C. macrophylla* and *C. ovata.*

Botany

Karonda is small to big shrub, usually 2-4 m tall. The stem is rich in white latex and the branches contain sharp spines. The leaves are oblong and conical, 10-15cm long and 5-8 cm wide, green above and brown below. The flowers are small, white or yellowish and found in groups and are 3-5 cm in diameter and have 5 pink sepals, 5 white-pink petals 5 stamens, Ovary superior, 2 celled and ovules 4 in each cell. The fruit is a berry, which is formed in clusters of 3-10, 1.8 to 2.5 cm long, green in colour when unripe and turns red and finally black on maturity.

Floral Biology

Flowering and fruiting takes place twice a year. Variable percentage of anthesis has been registered in different genotypes. Peak period of anthesis is recorded from 3-6 to 8-10 PM, depending on the locality, cultivar and atmospheric temperature. Dehiscence of anthers takes place before anthesis between 12 noon and 4 PM. The dehiscence takes place in the longitudinal fashion and completes within an hour. Atmospheric temperature and humidity influence the time and duration of dehiscence. The maximum receptivity of stigma is found from two days before anthesis and to two days after anthesis. Mostly self pollination is evidenced.

Breeding Objectives

(i) To develop a cultivar with high yield potential and producing big size fruits with attractive skin colour.

(ii) To develop a cultivar producing fruits of good quality having higher pulp content, less number of seeds, high TSS, low acidity and high ascorbic acid.

(iii) To develop a cultivar of dwarf stature with less spine density.

(iv) To develop a cultivar having drought resistance and free from major diseases and pests.

Limitations in Breeding

1. Heterozygous.
2. Long juvenile phase.
3. Lack of information on inheritance pattern.

Germplasm Resources

Germplasm collection consisting 3 *Carissa* species and 24 cultivars have been maintainted RFS, Vengarula, CHES, Chethalli (45), CIAH, Bikaner, CISH, Lucknow, NDUAT, Faizbad, MPKV, Rahuri and IIHR, Bengaluru.

Sources of useful traits: Not available.

Inheritance Pattern

The inheritance linkage of different economical characters has been studied. Based on work done so far, following conclusions can be drawn on the inheritance pattern of certain characters.

(i) The magnitude of phenotypic coefficient of variation (PCV) was slightly higher than the genotypic coefficient of variation (GCV) for all characters. The PCV and GCV was more than 100% for pomace per fruit while it ranged from 60 to 90% for seed weight per fruit, acidity, non-reducing sugar and average fruit weight. Heritability was above 90% for average fruit weight, bearing potential, non-reducing sugar, reducing sugar, acidity, total soluble solids and pomace content, while genetic advance in terms of percentage mean was high for pomace, fruit weight, acidity, seed weight per fruit and non-reducing sugar. The magnitude of variation was maximum for total soluble solids followed by pulp weight. The characters, equatorial diameter, polar diameter, individual fruit weight, pulp weight and total soluble solids were less affected by environment, which showed close correspondence between genotypic coefficient of variability and phenotypic coefficient of variability, while characters such as seed weight per fruit and number of seeds per fruit were most affected. High heritability estimates with low genetic advance was observed for TSS implying less genetic gain if selected on this basis while, selection for equatorial diameter, polar diameter, individual fruit weight and pulp weight had high heritability, high genotypic coefficient of variability and high genetic advance is expected to result in considerable genetic gains. The plant having higher number of umbel segments was associated with higher yields.

Breeding Methods

1. Selection: From the considerable variabilty with respect to fruit size, shape and colour, desirble types having green, pink and white fruits have been identified based on fruit colour. Some superior types have been identified:

Pant Suvarna: It was selected and released from G.B. Pant University of Agriculture and Technology, Pantnagar in 1986.

Pant Manohar: It was developed and released from G.B. Pant University of Agriculture and Technology, Pantnagar in 1991.

Pant Sudarshan: It was developed and released from G.B. Pant University of Agriculture and Technology, Pantnagar

Pant Suvarna: It was developed by selection from G.B.Pant UA&T, Pantnagar.

Konkan: This sweet type was developed by Fruit Research Station, Vengurla, Maharashtra.

Thar Kamal: Released by CIAH, Bikaner for semi- arid regions.

CHES-K- 2: It is a promising genotype developed at Central Horticultural Experiment Station, Godhra.

CZK-2011: It is a selection made at Central Arid Zone Research Institute, Jodhpur, Rajasthan.

CZK - 2022: It is a selection made at Central Arid Zone Research Institute, Jodhpur, Rajasthan.

CZK – 2031: It is a selection made at Central Arid Zone Research Institute, Jodhpur, Rajasthan.

CISH Kr-11: It is a superior genotype selected at Central Institute of Subtropical Horticulture, Lucknow.

At Rahuri some promising types No. 3, No.12, No.13 and No.16 have been identified.

Hybridization: Crossing between the varieties like Konkan Bold and Pant Manohar may result some positive traits in the progenies.

Micropropagation

Shoot tips from mature plants of cv. Pant Sudarshan were cultured on Murashige and Skoog's (MS) basal medium supplemented with benzyladenine (BA) and indolebutyric acid (IBA) during different seasons. Shoot proliferation was highest on MS basal media supplemented with 3.0 mg l^{-1} BA. Rooting of microshoots was noted to be the best in 1/2 MS plus 0.8 mg l^{-1} IBA and 0.2 mg l^{-1} NAA. The rooted plantlets were successfully acclimatized in vermiculite:sand:soil potting mixture

MULBERRY

Origin: China, Korea, Japan and in the Himalayan foothills.

Taxonomy status:

Order: Rosales

Family: Moraceae

Botanical name: *Morus alba* L. (White mulberry)

M. nigra L (Black mulberry)

Chromosomes number: x=14, 2n=28 (*M. alba, M. indica*); x=22, 2n = 308 (dexoploid) *M. nigra.*

Genome size: 357 Mb (*M. notabilis*)

The species are *M. nigra*, black mulberry (Iran), *M. indica* (India), *M. rubra*, Red mulberry (North America) etc. *Morus alba* L. (white mulberry), *Morus macroura* Miq. (Himalayan mulberry), *M. cathayana* Hemsl. (Chinese mulberry) and *M. atropurpurea* Roxb. (Guangdong mulberry) are native of China. *M. alba* and *M. rubra* (red mulberry) have also been naturalized in Europe and North America. *M. nigra* L. (Black, English or Persian mulberry) is native of North America. *M. indica* L. syn *M. australis* Poi. (Korean mulberry or Aino mulberry), *M. laevigata* Wall. ex Brandis (tut) and *M. serrata* Roxb. (Himalayan mulberry) are found wild in Himalayan region of India.

Botany

Morus alba is a deciduous, monoecious or dioecious tree of moderate size growing up to 15 m in height having a spreading and umbrella shaped crown, grey to yellow branches, smooth bark and slender young twigs. Leaves of white mulberry are simple, alternate, stipulate, petiolate, entire or lobed. The number of lobes varies from one to five. Leaves of the red mulberry are larger, thicker, blunt toothed and often lobed. They are rough on their upper surfaces and pubescent underneath. The smaller black mulberry leaves are similar to those of the red mulberry morphologically, but with sturdier twigs and fatter buds. The leaves are thin, small, 5-15 cm in size, broadly ovate, sometime almost orbicular or triangular, lobed and with coarsely toothed margins. The inflorescence in mulberry is a catkin with pendent or drooping peduncle bearing unisexual flowers. Male catkins are usually longer than the female catkins. Flower is small, inconspicuous, incomplete, unisexual and hypogenous. Male flower has 4 sepals, which are imbricate, and four inflexed stamens in bud. Female flower has 4 decussate, acrescent and succulent sepals and one celled and two partite or bifid ovary with pendulous ovule. The fruit is a sorosis that is composed of a collection of individual fruits (achens). The fruits are borne on short stalks as long as the fruit and are white to pinkish purple but usually red when ripe, globose and 2 cm in diameter. Seed is subglobose with copious and fleshy albumen and the embryo is incurved.

M. alba has several botanical varieties named on the basis of mode branching and shape of leaves. These are *M. alba var. pendula* Dipp. which generally has slender and drooping branches; *M. alba* var. *pyramidalis* Ser. which has branches giving the tree a pyramidal form; *M. alba* var. *tartarica* Loudon (Russian mulberry) which is very ornamental, dwarf having bushy top and black fruit and can be used as a very hardy rootstock; *M. alba* var. *japonica* Audib. which has large and pointed leaves and oval and red fruits; *M. alba* var. *multicaulis* (White mulberry) is vigorous in growth, has large pointed unlobed leaves, bears sweet fruit and grown in China for silkworms; *M. alba* var. *celtididifolia* HBK. is a medium sized tree of 7m height having small smooth and less lobed leaves and globular fruit.

M. nigra L. Black mulberry is deciduous small, spreading tree growing up to 10 m height, having thick, dark brown young twigs with a velvet-like hairy covering, thick alternate 5-20 cm long leaves with ovate deeply cordate base and abruptly acute apex, dull surface, dark green and rough above and hairy beneath, coarsely and sharply roothed margin which may sometimes be lobed. Flowers are similar to those of *M. alba*. Fruits are more or less sessile, dark purple, about 2 cm in diameter, acidic, becoming sweet only when completely ripe and are of good quality.

Floral Biology

The blooming starts from middle of January. Anther becomes mature within 2-3 days after the opening of flower and dehiscence a day after, starting from early morning and maximum during 11 AM to 1 PM and stigma remains receptative for 7-10 days. Mulberry plants having anemophilous flowers cross each other easily and naturally.

Breeding Objectives

(a) Higher leaf yield per unit area.
(b) Early sprouting behaviour.
(c) Easy to root when propagated through stem cuttings.
(d) Height and branching habit.
(e) Palatable leaf to rear silkworms.
(f) High fruit yield, high nutritional value.
(g) Resistance against cold, diseases and pests.

Limitations in Breeding

1. Long juvenile period.
2. High heterozygosity.
3. Dioecious nature.

Germplasm Resources

Large numbers of germplasm accessions are available in China, India, Japan, Korea and Vietnam There are more than 2600 mulberry accessions maintained in China. Japan has 1375 germplasm accessions and Korea has 614 accessions.

The Central Sericultural Germplasm Resources Centre, Hosur, Tamil Nadu is the national repository for mulberry. It has the largest collection of 978 accessions. Besides this, several small-scale germplasm collections are also available at different Central Sericultural Research and Training Institutes located in Mysore (Karnataka), Berhampore (West Bengal), and Pampore (Jammu and Kashmir). CISH, Lucknow (15), NDUAT, Faizbad, MPKV, Rahuri and IIHR, Bengaluru are also maintaing the germplasm of mulberry.

Sources of useful traits:

Higher yield: Mysore local, Bombay piasbari, Kanca-2, Bilidevalaya, kajli, s1, Bc259, C776 and S-36.

Suitable for hilly region: $BC_2$59.

Frost and high temperature tolerant: Thar Lohit and Thar Harit.

Salt and drought: C776.

Suitable for irrigated area: Victoty-1.

Suitable for rainfed region: RFS-175 and S1.

Inheritance pattern: There is no information available.

Breeding Methods

1. Introducation: The varieties like Himalyan DMOR9 (Himalayan mulberry), Pakistan White, Sahranpur Local, Australian Green, Oscar, Maple Leaf, Bukuklu, Big Leaf FL, Illinois Everbearing and Black Prince should be introduced.

2. Selection: In India, Local selections have been made based on fruit size and quality e.g. Thar Lohit (selected at CIAH, Bikaner), Thar Harit (Pushkar (Rajasthan) and CIAH-3 (Hisar, Haryana).Types having elongated fruits occur in Punjab. Local selections have been made based on fruit size and quality. Types having elongated fruits occur in Punjab. Collections of *M. alba* types have been maintained at Saharanpur, Sardar Krushinagar and Varanasi and of *M. nigra* at Varanasi and Ludhiana.

Most of the white mulberry types are grown as ornamental trees or for rearing silk worms. Varieties selected in USA like Downings Everbearing, New American, Wellington, Trowbridge, Thorburn, Ramsey White, Victoria, Munson and Merrit are grown for fruit. Black Persian is an important black mulberry variety. It has plump, black juicy fruits of 4x 2.5 cm size with a rich and subacid flavour. Shangri La is another black cultivar of Japan. Superior selections of red mulberry made in USA are Johnson, Hicks Everbearing, Stubbs, Lampasab, Townsend, Tehama Riveria, Travis and Illinois Everbearing. Mahi and Thai are cultivars respectively from Pakistan and Thailand.

Out of more than 20 species, *Morus alba, M. serrata, M. indica* and *M. latifolia* are more useful. Central Silk Board has developed high yielding variety Kanva-2, S-30 and S- 54 suited for Karnataka, AP and Tamil Nadu; and S-162, S-519, S-633 for Punjab, J&K, UP and West Bengal.

3. Hybridization: Interspecific hybridizations among *M. alba, M. australis, M. cathayana,M. latifolia, M. nigra and M. sinensis* revealed varied results as in some crosses fertility was very high, whereas in some others it was very poor.

Hybrid forms between *M. alba and M. rubra* also exist, e.g. Collier, Ilinois and Everbearing. Varieties released by Central Sericultural Research and Training Institute, Central Silk Board, Srirampura, Mysore are Mysore local, Bombaypiasbari, Kanca-2, Bilidevalaya, Kajli, S1, Bc259, C776, S-36, Victory-1 and RFS-175.

4. Polyploidy: Triploids are developed by crossing diploids and artificially induced tetraploids. Triploid varieties are cold and disese resistant and are extensively cultivated in Japan for their superior quality leaves. 122 triploid varieties are reported in Japan.

5. Biotechnology: In the past two decades, research on mulberry has enormously benefited from molecular markers. Application of DNA fingerprinting in mulberry was initiated in Japan. Random amplified polymorphic DNA (RAPD) was the first DNA based marker used in mulberry.

6. Germplasm conservation: In India, mulberry genetic resources are conserved in four different ways such as *in situ* conservation, *ex situ* conservation, *in vitro* conservation and DNA banking.

MANGOSTEEN

Origin: Southeast Asia

Taxonomy status:

Order: Malpighiales

Family: Clusiaceae

Genus: *Garcinia*

Species: *mangostana*

Botanical Name: *Garcinia mangostana* L.

Chromosome number: 2n = 4x = 56–76, 88-90, 96 and 120-130.

Genome size: 6.00 - 6.05 pg.

The related species are *G. dulcis, G. xanthochymus, G. prainiana, G. cowa, G. hanburyi, G. cambogia* and *G. hombroniana.* Mangosteen is believed to have arisen from natural hybridization between *G. hombroniana* and *G. malaccensis.*

Botany

The mangosteen is evergreen, dioecious, slow growing and erect tree reaching up to 6-25m in height, with a pyramidal crown. It has dark brown or nearly black flaking bark, the inner bark containing much yellow, gummy and bitter latex. The leaves are simple, ovate oblong or elliptic, opposite, short stalked, leathery and thick, dark green, slightly glossy above and yellowish green and dull beneath, 9-10 cm long and 4.5-10 cm wide with a conspicuous pale midrib. The flower is 4-5 cm wide and fleshy, male and hermaphrodite on the same tree. The flowers are borne in clusters of 3-9 at the branch tips. There are 4 sepals and 4 ovate, thick, and fleshy petals, green with red spot on the outside, yellowish red inside and many stamens, though the aborted anthers bear no pollen. The hermaphrodite flowers are borne singly or in pairs at the tips of young branchlets, their petals may be yellowish green edged with red or mostly red and quickly shed.

The fruit capped by the prominent calyx at the stem end and with 4-8 triangular, flat remnant of the stigma in a rosette at the apex, is round, dark purple to red purple and smooth externally, 3.4 – 7.5 cm in diameter. The rind is 6-10 mm thick, red in cross section, purplish white inside. It contains bitter yellow latex and a purple, staining juice. There are 4 to 8 triangular segments of snow white, juicy soft flesh (actually the aril of seeds). The fruit may be seedless or have 1-5 fully developed ovoid oblong, somewhat flattened seeds, 2.5 cm long and 1.6 cm wide that cling to the flesh. The flesh is slightly acid and mild to distinctly acid in flavour and is acclaimed and exquisitely luscious and delicious. The mangosteen is reproduced by parthenogenesis.

Floral Biology

Flowering in India occurs in April-May and again in October- November. The flowers are short lived; they open in the late afternoon and the petals fall shortly thereafter. The yellowish green petals of hermaphrodite flowers quickly shed.

Breeding Objectives

1. To develop cultivars having wide adaptability.
2. To develop cultivars having short juvenile period.
3. To develop early maturing, high yielding cultivars bearing fruits of good keeping quality.
4. To develop cultivars having longer seed viability.
5. To develop cultivars free from fruit cracking.
6. To develop cultivars free from gamboge disorder.
7. To develop cultivars having high tolerance to drought.

Limitations in Breeding

1. Long juvenile phase.
2. Narrow genetic base.
3. Slow growth.
4. Homogeneous and has a single genotype.
5. No male plant (absence of a pollen source).
6. Low seed production.
7. Seed is apomictic.
8. Lack of information on inheritance pattern.

Germplasm Resources

Collections of selected, apparently distinct, trees are kept in Brunei, Indonessia, Malaysia, Papua New Guinea, the Philippines, Thailand and USA. The Malaysian Department of agriculture has identified 16 accessions which show variation for fruit size and shape, seed number, shelf life, fruiting precocity and external colouration. In India, the institutes like IIHR, Bengaluru and TNAU, Coimbatore are maintaing the germplasm of mangosteen.

Genetic resources/sources of useful traits and Inheritance pattern: No information is available.

Breeding Methods

1. Selection: Breeding of mangosteen plants are propagated clonally by seed and does not need sexual fertilization to produce fruit and hence genetic combination resulting from the seeds, show little genotypic variation to enable selection. Thus there are few varieties or cultivars, but all mangosteen tree population is not genetically identical apparently due to infrequent natural mutation. The Malaysian Department of Agriculture has released the clones GA1 and GA2. The commercially available Malaysian cultivar Mesta is seedless and has a small fruit size than the typical mangosteen. According to Corner (1952), the fruits from seedling tree are fairly uniform; only one distinct variation is known and that is in the Sulu Islands. The fruit is larger having thicker rind than normal and the flesh more acid with more pronounced flavour. In North Bormeo, a seemingly wild form has only 4 carpels, each containing a fully developed seed which probably is not unique.

2. Hybridization: Crossing mangosteen with the closely related species *G. hombroniana* to create new varieties may be possible to develop cultivars having desirable traits like high fruit yield, long seed viability, easy propagation, faster growth rate in nursery, short juvenile period and high tolerance to low relative humidity.

3. Biotechnology:

There are no reports on micropropagation of *Garcinia mangostana* from bud explants. However, multiple plantlets have been produced in vitro via organogenesis from leaf and seed explants. Study of genetic variability existing in different countries using molecular markers would be very useful. Gene transfer could be useful to reduce the juvenile phase and extend shelf life of fruits.

KOKUM

Origin: Western Ghats of India.

Taxonomy status:

Order: Malpighiales

Family: Clusiaceae

Subfamily: Clusiodieae

Tribe: Garcinieae

Genus: *Garcinia*

Species: *indica*

Botanical name: *Garcina indica* L.

Chromosome number: 2n= 54

Genome size:?

Related species are *G. gummigatta* syn. *G. cambogia*, *G. mangostana*, *G. tinctoria*, *G. xanthochymus*, *G. morella*, *G. cowa* and *G. hombroniana*.

Botany

Kokum is an evergreen, dioecious tree and can grow as high as 18m with a pyramid canopy. The simple and opposite leaves, about 10 x 5 cm, glabrous above and paler beneath, are red when young and dark green when mature. The flowers are small and unisexual; the male and female flowers are found on separate trees. The calyx is of four free sepals and corolla of four to five imbricate petals. The male flowers have 10 to 20 stamens. The female flowers have ovary of 4 to 8 chambers which is topped with a lobed and sessile stigma. The fruit is usually deep purple to pink berry. The fleshy rind of the fruit is juicy and acidic. The fruit contains five to eight large seeds embedded in an acidic pulp. The kernels account for 61% of the weight of the seed, while the oil content of the kernel accounts for about 44%.

Floral Biology

The tree comes to bloom from November to February and the fruits ripen in April-May. The trees could be designated into the following types on the basis of preponderance of particular type of flowers and the bearing tendency of individual tree:

Type-I: Staminate or male tree: The flowers mostly have long pedicels, mass of stamens crowded on receptacle and sometimes rudimentary; pistil is with pointed apex. They are incapable of producing any fruit and serve only as pollinators.

Type-II: Hermaphrodite or bisexual tree.

Type-III: Pistillate or female tree: Flower is identified by short pedicel, well developed pistil and two or four tufts of staminodes below.

The maximum anthesis of male (59.2%) and female (58.7%) flowers occurs between 0600 and 0800 hrs, and 21.9 and 24.0% between 1600 and 1800 hrs. The dehiscence of anthers takes place 15-20 minutes prior to anthesis. Pollen viability in male and bisexual flowers is 96.29 and 95.75%, respectively. Wind appears to be an important agent of pollination in kokum since 11.4-23.1 atmospheric pollen grains were observed. The maximum stigma receptivity (80%) was observed on the day of anthesis followed by 68, 64 and 44%, respectivity after 1, 2 and 3 days of anthesis. Kokum is predominantly a dioecious plant and therefore cross pollination is the rule. Very high percentage (68-78%) of female flowers of kokum set fruits.

Germplasm Resources

The Institutes like CCARI, Goa, ICAR NBPGR Research Centre, Thrissur, IIHR, Bengaluru and Dr. BSKKV, Dapoli conserve germplasm of kokum.

Breeding Methods

1. **Selection:** Fourteen kokum types have been identified at KKV, Dapoli from the natural variability on the basis of traits such as early maturity and fruit size and yield. Khanvilkar *et al.* (1987) reported that genotypes KK-87, KK- 155, KK-76, KK-153 and K 149 as promising in respect of fruit yield and number. Konkan Amrit has been released from Dr. B.S. Konkan Krishi Vidyapeeth. It bears big size fruits weighing about 30g. A unique variety in Uttar Kannada locally known as 'bili murugalu' bears yellow coloured fruits. It is believed to posses medicinal properties. Variety Konkan Amritha has been developed by clonal selection. This variety is considerably early having short harvesting period (78 days) in few pluckings.

5 Temperate Fruits

APPLE

Origin: The regions within the Tian Shan forest regions of Central Asia, including western mountains, Kazakhstan, Kyrgyzstan and Tajikistan.

Taxonomy status:

Order: Rosales

Family: Rosaceae*

Subfamily: Rosoideae

Tribe: Maleae

Section: *Malus*

Series: Sievrsinae

Botanical name: *Malus* x *domestica* Borkh.

Chromosomes number: x=8 or 9; n= 17; 2n=34=2n=3n=51

Genome size: 750 Mbp (Golden Delicious)

The genome of modern domestic apple is 46% of *M. sieversii,* 21% of European crab apple and 33% of uncertain sources.

* Rosaceae family is classified into three subfamilies (Potter *et al.*, 2007); the large ones are Rosoideae and Amygdaloideae, and the small one is Dryadoideae with fewer than 30 species.

The related species are *M. sieversii, M. sylvestris, M. angustifolia, M. ioensis, M. coronaria*, and *M. baccata, M. spectabilis, M. micromalus, M. zumi, M. sargentii, M. sikkimensis, M. mandshurica, Malus x scheideckeri, M. orientalis and M. prunifolia.*

Botany

Apple is small to medium sized, branched, deciduous or rarely evergreen tree or shrub, can reach a height of 10 m with a single trunk and a broadly spreading canopy. Young stems and twigs are somewhat tomentose (hairy), while older branches are glabrous (smooth); buds ovoid, with several imbricate scales. Leaves are alternately arranged, dark green, simple oval-shaped with a serrated edge, 4-13 cm long x 3-7 cm wide, with irregularly saw-toothed margins, and usually hairy beneath. Apple flower buds (FB) are often initiated on terminal buds of short shoots called spurs; the axis of the spur is called the bourse. The white to pink flowers appear in cymes. Flower is hermphrodite, epigynous, consists of 5 sepals, 5 petals usually suborbicular or obovate; stamens 15-

20, with usually yellow anthers; ovary inferior, 5 carpels; 5 styles, connate, 2 ovules per carpel. Fruit is a pome, usually without grit cells, persistent or deciduous calyx. Seeds are smooth, shiny and chestnut brown.

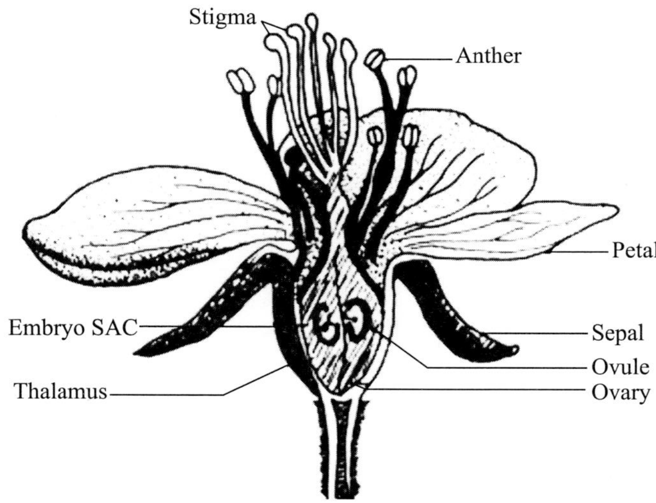

Fig. 16. L.S. of apple flower showing different parts.

Floral Biology

The flower development takes roughly 10 months, beginning with the transition from vegetative to reproductive phase (late June in the northern hemisphere and ending with anthesis (late April to the beginning of May) in the subsequent year. A greater percentage of the larger basal flowers set fruit compared with the smaller terminal ones and the king blossom. The king blossom is usually allowed to open and perhaps even the second blossom; then one or more of the remaining blossoms are emasculated when they are at the balloon stage of development. The remaining flowers are removed at the time of emasculation.

Flowering in India starts in the second to third week of March. The time of anthesis is 6-9 AM and dehiscence of anthers occur simultaneously and stigma receptivity starts at the time of anthesis and continues for 2-3 days. Mechanical transfer of pollen from anthers to stigmas is a pre-requisite to fertilization of ovules and development of fruits and seeds. Pollination is done by insects, particularly bees. Most flowers are admirably designed to facilitate pollination, but in a few genotypes such as Delicious, the stamens are longer than styles, which allow the bees to visit flowers without making contact with stigmas resulting in considerable loss of crop.

Breeding objectives for scion

(i) To develop precocious and regular bearing cultivars having wide range of environmenal adaptability.

(ii) To develop compact, well-branched and spur-type trees.

(iii) To develop cultivars for low/medium and high chilling.

(iv) To develop early ripening, high yielding cultivars having better adaptability.

(v) To develop cultivars with good fruit qualty (large size > 70 mm in diameter, ovate or conical shaped, free from blemishes and good taste and texture) and storage quality.

(vi) To develop cultivars having market and processing stability.

(vii) To develop cultivars having resistance to insect pests (wooly apple aphid and rosy apple aphid) and diseases (fire blight, alternaria blotch, blotch, canker, scab, crown rot, powdery mildew.

(viii) To develop cultivars tolerant to cold and drought.

Breeding objectives for rootstock

(i) Focus on adequate anchorage and rooting ability and productivity and control of size of scion.

(ii) Compatibility with scion and induce precocity, productivity and fruit quality in scion.

(iii) To develop rootstock for wide range of environment viz., winter hardiness and drought tolerance.

(iv) To develop rootstocks resistant to biotic stresses.

Limitations in Breeding

1. Long juvenile phase.
2. Heterozygosity.
3. Narrow genetic base.
4. Self incompatibility.
5. Sterility.
6. Polygenic control of most horticultural traits.
7. Polyploidy.
8. Inbreeding depression.
9. Parthenocarpy.
10. Apomixis.

Germplasm Resources

In the world, twenty institutions held 67 wild species taxa, with between 1 and 2,300 accessions for each species. Seventeen species hybrids were also documented in the collections, with a total of 243 accessions (Bramel and Volk, 2019). *Malus niedzwetzkyana* and *M. sieversii* are now on the Internation Red List threatened species (IUCN; East Wood et al., 2009).

The largest collection of apple germplasm is held at the Plant Genetics Resource Unit at Cornell University, Ithaca, New York, where about 4,000 accessions are being maintained.

In India, germplasm of apple (204 accession) was made through introductions and indigenous collections and these have been conserved *ex-situ* at national field gene bank at CITH, Srinagar. Field gene banks have been maintained at Regional Horticultural Research and Training Insitute, Dr YSPUHF, Mashobra (156 accessions); Temperate Horticultural Research Station Dr YSPUHF, Kotkhai, Shimla (215); NBPGR Research Station, Phagali; IARI RRS, Amartara, Shimla (38 *Malus* species and 69 accessions), SKUAST, Srinagar; Hill Fruit Research Centre, Chaubattia and CITH, Srinagar (80 accessions).

Genetic sources of useful traits

Low chilling: Dorsett Golden, Anna, Tropic Mac, Tropic Sweet, 88-20, Ein Schemer, 60-39, Tamma, Neomi, Tropical Beauty, Gallia Beauty, Winter Banana, Tame, Vered, Rome Beauty, Winter Banana, Schlomit, HRMN-99, Michel Molies, King David and *Malus sieboldii.*

Columnar : Wijicik Mclntosh.

Self fruitful: Golden Delicious.

Parthenocarpy: Spencer Seedless (apetalous).

Apomixis: *M. hupehensis* and *M sikkimensis.*

Early ripening: Yandik-Ovske, Papisovka Canniage**,** Benoni, Irish Peach, Early Shanburry and Fanny.

Red Fleshed: Niedwetzkyana, Estonia, Pendragon, Derek Mills, Pink Pearl, Pink Sparkle and Red Wonder.

Keeping Quality: Ambred, Ambstarking, Ambroyal, Ambrich, Chaubattia Princess, Chaubattia Anupam and Ambri.

Canning: York, Stayman, Golden Delicious, Northern Spy, Rhode Island Greening, Yellow Newton and Jonathan.

Juice: James Grieves, Howgate Wonder, Granny Smith, Jonathan, Golden Delicious, Gala Mast, Fuji, Gala, Lord Lambourne, and Allington.

Cider: Esopus Spitzenburg, Tremblets Bitter, Bramley's Seedling, James Grieves, Baldwin, McIntosh, Northern Spy and Russet.

Sauce: Cortland, Granny Smith, Jonathan, McIntosh, Crispin, Empire, Russet, Golden Delicious, Arkansas Black, Gala, Liberty, Lodi, Mutsu, Northern Spy, Rome Beauty, Rhode Island Greening, Twenty Ounce, Monroe, Spartan, Winesap and York.

Pies: Baldwin, Jonathan, Northern Spy, Rhode Island Greening, Rome, Cortland, Crispin, Empire, Golden Delicious, Spartan, Russet, Arkansas Black, Gala, Granny Smith, Lodi, McIntosh, Mutsu, Rome Beauty, Spartan, Winesap and York.

Baking: Granny Smith, Northern Spy, Rome, Arkansas Black, Cortland, Empire, Gala, Golden Delicious, Jonathan, Liberty, Lodi, McIntosh, Mutsu, Rome Beauty, Spartan, Winesap, Bramley's Seedling and York.

Freezing: Empire, Golden Delicious, Jonathan, Liberty and Mutsu.

Salad purpose: Baldwin, Cortland, Granny Smith, Red Delicious, Golden Delicious and Mutsu.

Non-browning: Arctic Granny, Arctic Golden.

Bitter pit: Golden Delicious, Mclntosh.

Early bearing: Chaubattia Swarnima (Benoni x Red Delicious), Chaubattia Anurag (Esopus Spitzenberg x Red Delicious), Early Shanburry, Fenny, Red June.

Scab: *Malus sieversii*, Borovinka, Hagloe Crab, Tita Zetei, Braeburn, Britegold, Delicious, Empire Red, Enterprise, Fuji, Goldstar, Jonafree, Nittany and Wealthy Red, Prima, Primula, Sir Prize, Jonafree, Dayton, Priscilla, Sir Prize, MacFree, Freedom, Coop-12, Coop-13, Firdous and Shireen, *Malus floribunda,* Prima, Liberty, Florina, Cox's Orange Pippin, Florina, Firdous and Shireen.

Fire blight: *Malus sieversii*, G-41.

Collar rot (*Phytophthora cactorum*): *Malus sikkimensis*, *M.orientale*, *M.purpurea*, *M.floribunda Malus sieversii, Docynia hookeriana.*

Root rot: *Malus baccata*, *M. floribunda*,

Replant disease: *Malus sieversii.*

Codling moth: *M. zumi.*

Woolly aphid: Northern Spy, MM-104 (M-2 x Northern Spy) and *M. hupehensis.*

Powdery Mildew: Borovinka, Hagloe Crab, Tita Zetei, Braeburn, *M. robusta, Docynia hookeriana*, *Malus sikkimensis* and *M. zumi.*

Anchorage: MM-111.

Drought: MM-111.

Low temperature: McIntosh, Summerred, Caravelle, Lobo,Quinte, Norland, Westiand, Parkland, Ottawa 3 and Ottawa 8.

Genetics of important traits:

Alston *et al.* (2000) updated the *Malus* gene list to include 145 genes, with 29 involved in pest or disease resistance. Gene discovery is increasing as sequences from other species are being used to investigate *Malus*.

Main characteristic genes have been identified in different cultivars. The genes related with ethylene biosynthesis (ACS, ACO and ACC) regulate fruit softening. Albinism (al gene), pale green lethal seedlings (l gene), and color due to anthocyanin genes, as well as greasy skin (Gr gene), have been studied by several researchers. Genes affecting petals have been found to control apetaly (ape) or double petals (Pd). Genes related with apple fertility are MADS-box genes associated with the development of floral meristems and organ identity, MDH1 (apple homoeobox gene) involved in the control of plant fertility, pollen lethal (P-1, P-2, P-3, P-4 and P-5), and pollen incompatibility (S-alleles). Some genes have been found affecting fruit quality, such as aroma (Ar), malic acid (Ma) or bitter pit (Bp-1 and Bp-2). Sensibility to russeting is attributed to the Ru gene. Studies in growth regulation have provided deep knowledge in apple dwarfing (genes st-1 and st-2 for sturdy dwarf; genes dw-1, dw-3 and dw-4 for early dwarf; and cr for crinkle dwarf), regrowth promoter (G), Gibberellin gene, Knotted 1-like homoeobox expressed during growth and development, gene MdPIP1 controlling fruit expansion and in plants under osmotic stress and DAD1 as inhibitor of programmed cell death. Different pest and disease resisting genes have been localized, such as genes for curling aphid (*Dysaphis devecta*Wlk.) resistance (genes Sd-1 to Sd-3 and the precursor Pr-Sd), for WAA resistance (*Eriosoma lanigerum*) (Er-1), for yellow mottle (ym-1, ym-2, ym-3), for hypersensitivity to *D. plantaginea* (Sm-h), for fire blight resistance (Alston and Briggs, 1970), for *Glomerella cingulata* susceptibility (Gb), for *Gymnosporangium* resistance (Gy-a and Gy-b), for *Phyllosticta solitaria* susceptibility (Ps-1 and Ps-2), for *Phytophthora cactorum* resistance (Pc), for *P. leucotricha* resistance (Pl-1, Pl-2, Pl-w and Pl-d) and the precursor Pr-Pl-1, and for scab (*Venturia inaequalis*) resistance (Va, Vb, Vbj, Vf, Vfn, Vm, Vr, Vr2).

Inheritance pattern of economically important traits

Chilling requirement: Generally quantitatively inherited, although the low chill requirement of Anna is thought to be controlled by one major and a number of minor genes.

Cold hardiness: Quantitatively inherited, largely additive.

Growth rate: Quantitatively inherited; QTL identified.

Vigour: Quantitatively inherited, largely additive.

Season of flowering: Quantitatively inherited, largely.

Harvest date: Quantitatively inherited; QTL identified.

Juvenile phase length: Quantitatively inherited; QTL identified (Liebhard *et al.* 2003).

Incompatibility: Numerous S-alleles exist, semi-compatible combinations.

Plant habit: Compact Single dominant gene (Co) identified for compact columnar habit in Wijcik clone of McIntosh; genetics more complex in other sources of the compact branching habit of dwarfing exist that are regulated by recessive alleles (d1 – d4).

Spur bearing habit: Generally quantitatively inherited, but spur habit in sports of Golden Delicious, Redspur and Starkrimson result from a single recessive gene.

Russett: Single dominant gene identified with many modifying polygenes; some russeted clones do not transmit russet to offsprings.

Acidity: Quantitatively inherited; QTL identified for malic acid gene, Ma.

Firmness: Quantitatively inherited; QTL identified.

Skin colour: Quantitative but may be regulated by a few major genes; one possibility is that three dominant, major genes produce colour; marker identified for a fruit colour gene (Rf) regulating the red/yellow dimorphism; EST research indicated that myb transcription factors are important.

Fruit Shape: Quantitatively inherited with a low genotype by environment.

Fruit Size/weight: Quantitatively inherited; highly heritable; QTL identified.

Red flesh: Dominant gene.

Storage disorders: High heritability for soft scald and superficial scald; moderate heritability for water core; low heritability for external pit, internal pit, brown heart, breakdown and chilling injury.

Sugar content: Quantitatively inherited; QTL identified.

Texture: Quantitatively inherited; QTL identified ; Md-ACS1 found to be closely associated with fruit softening.

Fruit colour and spineness: Polygenes.

Alternaria blotch, crown rot, wooly apple aphid, rosy apple aphid and rosy curling aphid: Single dominant gene.

Disease resistance: Polygenes.

Pre-selection criteria:

Breeding new apples is a long and costly process as apples have an extended juvenile period from seed to flowering that can last five or more years. Apple breeders often propagate seedlings on precocious rootstocks to shorten the cycle, but this only reduces the cycle by approximately two years. This long duration from seed to flowering is especially problematic for breeders who wish to introgress alleles from wild apple species or conduct parallel crosses over two or more generations to pyramid and combine multiple alleles for disease resistances. One relatively fast strategy to reduce the generation time is the use of transgenic stocks that carry the early-flowering gene BpMADS4, originally identified in birch trees. In apple, T1190 is a widely used BpMADS4-trangenic stock developed by Henryk Flachowsky (Julius Kühn-Institut, Germany) that can reduce the juvenility period to less than 10 months (Flachowsky *et al.*, 2007, 2011). In RosBREED, Jay Norelli used T1190 to pyramid and combine fire blight and scab resistance from apple cultivars Splendour and Enterprise into a single individual that can be used as a disease-resistant donor parent.

Beakbane *et al.* (1939, 1941, 1947, 1975) reported correlations between dwarfing propensity and various structural features. These include the percentage of live tissue of the root cross section; the bark/wood ratio, the percentage of medullary ray tissue in root transection; and high stomatal density of leaves. Some rootstock breeding programs have employed the bark/wood ratio to reduce populations or to select a reduced number of individuals most likely to be dwarfing. Using this rapid flowering trait, breeders are cycling through a generation each year, crossing early flowering transgenic offspring with elite cultivars or wild relatives with desirable traits like resistance to apples primary pathogens, apple scab and fire blight (Flachowsky *et al.*, 2011). In apple, lateral branching has high heritability and is correlated with topological traits. A correlation between the pH of the leaf sap and the pH of the fruit juice, sweet type has pH of leaves 5.7 and the acid types between 5.5-5.6. Useful correlations have been established between late leafing and late flowering. Another correlation that is often claimed is between red fruit color and anthocyanin pigmentation of other parts of the tree such as one year old shoots or leaf petioles. Selection techniques are used for canker disease resistance involving wound (scar on leaves) inoculation on seedling, cut seedling test or wounds made at the base of shoot of one year old wood in glass house for screening for resistance to collar rot.

RosBREED has developed DNA-information on rosaceae crops by which that can enhance the efficiency, accuracy, pace, and creativity of breeding programs, enabling breeders to pursue new cross combinations and use their resources effectively by saving thousands of dollars and hundreds of hours wasted on caring for inferior offspring that will never have commercial value (Peace, 2017).

Breeding Methods

1. Introduction: The cultivars like Skovfoged, Dorsett Golden, Red Spur, Well Spur, Vermont Spur, Strude Spur, Ace Spur, Royal Delicious, Vance Delicious, Top Red, Scarlet Gala, Skyline Superme, Bright-N-Early, Red Chief, Oregon Spur, Silver Spur, Scarlet Spur, Starking Delicious, Rich-a-Red, Mollies Delicious, Mayan, American Apirouge, Michael Schlomit, Dark Baron Gala, Gale Gala, Red Fuji, Golden Spur, Prima, Liberty, Super Chief, Sandidge, Red Delicious, Jonathan, Gala Redlum, Golden Delicious, Red Velox, Mich Gala, Wilton Star, Erosa, Gala Mast, Golden Clone B, Fuji, Zehn and Aztec have been introduced from USA, Russia, UK, Canada, Germany, Israel, Netherlands, Australia, Italy and Denmark.

Varieties to be introdued: Jonagold Decoster, Sun Fuji, New Adams Apple, Auvli Early Fuji, Sansa, Ruby Jon, Early Red One Delicious, September Wonder Fuji, Red Fuji, Top Export Fuji, Hillwell and Buckeye Gala.

Clonal rootstocks: M-9 (dwarf), M-7, M-26 and M-106, (semi dwarf), MM-111, (semi vigorous to vigorous) and MM-111 (drought tolerant).

2. Selection: Selective breeding practiced for thousands of years has been used to produce tastier fruits with greater resistance to biotic and abiotic stress. Several varieties have been selected by growers.

Dorsett Golden: It was first selected in 1959 by Mrs. Irene Dorsett of Nassau, Bahamas.

Rome Beauty: Chance seedling originated in Ohio, USA, by H. N. Gillett in 1848.

Winter Banana: Originated at Cass County, Indiana, USA by D. Flory.

Red Delicious: Seedling selection in 1887 at Peru, Iowa, USA.

Golden Delicious: Seedling selection in 1890 at West Virginia. It was used as a parent in hybridization programme.

Jonathan: It was discovered in 1826 as chance seedling on the farm of Philip Rick in Woodstock, New York.

McIntosh: Seedling discovered by John McIntosh in 1811 in Dundas County, Onatario, Canada.

Granny Smith: Originated in Australia in 1868 and named after Maria Ann Smith, who propagated the cultivar from a chance seedling.

Braeburn: Originated in New Zealand and was named after Braeburn Orchards where it was first grown.

Bountiful (Marring Bountiful): Selected from open-pollinated seedling Cox's Orange Pippin by F.H. Alston at East Malling Research Station Station, East Malling.

Baldwin: Originated about 1740 as a chance seedling in Wilmington, MA.

Ambri: It is a seedling selection indigenous to Kashmir.

CITH Lodha Apple: Selection at CITH, Srinagar.

HRMN-99: Selected by Mr Hariman Sharma at Paniala, Bilaspur, H.P. Requires very low chilling.

3. Hybridization: In India, hybridization was initiated in SKUAST, Srinagar, Kashmir, Dr YSPUH & F, Nauni (Mashobra), HP, HETC, Ranikhet, Uttarakhand, and CITH Res. Stat., Mukteshwar, Nainital, Uttarakhand.

Table 16. Apple hybrids developed in India.

Hybrids	Parentage	Important characters	Origin
Lal Ambri	Red Delicious x Ambri	Good yield and fruit quality; excellent keeping quality. Free from scab.	SKUAST (K), Srinagar, J&K
Sunehari	Ambri x Golden Delicious	Good yield and fruit quality. Free from scab.	SKUAST (K), Srinagar, J&K
Akbar	Golden Delicious x Rome Beauty x *M. floribunda*	Yield, scab resistance and fruit quality.	SKUAST (K), Srinagar, J&K
Firdous	Ambri x Cox's Orange Pippin	Yield and fruit quality.	SKUAST (K), Srinagar, J&K
Shreen	Lord Lambourne x Melba x R-12740-7A	Yield, scab resistance and fruit quality.	SKUAST (K), Srinagar, J&K
Ambred	Red Delicious x Ambri	Good yield and fruit quality.	YSPUH & F, Nauni (Mashobra), HP
Ambrich	Rich-a-Red x Ambri	Good yield and fruit quality. Free from scab.	YSPUH & F, Nauni (Mashobra), HP
Ambstarking	Starking Delicious x Ambri	Good yield and fruit quality.	YSPUH & F, Nauni (Mashobra), HP
Ambroyal	Starking Delicious x Ambri	Good yield and fruit quality.	YSPUH & F, Nauni (Mashobra), HP
Chaubattia Anupam	Early Shanburry x Red Delicious	Early maturity, yield and fruit quality.	YSPUH & F, Nauni (Mashobra), HP
Chaubattia Princess	Early Shanburry x Red Delicious	Early maturity, yield and fruit quality.	HETC, Ranikhet, Uttarakhand
Swarnima	Benoni x Red Delicious	Early maturity, yield and fruit quality.	HETC, Ranikhet, Uttarakhand
CITH Lodh Apple 1	Bud sport of Red Delicious	Good yield and fruit quality with improved over colour.	CITH Res. Stat., Mukteshwar, Nainital, Uttarakhand
Pusa Gold	Golden Delicious x Tydeman's Early Worcestor	Resist. to Powdery mildew and apple scab	IARI, RS, Shimla
Pusa Amartara Pride	Royal Delicious x Prima	Resist. to Powdery mildew and apple scab	IARI, RS, Shimla

***Advances made in USA, Canada, European countries, Australia, Brazil, New Zealand, Russia, China, Korea and Japan*:**

The cooperative disease resistance breeding program (between Purdue University, Rutgers University, and the University of Illinois) has been active for over 40 years to develop scab resistant cultivars. About 70 cultivars from various programs around the world have been introduced based on the *Vf* gene and a number show promise as potentially important cultivars, including Enterprise, GoldRush, Jonafree, Liberty and Pristine.

Anna: It is cross between Red Hadassiya and Golden Delicious and was bred by Abba Stein from Ein Shemer, Israel.

Ariane: Hybrid of Florina x Prima with pollen from Golden Delicious. Developed by F. Laurens,Y. Lespinasse and Fouillet, A at Institut national de la recherche agronomique (INRA).

Akita Gold: This is a hybrid between Golden Delicious and Fuji.

Ambrosia: This is a cross between Starking Delicious and Golden Delicious.

Arlet: This is a cross between Golden Delicious and Idared.

Autumn Crisp: Developed from a cross of Golden Delicious and Monroe.

Bohemia: This is a cross between Lord Lambourne and Golden Delicious.

Candel: This is a cross between Jonathan and Golden Delicious.

Cameo: This is a cross between Golden Delicious and Red Delicious.

Caudle: Golden Delicious × Red Delicious.

Champion: Golden Delicious × Cox's Orange Pippin.

Chantecler: Golden Delicious × Reinette Clochard.

Civni: It was first developed in 1985 as a cross of 'Gala' and Elstar apples by the Consorzio Italiano Vivaisti, an Italian apple growers' consortium from Ferrara.

Pinova: Hybrid between Clivia and Golden Delicious and was created in 1965 at the Institut für Obstforschung of Dresden.

Pillnitz: Hybrid between Clivia and Golden Delicious.

Cortland: Cross between McIntosh and Ben Davis.

Cosmic Crisp: Enterprise x Honey Crisp.

Cripps Pink (Pink Lady®): Golden Delicious × Lady Williams.

Tentation ® (Delblush): It is a cross of Grifer (Blushing Golden) and Golden Delicious. It was created in France in 1979 by Georges Delbard.

Elstar: Ingrid Marie × Golden Delicious.

Empire: Cross between Mclntosh and Delicious.

Ein Shemer: Cross between Zabidani and Golden Delicious.

Falstaff: Cross between James Grieve and Golden Delicious. Originated in East Malling Research Station, East Mallimg, England by F.H. Alston.

Fiesta™ (Mulling Fiesta): Cross between Cox's Orange Pippin and Idared, originated by F.H. Alston, J.R. Stow, J.W. Palmer, L.H. Clark, R.F.V. Williams, and H.F. Ermen, East Malling Research Station, East Malling.

Firm Gold: Cross between U.S. PP 2024 and Starkrimson Red Delicious.

Fuji: Ralls Janet x Red Delicious cross developed at Tohoku Research Station in Fujisaki, Aomori, Japan.

Gala: Cross between Kidds Orange and Golden Delicious. It was discovered by orchardist J.H. Kidd in New Zealand in 1934.

Ginger Gold: Cross between Albemarle Pippin and Golden Delicious.

Gloster (Gloster 69): Glockenapfel × Richared, originated at Jork Fruit Res. Sta., Jerk, Germany, introduced in 1969.

Greensleeves (Malling Greensleeves): James Grieve × Golden Delicious. Originated by F.H. Alston, R. Watkins, and J.W. Bates, East Malling Research Station, East Malling, England.

Honeycrisp: Macoun x Honeygold. Developed by J.J. Luby and D.S. Bedford, Univ. of Minnesota, St. Paul, MN during 2007 and sold in Europe as Honeycrunch®.

Iduna: Golden Delicious × Glockenapfel.

Jester (Malling Jester): Worcester Pearmain × Starkspur Golden Delicious. Invented by F.H. Alston in East Malling, Kent, England.

Jonagold: A cross between Golden Delicious and Jonathan. It was developed by R.Way, H. Bennett, L. Klein, J. Einsett, and K. Livermore in 1953 at the New York State Agricultural Experiment Station in Geneva, New York.

Jupiter (Malling Jupiter): Cox's Orange Pippin × Starking Delicious. Originated in East Malling, Kent, England by F.H. Alston.

Kent (Mulling Kent): Cross of Cox's Orange Pippin with Jonathan. Originated in East Malling, England, by H. M. Tydeman.

Maayan: F2 selection from the cross (Calville St. Sauveur x Damascus) x Delicious. Developed by C. Oppenheimer at Rehovot, Israel.

Michal: F2 selection (Calville St. Sauveur x Damascus) x Delicious. Originated at Rehovot, Israel, by C. Oppenheimer.

Maigold: Fraurotacher × Golden Delicious.

Melrose: Cross between Jonathan and Delicious. Originated at Wooster, Ohio, by F. S. Howlett in 1944.

Mollie's Delicious: Cross of (Golden Delicious x Edgewood) x (Red Gravestein x Close). Originated at New Brunswick, N. J. by L. F. Hough and C. H. Bailey in 1966.

Monroe: Cross between Jonathan and Rome Beauty.

Mutsu (syn. **Crispin**): Cross between Golden Delicious and Indo Cross, developed at Aamori Apple Expt.Station, Aamori, Japan.

Opal: Cross between Topaz and Golden Delicious.

Sunburst ®: Gala × Falstaff. Released by EMRS, East Malling.

Pink Lady: Cross between Golden Delicious and Lady Williams. Originated from the breeding program of the Western Australian Dept. Agr. Selected in 1979 by John Cripps and introduced in 1989.

Pinova: Cross between Clivia and Golden Delicious.

Prima: It was derived from a cross between PRI 14-510, a seedling grown at the University of Illinois derived from *Malus floribunda* 821 as the seed parent carrying the Vf gene for resistance to apple scab,with pollen from a New Jersey seedling, NJ 123249. It was released from the cooperative program involving Purdue University, Rutgers University, the State University of New Jersey, and the University of Illinois.

Rave®: Natural cross of Honeycrisp and MonArk (AA44). Developed by David Bedford and James Luby at the University of Minnesota, Minnesota. It is also known as MN55 or First Kiss®.

Redsleeves (Malling Redsleeves): Cross between Exeter Cross and TSR15T3 (a PRI scab-resistant selection). Originated at East Malling Res. Sta., East Malling, Kent, England by F. H. Alston.

Rubinette: Cross between Golden Delicious and Cox's Orange Pippin.

RubyFrost: Cross between Braeburn and Autumn Crisp developed at Cornell University, Ithaca.

Scifresh (Jazz™): It is a cross between Royal Gala and Braeburn. It was developed in New Zealand in collaboration between apple marketer ENZA, orchardists and the Plant & Food Research Institute, New Zealand.

Sekai Ichi: Cross between Golden Delicious and Red Delicious.

Spigold: Cross between Northern Spy and Golden Delicious.

Sundowner: Cross between Golden Delicious and Lady Williams.

Suntan (Mulling Suntan): Cox's Orange Pippin × Court Pendu Plat. Orig. from breeding program at at East Malling Res. Sta., East Malling, Kent, England, by F. H. Alston and R. Watkins.

Tentation delblush (Grifer): Cross between Blushing Golden and Golden Delicious.

Tropic Sweet: Polycross from selections from Jerseymac x Anna, originated at Gainesville, Fla. at Univ. of Florida, by W. B. Sherman and P. M. Lyrene.

Triumph: Came from its parents, Liberty and Honeycrisp, at University of Minnesota, St. Paul, MN.

Lucy: Honey crisp x Airlie Red, created by Chelan Freash, Chelan, Washington.

Lucy: Honey crisp x Airlie Red, created by Chelan Freash, Chelan, Washington.

Redlove'® series: It was developed by a private breeder (Markus Kobelt, Lubera nursery in Switzerland. They are hybrids from cross-pollination of red fleshed and scab-resistant plants. These include **'Redlove Calypso'.**

Scab resistant cultivars: Redlove Circe, Redlove Era, Redlove Odysso and Redlove® Sirena.

Columnar apple: McIntosh Wijcik is heterozygous for a dominant gene, Co, imparting a compact growth form with short internodes and very few lateral branches. With minimal pruning, these trees can be maintained as a single column. At East Malling, Tobutt has combined this columnar habit with other characters to produce selections likely to be useful as garden trees, as pollinizers in conventional orchards, and for very high density plantings. The relative absence of branches and the well exposed fruit on these trees suggests that a system based on columnar trees may lend itself more readily to mechanized harvesting of dessert fruit than other growing systems in existence. Four columnar types named Maypole, Telamon, Trajan, and Tuscan have been selected for garden culture, described by Tobutt were used in the crosses.

Trajan: Produced by crossing Golden Delicious x Wijcik at East Malling Research Station, East Malling in 1986. The new variety yields medium sized asymmetric generally round-conical fruit having a yellow-green skin which is partially covered with a red flush having prominent lenticels. The unique growth habit makes extremely high density plantings possible with concomitant increased fruit yields.

Telamon: Originated from a cross Wijcik x Golden Delicious, created at East Malling Research Station, East Malling in 1986.

Tuscan: Originated from a cross between Wijcik variety of McIntosh apple tree and Greensleeves created at East Malling Research Station, East Malling in 1986.

Maypole: Produced by crossing Wijcik x *Malus Baskatoag* at East Malling Research Station, East Malling in 1986. A new and distinct variety of ornamental crab apple tree is provided which exhibits an extremely columnar growth habit which generally resembles that of a vertical pole.

Rosalie: Aneta x Maypole.

Obelisk: Produced by crossing Cox's Orange Pippin and Court Pendu Plat and Wijcik.

Smaragd: Originated from a cross between Granny Smith and McIntosh Wijcik, created at Faculty of Agriculture, Novi Sad, Serbia in 1992,

Rootstock Breeding: The main accomplishment in apple rootstock breeding was the development of dwarfing apple rootstocks, which started in the early 1920s, by the East Malling Research Station, Kent, England. Drs. Wellington and Hatton sorted out the incorrect naming and mixtures then widespread in apple rootstocks distributed throughout Europe. These verified and distinct apple rootstocks were then distributed throughout the world as Paradise (Malling) apple rootstock selections.

In collaboration with John Innes Institute, then based at Merton, a further series of rootstocks was produced, all exhibiting resistance to the woolly apple aphid (*Eriosoma lanigerum*). This was, and still remains, a very serious pest in many parts of the southern hemisphere, where it attacks the root systems of apple trees. Malling Merton (MM.) series of rootstocks gained resistance to woolly apple aphid from their Northern

Spy parent, are still used extensively throughout the world. The roostocks developed from this programme are

M. 116: MM. 106 × M.27.

M. 27: M.16 x M.9.

MM. 104: M 2 x Northern Spy.

MM 109: M 2 x M.9.

MM 111: Northern Spy x Merton 793.

Several promising new apple rootstocks have emerged from HRI-East Malling program of breeding and selection:

AR86-1-25: It is a selection from M.27 x MM.106 cross. It produces trees with similar in vigor, yield and fruit size to MM.106 but has additional great advantage compared with MM.106 in its strong resistance to collar rot.

AR86-1-20: It is a sibling from the same cross as in case of AR86-1-25.

In India, IARI Regional station, Shimla has developed **Pusa Seb Mulvrint –1** from *Malus baccata* (Shillong) with high grafting success, easy propagation and resistance to root rot, powdery mildew and wooly aphid.

New dwarfing rootstocks

BM 427: M.4 x Antonovka Kamensischka, originated at Balsgård, Sweden.

G.65: M.27 x Beauty, originated at Cornell University, Ithaca.

JM.1, 5 and 8: Marubakaido x M.9, originated at Apple Research Centre, NIFTS, Japan.

J-TE-G: M.9 x Croncels, Techobuzice, Czech

P.22: M.9 x Antonovka, originated at Skierniewice, Poland.

P.61: P.16 x M.26, originated at Skierniewice, Poland.

P.66: P.22 x M.26, originated at Skierniewice, Poland.

Voinesti 2: M.9 x Cretesc, originated at Voinesti, Romania.

Intergeneric and interspecific hybridizaion

Outside of the genus *Malus*, the potential for natural hybridisation with other genera appears limited. Breeding programs have produced intergeneric hybrids between apple and pear (*Malus* × *Pyrus*), and apple and hawthorn (*Malus* × *Crataegus*) as reported by Robertson *et al*. (1991). However, these relied on techniques such as embryo rescue (Banno *et al*., 2003). Forced intergeneric hybridisation of *Cydonia* (quince) with *Malus* resulted in fertile genotypes which have been identified as an artificial hybrid genus × *Cydomalus*. However, seedlings produced are generally weak and are of low viability and germinability.

Interspcific hybridization: Most species in the genus *Malus* can be readily hybridised after artificial cross pollination, but interspecific crossing can also occur under natural conditions. The capacity for inter-species hybridisation within the genus *Malus* is evident by the numerous hybrids among *Malus* spp. The domesticated apple is the result of an interspecific hybridization named *Malus* x *domestica* Borkh.

4. Backcrossing: In apple, backcrossing programmes have resulted in successful commercial selections (carrying disease resistance from wild species) by the second backcross generation.

5. Mutation: Natural mutations have played an important role in the evolution of new apple varieties.

Spur type mutants: Golden Spur, Red Spur, Starkrimson, Golden Spur, Red Chief, Oregon Spur, Well Spur and Silver Spur.

Colour mutants: Vance Delicious, Top Red, Skyline Supreme Scarlet Gala and Bright-N-Early.

Red Delicious sport: Red Spur, Rich-a-Red, Vance Delicious, Hardeman, Atwood, Common Earlired, Gardner (s), Hapke, Hardispur (s), Harrold, Hi-Early, Huebener, Imperial, Maehara, Red King, Red Prince, Red Queen, Redspur (s), Royal Red Starkrimson (s), Topred, Vance, Watson, Wellspur(s) and Woodspur (s). **Starking Delicious**: Also known as Royal Delicious, it is a Limb Sport of Red Delicious in Monsoville, New Jersey, USA (1921).

Top Red Delicious: Bud sport of Shotwell Delicious.

Red Chief: Bud sport of Starkrimson.

Skyline Supreme Delicious: Bud mutant of Starking Delicious.

Braeburn sports: Braestar™ (Brayleet), Joburn™, Braeburn (Aurora) and Eve™ Braeburn and Chinook.

Jonagold New Sports: Jonica, Rubinstar and DeCoster Jonagold and Nicolais King Jonagold, Jonagold Romagold, Morrens Jonagored and Red Jonaprince.

Fuji sport: Some of the older sports of Fuji include: BC #2, Naga-fu 2, 6, and 12, Akifu #1, Seiko-fu Red Fuji, Red Fuji T.A.C. 114 and Yataka. Newer sports include: Autumn Rose Fuji, Beni Shogun Fuji, Blaze™ Fuji, Blaze Fuji and Desert Rose Fuji.

Gala New Sports: The interest in new and improved sports of Gala is all too reminiscent of Delicious. New sports are Galaxy Gala (better coloured), Scarlet Gala (better coloured), Autumn Gala (Harry Black), Brookfield Gala (Baigent) and Buckeye® Gala.

Golden Delicious Strains/clones: Lucky Rose Golden A, Smoothie Golden Delicious (Gibson), Autumn Gold, Elliot, Golden Glory, Golden Supreme, Goldrush, Shizuka, Stark Ultragold and Suncrisp, Hampshire (Gould).

Jonagold new Sports: Jonagored, Jonica, Rubinstar, DeCoster Jonagold, Nicolais King Jonagold, Morrens Jonagored Supra and Red Jonaprince.

Several cultivars of apple have been developed in France by mutation. These are Belrene, Courtgold, and Courtavel. The others are Dover (Netherland), Golden Haidegg (Autralia), James Grieves Double Red (Czech Republic), Mclntosh 8-F-2-32 (Canada), Mori-hou-fu 3A (Japan), Natsu-hime (Japan) and Shamrock (Canada).

6. Polyploidy: The triploid cultivars are Wilhelm, Gravenstein, Jonagold, Schöner von Boskoop, Ribston Pippin, Blenheim Orange, Kaiser, Baldwin, Mutsu and Kanadarenette. A few are tetraploids, e.g., Antonovka Ploskaya, Wealthy Tetraploidnyi,

Papirovka Tetraploidnaya, McIntosh Tetraploidnyi. Triploids are not suitable as pollinators. There is report of tetraploid giant sport -Perrinae Transparent. Apomixis has been seen in triploid species of *Malus sikkimensis*, *M. sargenti and Malus sieboldii* (diploid, triploid).

7. Biotechnology:

Micropropagation: Micropropagation using shoot proliferation has been achieved by standard protocols. Shoot proliferation medium is based on Murashige and Skoog salts with minimal organics, and the addition of cytokinins such as 4.4 μM benzylamino purine (BA) or 0.1 μM thidiazuron, the inclusion of 0.5 μM indolebutyric acid (IBA) and 1.4 μM gibberellic acid (GA_3) and 87.6 mM sucrose using agar as a gelling agent. Protocols for micro propagation of cultivars (Golden Delicious, Tydeman's Early Worcester, and Red spur), clonal rootstocks (MM 106, M 7, MM 111, M26 and Merton 793) have been developed.

In apple, tissue culture has been seen as a means to propagate fruiting cultivars on their own roots but results have been variable. Own-rooted apples trees are vigorous and branch freely. Flowering may be delayed as compared to planting budded trees.

Somatic embryogenesis: Embryogenesis has been induced from nucellar tissues from seeds and calli from various *in vitro*-grown apple seedlings or from flower buds and excised petals from field grown trees. However, there have been no reports of complete development of viable seedlings from adventitious embryony.

Molecular characterization: Molecular chararterization using isozymes SCAR, SSR and RAPD markers in apple has been developed.

Genetic transformation: Transgene technology has the potential for transforming useful characteristics from one cultivar to another within a species or from one to another. The first apple transformation was done by James *et al.* (1989) using *Agrobacterium tumefaciens*. Particle bombardment of apple leaf explants has received much less emphasis.

Apple rootstock M 26 was transformed with *rol*A and *rol*B (Zhu and Welander, 2000) and M 9 rootstock with *rolB* (Zhu *et al.,* 2001).

Fruit trees typically have long breeding cycles. Successful reduction of the juvenile period of apple to one single year was achieved via constitutive expression of the Bp*MADS4* transcription factor gene from silver birch (*Betula pendula* Roth.) in early flowering apple cultivars Pinova, Gala, Mitchgla Gala and Santana (Weigl *et. al.*, 2015).

Scab-resistant transgenic apple cultivars were obtained via *Agrobacterium*-mediated transfer of scab resistance genes *Rvi6* from *M. floribunda*, a species of crab apple that shows natural resistance to some strains of the apple scab fungus. Further research resulted in the first intragenic and cisgenic apple lines of the cultivar Gala with scab resistance, containing the *Rvi6* gene from *M. floribunda* under the control of the promoter from the apple Rubisco gene or the native *Rvi6*-promoter, respectively. These plants were planted in a field trial to study scab resistance in an orchard situation.

In 2015, the development of the first cisgenic apple with increased resistance to fire blight was reported. A cisgenic apple line C44.4.146 of the susceptible apple cultivar Gala Galaxy was regenerated using the *FB_MR5* gene from the wild apple *M. robusta*. The BpMADS4-based breeding technology allows for more rapid introgression of agronomically relevant traits (e.g., disease resistances) from wild apples into domestic apple cultivars.

Apples have been engineered using a gene silencing technique called ribonucleic acid interference (RNAi), in order to reduce the production of polyphenol oxidases (PPO), which causes the apple flesh to brown when sliced or bitten (Xu, 2013). Transgenic versions of the varieties Golden Delicious, Granny Smith, and Fuji have been approved for marketing. Research work on the allergenicity of apples has been expanded. Puhringer *et al.* (2000) found that the promoter for the major allergen Mal d1 is stress and pathogen induced.

PEAR

Origin: Asia minor to southwest Asia and East Asia.

Taxonomy status:

Order: Rosales

Family: Rosaceae

Subfamily: Rosoideae

Tribe: Pyreae

Botanical name: *Pyrus communis* L.

European pear predominates in Europe. *P. pyrifolia* (Burm.) Nak., Sand or Japanese pear is the main cultivated species in southern and central China, Japan and Southeast Asia.

Chromsomes number: 2n=2x=34

Genome size: 577.3 Mb (European pear)

512.0 Mb (*Pyrus* x *bretscheideri* Rehd.)

Related edible species are *Pyrus nivalis, Pyrus x bretschneideri, P. ussuriensis* and *P.* x *sinkiangensis.*

Botany

Deciduous trees or large shrubs, sometimes thorny; leaves alternate simple, petiolate, serrate, crenate or entire, rarely lobed, involvute in scales. Fowers that appear with or before the leaves are white, rarely pinkish and and are borne in corymbs on short spurs or lateral branchlets. The flowers are perfect and consist of five sepals, reflxed or spreading, and petals 5 clawed, sub-orbicular to broad oblong and 20-30 stamens with anthers that are usually red or purple. Styles vary from two to five and are free but closely constricted at base. The ovary has five locules with two ovules per locule. The fruit is a pseudo-fruit composed of the receptacle or the calyx tube, greatly dilated, enclosing the true fruit, and consisting of five cartilaginous carpels known as the core. Fruit is globose or pyriform, calyx persistent or deciduous, buttery texture of European pears tend to have crisp-textured flesh, and have abundant stone cells in the flesh of *P. pyrifolia*. Seeds are black in colour.

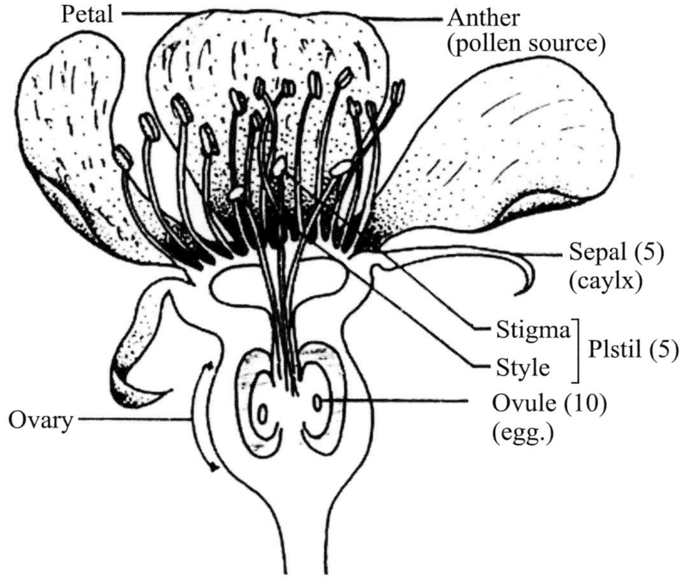

Fig. 17. Pear flower showing different parts.

Floral Biology

Pear blossoms open very quickly when weather is warm and dry. The anthesis takes place between 10 AM and 1.00 PM. The maximum anthers dehisce from 12 to 2.00 PM. Stigma receptivity occurs 2-3 days before anthesis.

Breeding Objectives

Scion breeding objectives

(i) Ecological adaptation.

(ii) Precocious, early ripening, self fertility, regular bearing.

(iii) Compact growth habit and high yield.

(iv) Superior fruit quality-red skin color, good fruit size and flavour, improved textural attributes (firm or soft buttery textures for European pear and sweet crisp juicy texture retaining good fruit firmness for Asian pear), good for processing and extended storage life.

(v) Resistance against diseases (scab, fire blight, black spot) and pests (pear psylla).

Limation in Breeding

1. Heterozygous.
2. Long juvenile phase.
3. Self incompatibility.
4. Sterility (Le Conte, Magness, Winter Nelis).
5. Parthenocarpy.
6. Inbreeding depression.
7. Lack of information on inheritance pattern.

Rootstock Breeding Objectives

1. Dwarfing.
2. Free standing of trees (good anchorage).
3. Precocity and productivity in scion varieties.
4. High yield efficiency.
5. Positive influence on fruit quality and size.
6. Efficient propagation ability and compatibility.
7. Drought and cold hardiness.
8. Tolerance to iron and calcium chlorosis.
9. Resistance to fire blight.
10. Resistance to pear decline.

Germplasm Resources

The USDA-ARS National Plant Germplasm System maintains clonal repository that includes more than 2500 unique clones or seedlots at Corvallis, Oregon representing world diversity for *Pyrus*. Large pear germplasm collections in western Europe are located at the National Fruit Collection at Brogdale Farm in Kent, England; Centre Wallon de Recherches Agronomiques in Gembloux, Belgium; Le Centre INRA Angers-Nantes in France; Federal Research Centre for Cultivated Plants in Dresden-Pillnitz, Germany and University of Bologna in Bologna, Italy.

Collections in pear were made through introductions and indigenous collections and have been conserved *ex situ* at national field gene bank at CITH, Srinagar and NBPGR Station. These accessions have been utilized for further varietal development program intented for Indian climatic conditions.

Field gene banks have been maintained at Regional Horticultural Research and Training Insitute, Dr YSPUHF, Mashobra (63 accessions),, Temperate Horticulturral Research Station Dr YSPUHF, Kotkhai, Shimla (56 accession), NBPGR Research Station, Phagli, IARI, RRS, Amartara, Shimla (7 *Pyrus* species and 15 accession), SKUAST, Srinagar, Hill Fruit Research Centre, Chaubattia and CITH, Srinagar (80 accessions).

Sources for economically important traits:

Low pH: *P. dimorphophylla, P. fauriei, P. mamorensis, P. betulifolia, P. pashia, P. calleryana.*

High pH: *P.cordata, P. amygdaliformis P. fauriei.*

Wet soil*:* *P. dimorphophylla, P. faurie, P. betulifolia, P. longipes, P. calleryana.*

Clay soil: *P. amygdaliformis*, *P. fauriei*, *P. betulifolia*, *P. pashia.*

Drought: *P. regelii, P. gharbiana, P. syriaca, P. salicifolia, P. cordata, P. amygdaliformis*, *P. longipes*, *P. pashia*, *P. calleryana.*

Winter hardiness: Paris, Verte, Nordha¨ user Winterforelle, *Pyrus betulifolia, P. pseudopashia, P. ussuriensis.*

Warm winters: *P. kawakamii, P.cordata, P. fauriei, P. amygdaliformis, P. betulifolia*, *P. pashia, P. calleryana.*

Cold hardiness: Seckel, Doyenne du Comice, Erika, Delta, *P. ussuriensis, P. longipes P. nivalis, P. elaeagrifolia, P. fauriei.*

Fire blight: *P. communis*, Seckel, Kieffer, Old Home, Harrow Delight, Sunrise, Blak'es Pride, Gem, Harrow Sweet, Harrow Gold, Shenandoah, Alexander Lucas (*P. communis*), Ya Li,Tzu Li (*P.* x *bretschneideri*), *P. hondoensis, P. fauriei, P. calleryana, P. ussuriensis.*

Blackspot: Kosui, Hosui.

Scab: *P. communis*, Bartlett (Williams), Beurre Hardy (Gellert), Kieffer, Dr. Jules Guyot, *P. pyrifolia.*

Powdery mildew: Doyenne du Comice, Winter Nelis, *P. dimorphophylla, Pyrus pashia* var. *Kumaonii* and *P. pashia* .

Pseudomonas blight: Beurre Hardy, Forelle.

Fruit rot: Passa Crassana, Clapp's Favourite, Louise Bonne dAvranches.

Pear psylla: Karamanka, Jerisbasma, Vodenjac, Monica, *P. betulifolia*, *P. dimorphophylla, P. nivalis, P. regelii, P. fauriei, P. calleryana, P. ussuriensis.*

Codling moth: *P.cordata, P. amygdaliformis, P. kawakamii*, *P. dimorphophylla*, *P. pseudopashia, P. fauriei, P. elaeagrifolia*, *P. betulifolia*, *P. calleryana.*

Low chilling cultivars: *P. pyrifolia,* Patharnakh, Gola, Le Conte, Keiffer, Smith, Baghugosha, China Pear, Pineapple, Baldwin, Tenn, Flordahome, Ayers Hood, Orient, Carnea, Tsu Li, Ya Li.

Dwarfness: Armida, David, Abate Light.

Evergreen: *Pyrus kawakamii.*

Precocious: *P. pyrifolia.*

Self-fertile: Bartlett, Nashpati, Beurre Hardy, Doyenne du Comice, Flemish Beauty, Osa-Nijisseiki, Akibae, Zuishyu and Shinjyu.

Red skin colour: Nordhaeuser Winterforelle, Rubia, Red Silk, Max Red Bartlett, Maxie™, PremP109, PREVAR™, Yanshan and Weinshan, Cheeky, Hortensia and Huobali.

Excellent fruit quality: Doyenne du Comice.

Good fruit quality in combination with high yield: Bosc (Kaiser Alexander), Dr. Jules Guyot, Pra¨ sident Drouard.

Drink: *P. nivalis.*

Ornamental: *P. calleryana, P. fauriei.*

Inheritance Pattern

Cold hardiness: Quantitatively inherited, largely additive.

Juvenile period: Positive correlation between length of the juvenility period and precocity; additive genetic control.

Pear/quince compatibility: Governed by single gene.

Plant habit

Dwarfing: Regulated by single genes.

Season of flowering: Quantitatively inherited.

Incompatibility: Numerous S-alleles exist, but there are compatible combinations.

Sterility: Governed by cytoplasmic and nuclear gametic factors.

Fruit quality

Firmness: Quantitatively inherited and highly heritable.

Flavor: Quantitatively inherited.

Juiciness: Regulated by a limited number of dominant genes.

Ripening: Polygenes.

Keeping quality: Quantitatively inherited (Zielinski *et al.* 1965).

Russeting: Quantitatively controlled in *P communis* with high heritability, wo genes regulate russetting in *P. pyrifolia.*

Skin color: Major genes determine – yellow dominant to green, blushed recessive to non-blushed; Deep red regulated by dominant gene; White flesh dominant to colored.

Shape: Quantitatively inherited with round and obovate dominant to pyriform and turbinate; moderate to high heritability for ratio of length to diameter.

Size: Quantitatively inherited; low heritability in some *P. pyrifolia* populations.

Sugar content: Heritability of soluble solids in *P. pyrifolia* is low to moderate depending on population.

Texture: Quantitatively inherited with relatively high heritability in *P. communis*; Presence of stone cells dominant to stoneless in most crosses of *Pyrus.*

Grit content: Regulated by at least four loci, acting independently and additively.

Pre-selection criteria

Stomata density, branch-cortex ratio, leaf thickness, palisade tissue thickness, palisade-spongy ratio and vessel density can be used as indices to identify growth potential at early growth period. A negative correlation between the thorns and the appearance of the first flowers was noticed for pears.

Breeding System

1. Introduction

Several varieties like William (Bartlett), Carmen, Max Red Bartlett, Z H. Copeace, Passe Crassane, Beurre Hardy, Clapp's Favorite, China Pear, Flemish Beauty, Devoe Early, Laxton's Superb, Red Bartlett, Starkrimson, Conference, Kashmiri Nakh, Bosc, Doyenne du Comice, Anjou, Fertility, Kieffer, Abate Fetel and Monarch have been introduced from USA, France, UK and Japan. Recently introduced varieties by NBPGR, New Delhi are Tsu Li, Hong Li, Worden, Seckle, Golden Russet, Bosc, Tailors Gold, Harrow Crisp and Harrow Gold.

Promising cultivars to be introduced: Columbia Red Anjou, Clapp's Favourite (Red), Packham's Triumph, Cascade, Sensation Red Bartlett, Nijisseiki, Kosui, Chorujo and Hosui (Nashi cultivars). The cultivars based on *P. ussuriensis* and *P. bretschneideri*, like Ya Li and Tsu Li of China should also be introduced.

Clonal Rootstocks: Quince-A (standard rootstock), Quince C (very dwarf) and Quince-B (semi vigorous), OG x F-51, Oregon-211, Oregon-249, OH x F-34, OH x F-69, OH x F-87, OH x F-230 and OH x F-233, *P. calleryana* (requires 400 chilling hours).

2. Selection: Jean Baptiste van Mons (1765–1842) in France did commendable job in selecting chance seedlings. Several selections have been done by nurserymen and growers:

Conference: Selected by Thomas Rivers, UK.

Abate Fetel: Chance seedling selected by Fetel, France.

Bartlett (Williams, Williams Bon Chretien): Discovered originally in 1765 by Mr. Stair in England.

Winter Nelis: Selected by Jean Charles Nelis, Beligian.

Bosc: Selected by Louis Bosc.

Doyenné du Comice: Open pollinated seedling selected in France.

Beurré dAnjou also known as Anjou, Winter Meuris, and Nec Plus Meuris: It is a Belgian cultivar developed/identified by Van Mons in 1823.

Rocha: Chance seedling selected in Portugal.

Beurre Bosc: Open pollinated selection.

Selection done in Kashmir: Kashmiri Nakh-1, Kashmiri Nakh -2 in CITH, Srinagar.

Selections done in Asian pear at PAU, Ludhiana: Punjab Red Blush, Punjab Gold, Punjab Nectar and Pathararnakh.

3. Hybridization:

Hybrid cultivars of European pear:

Ayers: It was developed by Brooks D. Drain of Tennessee Agriculture Experiment Station by crossing Garber with Anjou in 1937.

Aurora: Cross between Margherite Marillat and Bartlett.

Armida: Dr Jules Guyot x Comice.

Angelys: Winterdechant x Comice, developed by INRA, France.

Cascade: Max Red Bartlett × Comice, origiated by F. E. Reimer at Medford, Ore, USA.

Concorde: Comice x Conference, developed at the East Malling Research Station in England.

Condo: Conference x Comice, developed at CPRO, Wageningen.

Crispie: This hybrid combines the Asian Nijisseiki pear with the European Red Bartlett pear, bred by Crown Research Institute Plant & Food, New Zealand.

David: Dr. Jules Guyot x Comice.

Daves's Delight: Beierschmidt × Bartlett, originated at Kentville, Nova Scotia, by A. D. Crowe, Agriculture Canada Research Station, Kentville, Nova Scotia.

Eckehard™: Cross between Nordha user Winterforelle with Clapp's Favourite.

Gerburg: Cross between Clapp's Favourite with Nordha user Winterforelle.

Graf Dietrich: Clapp's Favourite x Nordha¨ user Winterforelle.

Graf Wilhelm: Comice x Nordha user Winterforelle.

Gra¨ fin Gepa: Cross between Nordha user Winterforelle with Baierschmidt.

Hermann: Cross between Dr. Jules Guyot and Bunte Juli.

Hortensia: Nordha¨ user Winterforelle x Clapp's Favourite.

Isolda: Cross between Dr. Jules Guyot with Bunte Juli.

Maxie™: This hybrid combines the Asian Nijisseiki pear with the European Red Bartlett pear.

Thimol: Nordha¨ user Winterforelle x Madame Verte.

Uta: Cross between Madame Verte with Beurre Bosc.

Butirra Precoce: Coscia x Bartlett.

Harrow Sweet: Bartlett x Purdue 80-51, developed at Agriculture and Agri-Food Canada (AAFC) Research Centre at Harrow Ontario, Canada.

Harrow Crisp: Bartlett x US56112-146, developed at Agriculture and Agri-Food Canada (AAFC) Research Centre at Agriculture Canada Research Branch, Research Station, Harrow, Ont. NOR 1G0, Canada.

Harrow Delight: Williams x Purdue 80-51 (Old Home x Early Sweet), Developed by R. E. C. Layne, Harrow, Ontario, Canada.

Harovin Sundown: Originated from a cross of Bartlett × US56112-146, made in 1972 by H. A. Quamme.

Clapp's Favourite: Flemish Beauty x Bartlett, developed in USA.

Cascade Oregon: Max Red Bartlett x Comice.

Delbard Premiere: Akca x Dr. Jules Guyot developed in France.

Packham's Triumph: Cross between Uvedale St. Germain and Bartlett, developed in Australia.

Jubileer DAr: Clapp's Favourite x Klementina, developed in Bulgaria.

Laxtons Superb: Marie Louise x Bartlett, developed in UK.

Asian or Nashi pear hybrids:

Kikusui: Taihaku x Nijisseiki.

Tsu-Li: Cross between *P. ussuriensis* and *P.* x *bretschneideri.*

Shinseiki: Cross between Nijiseiki and Chojuro.

Niitaka: It is a Japanese cultivar resulting from a cross between Amanogawa and Chojuro and produces large-sized fruits.

Kosui (syn. **Kousui):** Kikusu x Wasekozu.

Hosui: Kosui x Hiratsuka.

Hakko: Yakumo x Kosui.

Intergeneric hybrids: Intergeneric crosses between Japanese pear (*Pyrus serotina* var. *culta* Rehd.) and apple (*Malus pumila* Mill. var. *domestica* Schneid.) were tried with meagre success.

x *Sorbopyrus irregularis* (Münchh.) C. A. Wimm.: Hybrids of European Pear (*Pyrus communis*) and the Common Whitebeam (*Sorbus aria*). The original hybrid, known as the Bollwiller Pear, first arose at Bollwiller in Alsace, France, before 161. Shipova is important cultivar originates from Serbia/Montenegro. It is produce edible fruit with a sweet, yellowish flesh, which tastes similar to an Asian pear.

x *Pyronia veitchii* (Trab.) Guillaum: Hybrids between *Pyrus communis* and *Cydonia oblonga*. It has become a useful virus sensitive indicator plant. 'Luxemburgiana' is a form available from nurseries. Fruits round, to 8cm diameter, sweet and soft flesh, good flavour.

Interspecific hybrids:

Le Conte: *P. communis* x *P. pyrifolia* developed in USA.

Kieffer: *P. communis* x *P. pyrifolia.*

Orient: *P. communis* x *P. pyrifolia.*

Pineapple: *P. communis* x *P. pyrifolia.*

Chinese white pear *(Pyrus × bretschneideri*): *P. ussuriensis* x *P. betulifolia.*

Rootstocks: Fire blight resistant types have been found in progeny of the *P. communis* cross, Old Home x Farmingdale (OH x F rootstock series) and individuals from this family provide some size reduction. Clonal rootstocks derived from crosses between fire blight resistant pear cultivars Old Home (OH) and Farmingdale (F) are becoming popular in the USA, with selections OH x F 87 and OH x F 97 being the most widely used. Other dwarfing *P. communis* rootstocks include Pyrodwarf and BU 2-33 coming from Old Home x Bonne Louise dAvranches. These two selections are more dwarfing than OH x F selections, but may have other shortcomings that may limit their use.

4. Polyploidy: Naturally occurring polyploidy has been identified in both European and Asian cultivars, including that of Sha 01, a tetraploid ($2n = 4x = 68$) bud mutant of Korla Pear, a tetraploid Bartlett, and a triploid ($2n = 3x = 51$) Beurré Diel and Anli, a

P. ussuriensis cultivar. Triploids have been developed by crossing naturally occurring or induced tetraploids (following colchicine treatment) with diploid parents. Even though there is a range of available polyploids, pear breeding programs rarely use these to develop new cultivars. In crosses undertaken between Asian species, a range of tetraploid, triploid, and diploid combinations have been generated. For example, crosses between two tetraploids have yielded progeny of which 97% are tetraploid and 3% are aneuploid. Crosses of tetraploids with triploids, and reciprocal crosses, have yielded progenies with more or less equal numbers of triploids (34%), aneuploids (33%), and diploids (26%), while crosses between tetraploids and diploids have mostly produced diploids (61%) and triploids (36%).

5. Mutation: Several kinds of mutations have been identified after irradiation in *P. communis* including bloom time, blossom color, ripening time, fruit color and growth habit. Mutations have been induced that affected disease resistance and self compatibility in *P. pyrifolia*. The identification of Osa-Nijisseiki, which is a self-compatible mutation of Nijisseiki because of an S4 allele non-functional in the style but functional in the pollen has created an opportunity for pear breeders to incorporate this trait. Among these selected mutants in Asian pear, four cultivars were named and released, including Gold Nijisseiki, Osa Gold, Kotobuki Shinsui, and Shizukisui. Two self-compatible black spot resistant varieties show high promise.

Mutants in European pear:

Starkrimson: Bud mutant of Clapp's Favorite, discovered in Missouri, USA.

Red Bartlett, Max Red Bartlett: Bud sport of Bartlett.

Red Anjou: Bud mutant of Anjou.

Columbia Red: Bud mutation of Anjou.

Taylors Gold: A mutant clone of Comice.

6. Biotechnology:

***In Vitro* propagation**

Micropropagation consists of regeneration and multiplication of plants from cells and tissues, cultivated in nutrient medium. This method allows the mass production of plantlets, regardless of the time of year under excellent sanitary conditions. For pear culture, some limitations are experienced in micropropagation due to recalcitrant behaviour during the multiplication phase, difficulty in rooting induction and slow plant growth during acclimatization. Despite these issues, the CAV/UDESC micropropagation lab has been able to successfully micropropagate pear rootstocks since 2013, with high multiplication rates. The best micropropagation protocol for pear rootstocks, including the *Pyrus communis* selections, the OH x F series and Pyrodwarf has been briefly described.

Molecular markers and Quantitative Tait Loci (QTLs):

Molecular markers developed include isoenzymes, AFLPs, ISSRs, RAPDs and RFLPs. They have been used mostly for varietal identification of European and Asian pear. The molecular markers most useful for MAS are simple sequence repeats (SSRs), amplified fragment length polymorphism (AFLP), sequence related amplified polymorphism (SRAP), inter-simple sequence repeats (ISSR) and random amplified polymorphic DNA (RAPD). Yamamoto *et al.* (2002) identified 13 SSR markers from the Japanese pear (*Pyrus pyrifolia*). Of these, 10 could be amplified in the European pear (*P. communis* L.). Similar work identified 19 SSR markers from European pear. A genetic linkage map with 894 SSRs was developed to facilitate genetic and breeding studies. For black spot disease, two SSR markers were identified as tightly linked to the genes *Ani* and *Ana* in a Japanese cultivar. For psylla disease, 1 QTL was identified by SSR markers.

Markers have been developed recently for a few genes of interest, such as a black spot resistance gene and an ACC synthase gene or the S-alleles for self-incompatibility of pears. But no genetic map of pear as for apples is yet available. About 20 genes have already been cloned mostly from Asian varieties. They are involved in fruit ripening or quality and self-incompatibility.

The construction of genetic linkage maps with molecular markers has become essential to support genetic breeding programs, in order to identify quantitative trait loci (QTLs) associated with important characteristics. QTLs are regions responsible for the expression of phenotypic characteristics with continuous distribution. In the pear, the most important and studied characteristics are diseases such as black spot, the aphid *Dysaphis pyri*, fire blight and scab, rate of fruit growth and leaf traits.

Transformation

A major evolution of pear biotechnology is the development of *Agrobacterium tumefaciens* mediated transformation. About 10 genotypes were successfully transformed. Projects are in progress in several countries to express various transgenes in pear.

Matsuda *et al.* (2009) inserted the citrus *FLOWERING LOCUS T (CiFT)* gene in *P. communis* to induce early flowering. Another work aimed at the early flowering in *P. communis* was developed by Freiman *et al.* (2012), where the RNAi technique was used to silence the PcTFL1 gene, with promising results.

A dwarfed tree of Beurré Bosc was created by ARS horticulturalists Ralph Scorza and Richard Bell through genetic engineering inserting the rolC gene from *Agrobacterium rhizogenes* but the tree proved to be too small and slow growing. In recent years, research centers have invested heavily in the development of transgenic pear clones. These transgenic plants have being studied, analyzed and developed. Several centers are trying to obtain cultivars resistant to fire blight disease, and studies with genetic transformation for this purpose have been developed. This is the case of a study with the cv.

Passe Crassane that used the overexpression of the gene *Attacin E*, and thus it was possible to reduce the symptoms of fire blight disease in 6 genetic transformed lines.

Another study with *Pyrus communis* silenced the gene coding for ACC oxidase (ACO), and with this it was possible to decrease ethylene production by 85 %, which could increase shelf life. Another work focused on the overexpression of transcription factors (TFs), aiming at knowledge of the metabolic route of anthocyanin biosynthesis.

To date, there are no transgenic pear cultivars being marketed anywhere in the world. However, the Canadian company Okanagan Specialty Fruits, responsible for the development of Arctic® apple, is applying the same technology used in apple, to pear.

STONE FRUITS

Taxonomy status:

Order: Rosales

Family: Rosaceae

Subfamily: Amygdaloideae

The seven subgenera in *Prunus* are determined basically by how the leaves are rolled up in the bud, whether the flowers are organised in cymes or in racemes and finally by morphological characteristics of the generative organs, i.e. the size and colour of flowers, fruit, stone and seed traits.

AMYGDALUS (almonds): ***P. amygdalus***, *P. bucharica, P. fenzliana, P. kuramica, P. nana, P. orientalis, P. webbii.*

PERSICA (peaches): *P. davidiana, P. ferganensis, P. kansuensis, P. mira,* ***P. persica***.

ARMENIACA (apricots): *P. ansu,* ***P. armeniaca***, *P. brigantiaca, P.* x *dasycarpa, P. holosericea, P. mandshurica, P. mume, P. sibirica.*

PRUNUS (plums and prunes): *P. cerasifera, P. divaricata,* ***P. domestica***, *P. insititia, P.italica, P. spinosa, P. syriaca,* ***P. salicina***, *P. simonii, P. ussuriensis, P. americana, P.angustifolia, P. hortulana, P. maritima, P. mexicana, P. munsoniana, P. nigra, P. rivularis,P. subcordata.*

CERASUS (sweet and sour cherries): ***P. avium, P. cerasus***, *P. fruticosa, P. japonica, P.maackii, P. mahaleb, P. pseudocerasus, P. pumila, P. serrulata, P. tomentosa.*

PADUS (bird cherries): ***P. padus***, *P. serotina.*

LAUROCERASUS (bay cherries): ***Prunus laurocerasus***.

Prunus is a large diverse genus of woody plants which belongs to the subfamily Prunoideae of the family Rosaceae. The *Prunus* genus includes more than 400 species of flowering shrubs and trees, some of which are economically very important worldwide. Many species of this genus are economically important as 46 sources of edible fruits (e.g., apricots, cherries, almonds, peaches, and plums), oil, timber and 47 ornamentals. Other several species of *Prunus* or interspecific hybrids are used as rootstocks.

Botany

Deciduous or evergreen trees or shrubs, winter buds with many imbricate scales; leaves alternate, serrate, rarely entire stipulate; flowers perfect, solitary or in fascicles or racemes; sepals 5, usually white, sometimes pink or red; stamens numerous, perigynous, pistil 1, with elongated style, 2 ovuled; fruit is a drupe, usually 1-seeded. The large genus includes peach, nectarine, plum, cherry, apricot, almond, etc. About 200 species are mostly in the temperate zone, a few in the Andes of South America.

Germplasm Resources

Most of the European fruit producer countries also maintain extensive *Prunus* collections, gathering either original national genetic resources and introduced material. Since 1980, European countries rely on the European Cooperative Programme on Plant Genetic Resources (ECPGR), a network based on the concept that a better cooperation and coordination among genebanks and governments in Europe would make conservation and use of plant genetic resources more ECPGR operates through nine Networks, dealing with groups of crops or general themes related to plant genetic resources. The Prunus Working Group in the Fruit Network, was established in 1983. Thirty nine countries take part in the WG activities through their representatives, carrying out agreed work plans on GRs.

PEACH

Origin: China

Taxonomy status:

- **Order:** Rosales
- **Family:** Rosaceae
- **Subfamily:** Amygdaloideae
- **Tribe:** Amygdaleae
- **Genus:** *Prunus*
- **Subgenus:** Amygdalus

Botanical name: *Prunus persica* Batsch.

Chromsomes number: n= 8; 2n=2x=16

Genome size: 265 Mbp

Related species are *P. davidiana* (Carr.) Franch., Mountain peach, *P. kansuensis* Rehd., Wild peach and *P. mira* Koehne, Tibetan peach.

Botany

Peach tree is deciduous or evergreen and grows up to 7 m tall and wide. However, when pruned properly, trees are usually 3–4 m tall. The leaves are alternate, lanceolate, 7–16 cm long, 2–3 cm broad and pinnately veined. Flowers are borne singly on short peduncles (almost sessile), from lateral buds on 1 year-old wood; usually 1-2 flower buds/node. Flower is light pink, carmine to purplish; 2.5 cm in diameter, perfect. Single seed inside the ovary is surrounded by hypanthium. Colour of inner surface of hypanthium is indicative of flesh color; whitish-green = white, gold yellow. Flowers exhibit cleistogamy, self pollinating prior to opening. Petals can be large and showy, or small and curved on margins. Ornamental peaches contain fully double flowers, having many petals and a carnation-like appearance. Colours range from dark pink to white. Depending upon the variety the number of stamens varies from 33 to 59. The stamens are arranged in 2-3 whorls. The pistil is with single style and clavate stigma with monocarpellay ovary having one ovule.

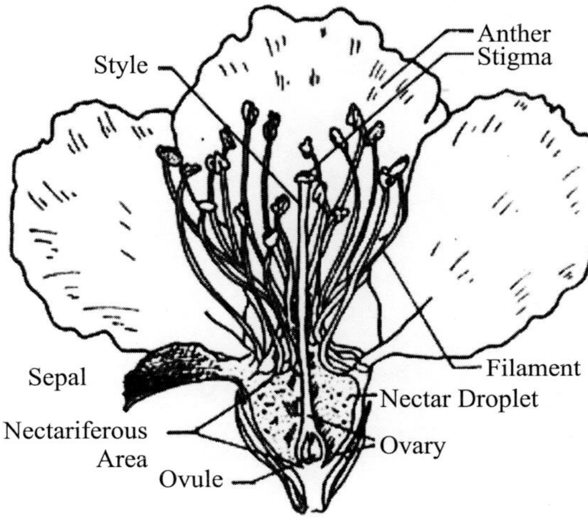

Fig. 18. L.S. of peach flower.

(Source: McGregor, 1976)

Floral Biology

Flowers take 5 hrs to complete anthesis from the dome shaped flower buds. Maximum number of flowers opens between 10 AM and 12 noon. Anther dehiscence starts between 8 and 10 AM and continues up to 12 noon. Dehiscence of all anthers is completed within a day and the remaining if any dehisce on the next day. The peak period of dehiscence is between 10 AM and 12 noon. Normally, the flowers are fully closed at 6 AM, but most of them open by 10 AM and all are open by noon. Most cultivars produce pollen at the time when the stigma is receptive that appears as yellow powdery mass at the dehisced anther lobes. Fertility varies from 0 to 90%. Stigma receptivity starts 4 days before anthesis and continues to be receptive till 3 days after anthesis. Flowers do not close at night; they may stay open and the stigma may be receptive for 3 days. Maximum fruit set is observed when the stigma was pollinated on the day of anthesis.

Scion Breeding Objectives:

1. To develop dwarf cultivars suitable for high density plantation and environmental adaptability (cold resistance and frost tolerance) for high elevation.
2. To breed cultivar for low chilling requirement for subtropical regions.

3. To breed cultivar to extend harvest season and for high yield and good quality (increased fruit size, improved skin colour, sweet crisp, increased flesh-to-pit ratio, fibreless, firm fresh, non- melting flesh, non-browning, free stone) and good storage life.
4. To extend seasonality by introducing early, late and very late ripening cultivars.
5. Resistance to diseases and insects (peach leaf curl, sharka (PPV), powdery mildew, brown rot, leaf curl, *Xanthomonas* spp. and green aphid (vector of PPV) and trunk borer.

Limitations in Breeding:

1. Long juvenile phase.
2. Flower and fruit drop.
3. Self incompatibility.
4. Gummosis.
5. Lack of genetic resources.
6. Lack of information on inheritance pattern.
7. Restricted adaptability.

Rootstock Breeding Objectives

1. Good in propagation capacity.
2. Tree size control.
3. Scion compatibility.
4. Resistance to biotic and abiotic stress.

Germplasm Resources

The National Clonal Germplasm Repository in Davis, CA is a United States Department of Agriculture (USDA), Agricultural Research Service (ARS) gene bank that houses national collections of various Mediterranean fruit and nut crops. The *Prunus* collection is the second largest in this gene bank. There are more than 90 taxa and more than 1600 accessions of *Prunus* spp. in the collection that includes almonds, apricots, cherries, peaches and plums.

The current database of peach contains 1479 accessions physically represented by 1452 individuals maintained by 17 Institutes from 12 European countries.

In India, large number of peach collections was made through introductions and indigenous collections and has been conserved *ex-situ* at national field gene bank CITH and at NBPGR. These accessions have been utilized for varietal development program intended for Indian climatic conditions. Germplasm accessions have been maintained at NBPGR Station, Shimla, NBPGR Station, Bhowali, Regional Station,

IARI, Shimla (14 *Prunus* species and 12 accessions), Dr YSPUHF, Solan, GBPUA&T, Pantnagar, SKUAST, Srinagar and CITH, Srinagar.

Genetic resource for useful traits

Root-knot nematodes (*M. incognita, M. javanica* and *M. arenaria*) **:** Nemaguard, *Prunus davidiana*, Okinawa, Nemared, Flordaguard.

Tree borer: Gold Ray, Elberta.

Green aphid: Selection S2678.

Canker: Cultivars derived from Elberta.

Mildew: *P. ferganensis, P. davidiana,* Oro A.

Leaf Curl: *P. amygdalus, P. davidiana.*

PPV: *P. davidiana.*

Low Chilling: Honey, Peento, Babocock, Earligrande, Flordaprince.

Cold: Bella, Calora, *Prunus mira.*

Frost: *P. davidiana.*

Drought: *Prunus mira, P. davidiana.*

Cold Hardiness: Bella, Calora, Redhaven, Harbrite and Hardired.

High Yield: Sunfire.

Earliness: Champion.

High Quality: Redhaven.

Flesh browning: Elberta, Biscoe, Emery, Redskin, Loring, Sunhigh Redglobe, Veteran, Early Elberta, Rochester, Hale, Alamar, Redhaven, and Sun Crest.

High Shelf Life: Parrot Deluxe.

Genetics of economically important traits:

Cold hardiness, chilling: Multigenes.

Tree habit, leaf characters, flower type: Governed by major genes.

Fruit size, shape, yellow/white flesh, melting flesh, free stone: Major genes.

Fruit maturity: Quantitatively regulated with high heritability.

Fruit Quality (Total soluble solids, sweetness, acidity and overall flavor): Quantitatively inherited.

Tree productivity and yield: Multigenes.

Pest and disease: Brown rot (*Monilinia fructicola*), Cytospora canker (*Leucotoma persoonii*), fungal gummosis (*Botryosphaeria dothidea*) and Leaf curl (*Taphrina deformans*): Dominant gene.

Root knot nematode: Two dominant resistance genes.

Green aphid: A single dominant resistance.

QTL have been identified for numerous horticulturally important traits including bloom, ripening time, fruit quality, storage life, freestone trait, internode length and pest resistance.

Pre-selection criteria

Branching index was developed for evaluation and selection of genotypes with reduced branching. Index values were highly correlated (r=2 0.7) with the total number of branches over two growing seasons and served as a good predictor of branching patterns observed in the third growing season, fruit flesh color associated with floral nectar color, fruit flesh color and foliage color and the color of senescent leaves, pollen sterility and late blooming, glandular foliage and resistance to leaf curl and early fruit maturity with solid red or variegated red pattern expressed on senescing leaves in late fall. The Solid or variegated red pigment in autumn leaves was a reliable marker for early ripening and that the intensity of red was inversely related to the length of the fruit development period instrumental in pre-selection for early ripening genotypes.

Breeding Methods

1. Introduction: A large number of varieties of peach have been introduced successfully. The cultivars Stark Early Glo, Candor, Red Gold, Glohaven, Cresthaven, Redhaven, Nimla, July Elberta, Sunhaven, Early Red June, J.H. Hale, World's Earliest, Peshwari, Quetta, Flordasun, Early Grande, Flordared, Matchless, Kanto 5, Shimizo Hakuto, Red June, Crawford Early, Paradel, Paradelux, Alexander, Florda Prince, Silver King, May Fire, Earligrande, Sun Crest, Florda Gold, Florda Queen, Elberta, Alton, Flordasun, May Fire, Snow Queen and Silver King are introductions from other parts of world. Recently introduced low chilling cultivars are Flordaprince, Tropic Beauty, Tropic Snow, Babcock, Bonita and Red Ceylon introdued in India from abroad.

The nectrine cultivars viz., Fantasia, Nectared, Snow Queen, Sunbest, Sunraycer, UF Royal, UF Queen, Sunmast, Sundollar, Suncoat, Sunred and Nectared have also been introduced from different parts of world.

Promising cultivars to be introduced: M.A.Black, Veteran, Chinese, Coral Star, Golden Monarch, Baby Gold 6, Brighton, Golden Jubilee, Kibt, Red Skin, Red Fremont, Fairhaven and Veteran.

Promising nectrine cultivars to be introduced: Red Gold, Supreme, Flavor Top, Arctic G Glo, Delicious Hardi Red, Meri Crest, Nectared-4 and Sunglo Harko.

Clonal rootstocks

GF-557 and GF-677 are drought tolerant. Siberian C, Rubira and Harrow Blood are dwarfing.

2. Selection: Most of peach cultivars were selected from chance seedlings and were named by growers in USA. Some examples included Early Crawford, Late Crawford, Reeves and Iron Mountain. Released cultivars from various places are:-

Hiley: Seedling selection of Belle.

J.H. Hale: Seedling selection from Elberta.

Elberta: Seedling selection from a Chinese Cling.

Georgia Belle: Selected from seeds of Chinese Cling.

Scarlet Prince: Originated from open pollinated seed of Blazeprince.

Scarlet Prince: Originated from open pollinated seed of Blazeprince.

In India, selection work was carried out at various places and following cultivars were released:-

Saharanpur: An early ripening, white fleshed peach hybrid was released from Saharanpur.

Sharbati: Selection released in 1950. It has chilling requirement of 30 to 40 hours.

Shan-e-Punjab: Selection done at PAU, Ludhiana from 16-33.

Pant Peach-1: Chance seedling from cultivar Sharbati.

Pratap: Selection done at PAU, Ludhiana from TA-170. Fruit matures one week earlier (3rd week of April) than that of Flordasun.

CITH, Srinagar released selections **CITH-P-1**, **CITH-P-2** and **CITH-P-3**.

The other selections made in India are Shaharanpur No. 6, Ranjit Bagh Early, Safeda (LR Brothers), Shaharanpur Hybrid 3 and China Flat.

3. Hybridization: Out crossing, inbreeding and back crossing are the principal breeding systems in peach. Breeding peach and nectarine for mild climates was initiated in USA and others countries of the world.

***Louisiana Agricultural Experiment Station*:**

Harvester: Resulted from a cross between Redskin and Southern Glow by J. C. Taylor and P. L. Hawthorne in 1998.

USDA-ARS, Fresno, California:

Springcrest: Resulted from a cross of FV89-14 3 × Springtime, made in 1958 by Victor E. Prince.

Spring Gem: Cross of P113-98[=PI00- 62 op=Redking op.] x FV9-164 [(=FV89-14 x Springtime)]; FV89-14 = [FV15-4 x Fireglow]; FV15-4 = Fireglow x Hiley. It was developed by D. W. Ramming and O. Tanner at USDA–ARS Fresno.

***ARS-USDA, Byron, GA*:**

Liberty Joy: BY99P1405 × BY02P3862 done by C. Chen and W. R. Okie.

Sureprince: F_2 of Dixiland x Durbin, developed by W.R. Okie.

Rubyprince: F_2 of Dixiland x Durbin, developed by W.R. Okie.

Autumnprince: F_2 of Dixiland x Durbin, developed by W. R. Okie, USDA-ARS Fruit and Tree Nut Research Unit in Byron, GA.

***Louisiana Agricultural Experiment Station, Calhoun, LA*:**

Harvester: Resulted from a cross between Redskin and Southern Glow made in 1960 andwas selected by J. C. Taylor in 1964.

***Michigan State University's South Haven Experiment Station*:**

Redhaven: Halehaven x Kalhaven developed by Professor Stanley Johnston.

Hermosa: J. H. Hale x Babcook.

Bonita: Prende x Sunghow.

Elberta: Peento x Lukens Honey.

Regal: Harvester x Surecrop.

July Prince: Resulted from a cross of L75-A50-20 x BY89P2787 in 1992.

Regal: Harvester x Surecrop.

***University of Arkansas, Fayetteville, Arkansas*:**

Amoore Sweet: Originated from a hand pollinated cross of Ark. 699N×Ark. 663CN. This new nectarine cultivar can be distinguished by its very firm, non-melting yellow flesh and resistance to bacterial spot disease.

Bowden: Released from a cross Ark. 699 × Ark. 663 nectarines.

***New Jersey Agricultural Experiment Station (NJAES), Rutgers University, Cream Ridge, NJ*:**

Raritan Rose: Cross between J.H Hale and Cumberland. A white peach variety released by the NJAES in 1936. It is a fairly attractive peach ripening in Redhaven season.

NJAES White Peach Selections:

K39-1 (D87-88 x J19-19): An attractive semi-freestone peach which ripens about 2 to 3 weeks before Redhaven.

D80-8 (NJ318 x H3-76): A firm, non-melting, clingstone peach ripening about 1 to 2 weeks before Redhaven.

NJ312 (NJ293 x Jing Yu): A semi-freestone peach ripening with Redhaven.

D89-9 (NJ318 x Flavorcrest): A freestone peach ripening just after Redhaven.

NJ318 (NJ256 x Biscoe): A freestone peach ripening about 3 weeks after Redhaven.

Saturn® (NJF2=NJ602903 x Pallas): A white-fleshed, semi-freestone, peen-tao type or doughnut-shaped peach which was released by the NJAES.

D33-1: (NJ291 x NJF8): A white-fleshed, clingstone, doughnut-shaped peach.

Nectrine hybrids:

***USDA–ARS, Byron, GA*:**

Juneprincess: F_2 seedling of Flamekist x P12-42 originated by W. R. Okie.

Mayglo: Developed by crossing Fayette x May Grand. Bred by Zaiger Genetics, Modesto, California, USA in 1984.

Fantasia: Resulted from a cross between Gold King and P101-24 (open-pollinated seedling of 'Red King') made in 1961. It was selected by J. H. Weinberger at the USDA-ARS, Fresno, CA, Station in 1964.

NJAES White Nectarine Selections:

K56-4 (B8-9-46 x Eastern Glo): A semi-freestone nectarine selection ripening 1 1/2 weeks before Redhaven.

K54-25 (B8-9-46 x Eastern Glo): A semi-freestone nectarine selection ripening 5 to 7 days before Redhaven.

Breeding peach and nectarine for hot climates

Low chill varieties available prior to breeding, lack size, colour, firmness and flavour (Sharpe *et al.*, 1954). These low to mid-chill varieties originated as seedling selections of germplasm from the peento group of South China or as possible hybrids of the peento group with the descendants of peaches introduced as seed by the Spainards through St. Augustine (Floyd, 1920). Peaches and nectarines are related to a native species of *Prunus* in China and the germplasm used in the breeding program had their low-chill origins from South China. Of these importations of germplasm, the most important became a seed imported from Okinawa. Crosses were made from which selections were again crossed to include desirable traits.

The Florida stone fruit breeding program began in 1952 under the direction of Dr. Ralph Sharpe. The major goal of the program was to take advantage of the climate in Florida to provide early, high quality fruit to the market (Sherman *et al.*, 1996).

In 1966, Dr. Wayne Sherman became the leader of the stone fruit breeding program, and he continued to focus on breeding for low-chill peach cultivars with non-melting flesh that perform well in Florida's subtropical climate. The University of Florida developed high quality, low-chilling, early-maturing peach and nectarine cultivars such as Flordastar, Flordaglo, Flordacrest, Tropic Beauty, and Tropic Snow. The last two were made jointly with Texas A&M University. These represent major achievements in developing a continuous ripening sequence in low chill peaches. Flordaglo and Tropic Snow are the first low chill white flesh cultivars recommended for growers.

The five cultivars developed for firm, non-browning flesh and short fuzz for fresh dessert quality are:

- **Flordastar:** Originated from a cross of Flordagold and Earli Grande peaches.
- **Flordaglo:** Originated from a cross of Sundowner* nectarine and Maravilha peach.
- **Flordacrest**: Originated from a cross of 5-13N and Tlordaking. 5-13N has complex parentage tracing back 5 generations to Okinawa and Hawaiian as sources for low-chill genes.
- **Tropic Beauty**: Originated from a cross of 3-2 and Florda prince peaches.
- **Tropic Snow**: Originated from a cross of 7-11 and Mara vilha peaches.
- **Flordacrest**: originated from a cross of 5-13N and Tlordaking. 5-13N has complex parentage tracing back 5 generations to Okinawa and Hawaiian as sources for low-chill genes.
- **Tropic Beauty**: Originated from a cross of 3-2 and Florda prince peaches.
- **Tropic Snow**: Originated from a cross of 7-11 and Mara vilha peaches
- **Flordaprince**: Resulted from a cross between Fla. 2-7 and Fla. 13-72. It is a low-chill-requiring peach that was released in 1981 by W. B. Sherman, P. M. Lyrene, R. H. Sharpe, and J.A.Mortensen.
- **July Princess:** Cross between 43P318 and Snow Princess developed by L. G. Bradford and J. M. Quisenberry at B Q Genetics, Le Grand, CA.
- **UF Gold**: Originated from the cross Fla. 84-8C (=Diamante op) x Fla 9-20C [=(Fla.K5W35 x NCA2219) op]; Fla.K5W35=Fla. 8-11 x Kaygold; NCA2219 = NC7130(=Halehaven x Redskin) x Candor.

***Hybrid developed at HRTC, Saharanpur in India*:**

Saharanpur Prabhat: Sharbati × Flordsun. An early ripening, white fleshed peach hybrid was released from Saharanpur.

Interspecific hybridization

***Prunus mira*:** A natural hybrid between peach and almond. Several rootstocks were developed for biotic and abiotic stress resistance in Europe.

Barrier-I: *Prunus persica* x *P. davidiana.*

Ishtara: Cross between *P. cerasifera* x *P. salicina* and *P. cerasifera* x *P. persica.*

Julior: *P. insititia* x *P. domestica.*

Jaspi: *P. salicina* x *P. spinosa*.

Citation: *Prunus persica* x *P. salicina.*

GF-677: *Prunus persica* x *P. amygdalus.*

Nemaguard: *Prunus persica* x *P.davidiana.* It is resistent to root knot nematode.

Myran: *Prunus belsiana* x *P.yunnan.*

Regal: Harvester x Surecrop.

Inerspecific Hybridization

The Prunus species viz., *P. americana, P. besseyi, P. brigantine*, *P. brigantine, P. cerasifera*, *P. cerasus, P. domestica, P. hortulana, P. japonica, P. munsoniana, P. nigra, P. pumila, P. salicina P. simmonii, P. spinosa, P. tenella, P. tomentosa* and *P. iger* produced sterile hybrids (Scorza and Okie, 1991).

Peacotum: It is a cross between peach, apricot and plum, developed at Zaiger's Genetics Inc.,

Nectaplum: Hybrids of nectarines and plums developed by Floyd Zaiger, Noted for sweetness and intense flavour, Spice Zee is only variety.

Peacherines: It is claimed to be cross between peach and nectarine and marketed in Australia and New Zealand. The fruit is intermediate in apppearance between peach and nectarine.

4. Mutation: Smooth skin mutant of peach is known as Nectrine. Early and late ripening forms of Fairhaven and Elberta are Compact sport of Redhaven**.** The cultivars of peach Shaji 1, Shaji 2 (China), Plovdiv 6 (Bulgaria)--yield, Hakutou Rs (Japan), Fuku-eknuba (Japan), Magnif 135 (Argentina) were developed through gamma radiation. Bud sports and early ripening mutants identified from Redhaven are Garnet Beauty and Stark Earli Glo

5. Biotechnology

Micropropagation: Peaches have been regenerated *in vitro* utilizing leaves, mature cotyledons, embryo-derived callus and immature zygotic embryos. Ramming has standardized the technique of *in ovulo* embryo culture which is used in breeding programmes for improvement of early maturing genotypes.

Molecular markers: Molecular markers have also been used to distinguish peach cultivars, measure their relatedness and determine their origins. RAPD has been found successful in identification of peach cultivars and comparison of genetic diversity. Linkage relationships with molecular markers have been described for 23 monogenic morphological traits associated with adaptation, flower color, fertility, leaf shape and color, plant habit, fruit quality and pest resistance.

Marker-assisted selection: Important characters and QTLs that are being mapped include:

- flower (bloom time, self-incompatibility, pollen-sterility and double-flowers),
- fruit (shape, pubescence, flesh colour, acidity and sweetness),
- leaf (red vs. green colour), and
- tree (pillar or weeping architectures) traits, and resistance to various pests and diseases including root-knot nematodes, powdery mildew and leaf curl.

Genetic mapping and QTL analysis: A growing number of molecular linkage maps have emerged of peach and its relatives; 5 are of pure *Prunus persica*, 2 of almond x *P. persica*, 2 of *P. persica* x *P. davidiana*, and one each of *P. persica* x nectarine, *P. persica* x *P. ferganensis* and myrobalan plum x an almond - *P. persica* hybrid. Map coverage ranges from 396 to 1300 cm, with 8 to 23 linkage groups being identified.

QTL have also been identified for 23 horticulturally important traits including bloom and ripening time, fruit quality, storage life, freestone trait, internode length and pest resistance.

Genetic Transformation

Transformations have been obtained via *Agrobacterium*-mediated transformation. The *gus* gene, which codes for β-glucuronidase, has been successfully utilized as a marker gene in genetic transformation peach. In addition, *npt*II (neomycin phosphotranspherase) gene, which confers kanamycin resistance, has been found quite useful. Peach has been transformed using microprojectile bombardment-based protocols with integration of *gus* and *npt*II genes. To manipulate tree architecture, transgenic peach seedlings expressing the *ipt* gene (which increases endogenous cytokinin levels and reduces apical dominance) were obtained.

Smigocki and Hammerschlag (1991) generated transgenic peach plants from embryogenic cultures of Redhaven using the sooty mutant strain of *A. tumefaciens, tms*:328::Tn5. This strain carries an octopine type Ti plasmid with a functional cytokinin gene and a mutated auxin gene. The transgenic plants with the cytokinin gene were dwarf, produced unusually high numbers of branches and had delayed leaf senescence. An efficient, repeatable peach transformation methodology awaits development.

ALMOND

Origin: West Asia.

Taxonomy status:

Order: Rosales

Family: Rosaceae

Subfamily: Amygdaloideae

Tribe: Amygdaleae

Subgenus: Amygdalus

Section: Euamygdalus

Botanical name: *Prunus dulcis* (Miller) D.A.Webb; syn. *Prunus amygdalus* (L.) Batsch

Chromosomes number: 2n=2x=16

Genome size: 238 Mb

The related *Prunus* species including *P. webbi, P. argentea, P. fenzliana, P. persica, P. bucharica, P. mira, P. kuramica, P. triloba* and *P. scoparia* are conferring several potentially useful traits.

Botany

The almond is a deciduous tree, growing 4–10 m in height, with a trunk of up to 30 cm in diameter. The young twigs are green at first, becoming purplish where exposed to sunlight, then grey in their second year. The leaves are 8–13 cm long, with a serrated margin and a 2.5 cm petiole. The flowers are white to pale pink, 3–5 cm diameter with five petals, produced singly or in pairs and appearing before the leaves in early spring. The number of stamens may vary from 20 to about 40 with the usual number being 30–33. Although the usual pistil number is one, some genotypes such as 'Eureka' tend to produce two, which may result in a double fruit. The fruit is 3.5–6 cm long. In botanical terms, fruit is a drupe; the outer covering or exocarp, fleshy but thin hull (mesocarp), and a distinct hardened shell. The hull undergoes limited enlargement during development, later becoming dry and leathery and dehiscing at maturity. The mature endocarp ranges from hard to soft and papery depending upon genotype. Inside the shell is the edible seed, commonly called a nut. Generally, one seed is present, but occasionally two occur. The pellicle of the kernel also changes colour from white to brown with maturity.

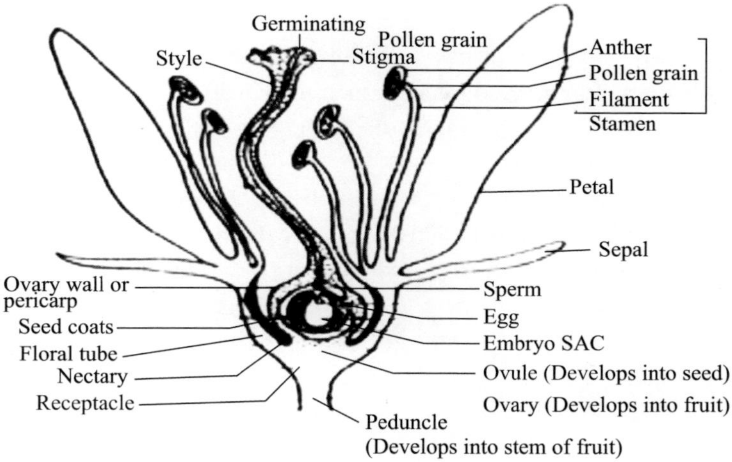

Fig. 19. L.S. of almond flower showing pollination and fertilization.

Floral Biology

Most of the flowers open between 10 AM and 12 noon. Anther dehiscence usually begins on the very day the flower opens albeit majority of anthers dehiscence a day after the anthesis. Stigma receptivity occurs immediately after anthers dehiscence. The stigmas and anthers are approximately at the same level, a condition associated with increased chances for self-pollination. The cultivated almond as well as most almond species express gametophytic self-incompatibility, which discourages self-fertilization, favors cross-pollination,and thus maintains genetic variability within seedling populations.

Genetic control of pollen–pistil: Self-incompatibility is through a single gene (S), which exists in a series of alleles including an allele for self-compatibility. Each diploid genotype carries two alleles of the series. Pollen grains, which have a haploid genome, are unable to fertilize a pistil possessing the same allele. Pollen genotypes with the same S-alleles as the pistil show self-incompatibility as well as cross incompatibility with other cultivars with the same S-allele. Cross incompatibility groups (CIGs) have been identified and incompatibility alleles have been assigned to many of them These important groups guide the selection of cultivar combinations used in orchard planting and provide important gene markers for pedigree studies. CIGs have now been identified for all major California cultivars. Self-fruitfulness requires a combination of both self compatibility and successful self pollination. Different degrees of self compatibility exist. Low pollen tube and ovule growth rates associated with certain genotypes can also decrease the probability of successful fertilization.

Breeding Objectives

Main aim is to develop cultivars having following characteristics:

1. Precocious and good vigour and branching habit.
2. Late blooming, self-fertility and self compatibilty, early ripening, regular bearing and higher productivity.
3. High kernal quality (bold size, paper shelled light in colour, shelling percentage).
4. Thin shelled, higher kernel yield and quality (bold size, free from double kernel, lighter in colour, sweet in taste, plumpy, tolerant to rancidity).
5. Regular bearing.
6. Tolerance to winter cold.
7. Resistance to pests like (Navel orange worm and white grub) and diseases like *Fusicoccum*, *Monilinia,* Stigmina blight, shot hole, *Pseudomonas amygdale,* etc.

Limitations in Breeding

1. Long juvenile phase.
2. Heterozygosity.
3. Self incompatibility.
4. Narrow Genetic diversity.
5. Information on inheritance pattern is not known.

Rootstock Breeding Objectives

1. Impart high vigour to the scion.
2. Adapted to cold and heat and resistance to diseases, insects and nematodes.
3. Compatible with most scion cultivars.

Germplasm Resources

The National Clonal Germplasm Repository in Davis, CA is a United States Department of Agriculture (USDA), Agricultural Research Service (ARS) gene bank that houses national collections of various Mediterranean fruit and nut crops. The *Prunus* collection is the second largest in this gene bank. There are more than 90 taxa and over 1600 accessins of *Prunus* spp. in the collection that includes almonds, apricots, cherries, peaches and plums.

The current database of almond contains 1003 accessions physically represented by 1003 individuals maintained by 17 Institutes from 12 countries of Europe.

In India, field gene banks have been maintained at Regional Horticultural Research and Training Insitute, Dr YSPUHF, Mashobra, Temperate Horticulturral Research Station Dr YSPUHF, Kotkhai, Shimla (31 accession), NBPGR Research Station, Phagali, IARI, RRS, Amartara, Shimla, SKUAST, Srinagar, Hill Fruit Research Centre, Chaubattia and CITH, Srinagar.

A list of sources of useful traits are given in Table 17.

Table 17. Genetic resources /sources of useful traits.

Trait	Cultivars/species
Low chilling	Achak, Desmayo Largueta, Ramillete, Marcona, Marta, Antoneta, Ferragnes, California Paper shell, Hybrid 15, Pathick's Wonder, JKS-55, H-98.
Late blooming	Cristomorto, Ferralise, Ferragnes, Fillippo, Ceo, Genco, Tuono, Texas, Tardy Large, Nonpareil.
Early maturing	Tuono, Tardy, Nonpareil, Pizzula d Avola, Fascionello, Fra Guillo Grande, Merced, Ne Plus Ultra.
Self-compatible	Fillippo Ceo, Genco, Tuono, Le Grand, AH-258, Moncayo, Ayles, Guara, *Prunus webbii*, Marcona, *P. bucharica.*
Cross compatible	Nonpareil, Mission, Ne Plus Ultra, Peerless.
High shelling (%)	Nonpareil, Ne Plus Ultra, Tardy Nonpareil, Vesta.
Completely sealed suture and cork retention	Merced, Kochba.
Very fertile cultivars	Marcona, Independance.
Less fertile cultivars	Bartre.
Compact cultivars	Tuono.
Unbranched and upright bearing cultivars	Bartre.
Dense foliage cultivars	Harriot, Jordanolo.
Hard shell cultivars	Cristomorto, Desmayo, Larqueta, Ferraduel, Marcona, Bartre, Tuono, Misssion, Pedre.
Large kernel cultivars	Bartre, Mollar de Tarragona, Marcona, Ferralise, Ferragnès, Verdal, Tardy Large.
Soft shell cultivars	Nonpareil, Texas, Ne Plus Ultra, Mollar de Tarragona, Carmel.
Small kernel cultivars	Texas, Ardechoice.
Sweet kernel cultivars	Ferragnes, Al, Jordan, Green, Bartre.
Double kernel cultivars	Texas, Ne Plus Ultra, Nonpareil, Sonora, Price and Mission.
Nematode resistance	Alnem 1, Alneum 88, Alnem 201, *Prunus davidiana.*

Inheritance Pattern

Chilling requirements: Multiple genes.

Blossom time: Major dominant gene.

Self fruitful: Dominant gene.

Tree characteristics: Additive gene action.

Time of ripening: Polygenic.

Hard Shell: Two major alleles.

Sweet kernel: Governed by qualitative dominant traits,

Bitterness: Single recessive genes.

Disease resistance: Qualitative genes.

Table 18. Degree of heritability of some traits in almond.

Trait	Established heritability
Bitterness	99
Shell texture	55
Suture opening of the shell	79
Double kernel	51
Length of kernel	77
Kernel color	42
Thickness of kernel	71
Shell retention	24
Kernel weight	64
Gumminess	15
Width of kernel	62
Retention of outer shell	34
Susceptibity to worm damage	30

Breeding Methods

1. Introduction: The cultivar California Paper Shell, Chellaston, Hardshell-5, Primorskhij, Nonpareil, Ne Plus Ultra, IXL, La Prima, Peerless, Drake, Pranyaj, Merced, Nikitskiy, Harriott, Jordanolo, Achak, Desmayo Largueta, Ramillete, Marcona, Marta, Antoneta, Ferragnes, Hybrid 15, Pathick's Wonder, JKS-55 and H-95 have been introdued in India from USA and other countres.

2. Selection:

Kester: It was originally identified as selection 2-19E at Univerity of California, Oakland, CA.

Nonpareil: Discovered as a chance seedling of unknown parentage.

Aldrich: Chance seedling selection by Clarence E. Aldrich and Maryanne Aldrich, at California.

Titan: Selection from Tardy Nonpareil open pollinated seedling; originated in Fresno, Calif., by the US Dept. of Agriculture in cooperation with Fresno State College.

Rosetta: Selection from Nonpareil open pollinated seedling.

***SKUAST, Srinagar*:**

Shalimar: It is a seedling selection and released at SKUAST, Srinagar. It has high shelling perentage and tolerance to leaf blight and dieback.

Makhdom: It is a seedling selection and released at SKUAST, Srinagar.

Parbat: It is a seedling selection and released at SKUAST, Srinagar having tolerance to leaf blight and gummosis.

Waris: It is seedling selection and released at SKUAST, Srinagar.

3. Hybridization: Improvement work was conducted at several **University/ Stations** in different parts of the world.

Institut de Recerca i Tecnologia Agroalimentaires (IRTA), Mas de Bover, Ctra. Reus-El Morell, km 3, 8. E-43120 Constant, Reus, Tarragona, Spain:

Tarraco: Bred from the cross FLTU18 x Anxaneta made in 1991. It blooms very late, it is self-incompatible, highly productive (heavy and consistent bearer).

Vayro: Bred from a 4-665 Lauranne cross made in 1991.

Constani: It is an open-pollinated seedling from the selection FGFD2, obtained in 1993.

Butte: Cross between Mission and Nonpareil.

Marinada: It is a seedling selection from a Lauranne x Glorieta.

Matan: Originating from a cross between Lauranne and Um.

ElFahem: It was bred in the Newe Yaar Research Center, located in the Yizreel Valley, Israel.

***University of California, Davis, CA*:** The cultivars Ne Plus Ultra, Merced, Kester Avalon, Carmel, Butte, Father, Price, Sonora, Sweetheart, Winter, Independence and Folsom were selected and developed.

Kester: Tardy Nonpareil × Arbuckle, developed by D. Kester, M. Viveros and T. Gradziel.

Ruby: Mission x Nonpareil.

Merced: Mission x Nonpareil.

Sonora: Nonpareil x Eureka (BC).

Thompson: Mission x Nonpareil.

Fritz: Mission x Drake, originated in USA.

Kapareil: Nonpareil x Eureka (BC), originated in USA.

Le Grand: Almond x Peach, originated in USA.

Carmel: Mission x Nonpareil, originated in USA.

Monterey: Mission x Nonpareil, originated in USA.

Padre: Mission x Swanson.

Ferragnes: Cristomorto x Ali, originated in France.

Golden Prince: Almond x Peach, originated in USA.

All in one: Almond x peach.

U.S. Department of Agriculture, Agricultural Research Service, Crop Diseases, Pests & Genetics Research Unit, Parlier, CA:

Yorizane: Cross between Tuono with selection 74-48 . It is a new self-pollinating almond variety with superb consumer traits such as size, colour, and flavor.

Adelaide Research & Innovation PTY LTD. (Adelaide, AU), Horticulture Innovation Australia Limited (Sydney, AU), Adelaide, Australia:

Capella, Carina and Maxima: Originated from a cross of Nonpareil and Lauranne done by Michelle Wirthensohn and Andrew Granger.

***University of Adelaide, Adelaide, South Australia, Australia*:**

Mira: derived from a cross between Nonpareil and Lauranne.

***Queensland Government Department of Agriculture and Fisheries, Australia*:**

Rhea: was derived from a cross between LeGrand and Keane.

***PAU, Ludhiana*:**

Hybrid-15: Thin Shelled x SIoh are developed by Department of Horticulture, PAU, Ludhiana.

Interspecific Hybridization

Interspecific crosses between related species (mainly *P. persica* x almond but also *P. webbii* almond) have been used for almond rootstock breeding in France, USA, Spain, and Yugoslavia. Sources of new genes may be the more developmentally distinct peach group including *P. persica*, *P. mira*, and *P. davidiana*.

Independence® (Alm-21 cv.): The new and distinct interspecific almond tree (*Prunus dulcis x Prunus persica*) x *Prunus dulcis*. Developed by Gary Neil Zaiger, 1908 Elm Ave., Modesto, CA.

Sweetheart: Originated from a cross at University of California, Davis (UCD). Selections SB3, 54-39E, and 25-26 were initially evaluated as selection SB13, 36-52 by University of California Davis, CA.

Rootstock Breeding

GF 677: It is a natural hybrid of peach and almond and was selected at the Institut national de la recherche agronomique (INRA), France.

GF 557: It was produced as a nematode-resistant hybrid of Shalil x almond, which is also resistant to chlorosis but sensitive to wet soils.

Three clones (**Felinem, Garnem and Monegro)** were selected from the progeny obtained in the cross between the Spanish almond Garfi as the female parent and North American peach Nemared as the pollen donor. These new rootstocks are characterized by red leaves, good vigor and easy clonal propagation, resistance to root-knot nematodes, adaptation to calcareous soils and graft compatibility with the whole range of peach and almond cultivars as well as some plum and apricot cultivars.

3. Mutation: Tardy Nonpareil has been patented as bud sport of Nonpareil. Supernova was selected from mutation in Italy. Carmel has been reported to be a bud sport of Nonpareil.

4. Biotechnology: An efficient *in vitro* protocol was standardized for micro propagation using dormant axillary buds as explants.

Molecular markers: Molecular markers have become an essential tool in breeding. Different types of molecular markers (isozymes, PCR, SSRs and AFLPs) have been employed for genetic characterization of plant material and the linkage of traits with agronomic interest, and the establishment of genetic relationships between cultivars and species.

SSR analysis confirmed previous isozymes studies which identified the almond as the most polymorphic species within the major *Prunus* tree crop species making it an ideal candidate for map construction. Marker assisted selection PCR-based markers of almond self-incompatibility S-alleles have been successfully used to identify different self-incompatibility genotypes. Similar results were obtained by Bosˇkovic *et al.* (2003) who identified major almond cultivar stylar S-RNase by electrophoresis in vertical polyacrylamide gels. PCR-based markers of almond self-incompatibility S-alleles have been employed to facilitate the integration of self-compatible S-alleles from related species. Screening efficiency and flexibility has been greatly increased with the development of successful multiplex PCR techniques. Using advanced cloning strategies, Ushijima *et al.* (2003) have described the structural and transcriptional analysis of a pollen-expressed F-box gene with haplotype-specific polymorphism strongly associated with self incompatibility. Molecular markers are currently being employed to elucidate the genetic basis of plant processes controlled by multiple genes. Campalans *et al.* (2001) have described a differential expression technique based on cDNA -AFLP (amplified restriction fragment polymorphism derived technique for RNA fingerprinting) to characterize genes involved in drought tolerance in almond. Results identified increased drought tolerance in specific genes associated with leaf function.

Genetic Transformation

Almond leaves were transformed with the marker genes gusA (β-glucuronidase) and nptII (neomycin phosphotransferase II) via Agrobacterium-mediated transformation. Following transformation, shoots were induced from primary explants on medium without kanamycin and exposed to selection 20 days after cocultivation.

PLUM

Origin: Caucasian region and China (Yangtze River basin).

Taxonomy status:

Order: Rosales

Family: Rosaceae

Subfamily: Amygdaloideae

Genus: *Prunus*

Section: Prunocerasus

The North American species bearing small fruits such as *P. americana*, *P. angustifolia*, *P. hortulana*, *P. munsoniana* and *P. maritima*.

Section: Euprunus

Botanical name: *Prunus domestica* L. European plum 2n=6x=48 (allopolyploid). *Prunus salicina* Lindl. Asian or Japanese plum 2n=2x=16.

Chromsomes number: 2n=16, 32, 48

Genome size: 308. 06 Mb (*Prunus salicina*)

: 0.61 pg (*P. angustifoila*)

Genome formula:

:CCSSSS (*Prunus domestica*)

:CC (*Prunus cerasifera*)

:SSSS (*Prunus spinosa*)

Prunus domestica is a hybrid of *P. cerasifera* (cherry plum, diploid) and *P. spinosa* (sloe, autopolyploid tetraploid).

Botany

Plums are small to medium sized, deciduous trees or shrubs. Leaves are alternate, serrate, glabrous to pubescent. Flowers are perfect, solitary or in umbel-like clusters, appearing before or rarely with the leaves with five sepals, petals 5, white in colour, attached to rim of the calyx, numerous stamens, single pistil, single carpel. Fruits are drupe, sulcate, glabrous, usually with bloom on epidermis, black, yellow, red and green in colour. Stone is flatter pit, compresed, usually longer than broad and smooth or nearly so.

The plums may be grouped into sections by inflorescence and place of origin. Section 1 has 1-2 flowers, rarely 3, while section 2 has flowers in clusters of 2-5. The diversity of plums occurs in nature and can be classified into several groups.

European plum (*P. domestica*)

It is the most important plum in Europe. It is also grown in other continents. The species can be divided in several groups considering the fruit characters. The fruits are round to oval in different sizes and colours. The flesh is juicy, soft and mostly clingstone. Usually, the fruits ripen earlier than those of prunes, although there are some exceptions.

Prunes: The fruits are oval to elongated, mostly smaller than ondaplums and generally high in sugar content. They can, therefore, be well used for drying. The colour is mostly dark blue to purple, but there are some cultivars with red, pink, yellow or bright colour. There are differences in the pubescence of shoots and leaves. The shoots of prunes are never pubescent. All prunes are plums but all plums are not prunes. Typical prunes are Prune dAgen and German Prune, which is the most widespread prune in Europe, called Hauszwetsche in Germany, Pozegacˇa in Yugoslavia, Beszterci in Hungary, Casalinga or Dro-Zwetsche in Italy, Quetsche Commune in France, Vinete romanesti in Romania and Kustandilska in Bulgaria, and in some countries known as Commun Plum.

Mirabelles: have often been classified to *P. insititia*. Although controversial, now they are regarded as a subspecies of *P. domestica*. Mirabelles have small round fruits (8–12 g) and 22–28 mm in diameter, mostly yellow coloured, often with red spots.

There are also green and dark purple varieties. The fruit is freestone, juicy, very sweet (18–20% Brix), full of aroma and of high quality.

Gage Plums, also called reineclaudes, have small to mediun round fruits in different colours from green (Green Gage) to yellow (Oullins) and purple (Graf Althans). The flesh is juicy, sweet, with aroma, very tasty and of high quality. The fruits are used mostly for fresh consumption.

European wild plums

Cherry plum or myrobalan (*P. cerasifera*): It shows the greatest diversity among all *Prunus* species both in morphology and in ecological adaptation. The highest incidences the centres in nature and of diversity were found near the coast of the Caspian Sea. The fruits found there vary in diameter from 17 to 37 mm and the fruit shape from elongated to round. The species occurs in this region up to 1800 m above sea level, adapted to widely different types of climate and growth conditions. *P. cerasifera* is diploid (2n = 2 = 16) and of interest because of its good productivity, resistance to diseases, drought and heat, its early maturity and tolerance to unfavourable conditions. It has small round fruits, with a diameter of 15–20 mm. Fruits of selected and cultivated types are bigger; some may reach nearly 40 mm. The fruit weight is between 9 and 15 g. The fruit colour ranges from yellow to red and dark violet. Most of them are yellow and then the fruits are frequently confused with mirabelles, but the fruit quality of myrobalan is much lower. The flesh is soft, juicy, sweet to sub-acid and of poor quality. The skin is tenacious and acid, and the stone is cling to semi-free. The fruits are not very suitable for fresh consumption, but favoured because of their early ripening.

For a long time, ***Prunus insititia*** (*P. domestica* spp. *insititia*) has been regarded as a species; but now it is classified as a subspecies of *P. domestica*. Members of the subspecies are used as rootstocks. They are of interest for hybridization in rootstock breeding. The subspecies was also used in Russian breeding programmes to get varieties with better winter hardiness. It can be divided into three groups (St.Julien A, Damson and Bullace). A differentiation between these groups is not always easy as natural hybridization occurs.

In 1754, Miller described a plum called St. Julien. Its rootstock characteristics were discussed throughout the old literature. Whether the name St. Julien was ever applied to a special cultivar is unknown. After studying the St. Julien plum over 40 years, Ku¨ppers (1976) came to the conclusion that this plum type range in appearance from small fruited as in St. Julien A, with green yellow colour, to wild bullaces like the rootstock GF 655/2 with black fruits. For a long time, seedlings have been used as rootstocks. Different types of St. Julien plum were selected in East Malling and propagated vegetatively. The most important are St. Julien A and Pixy. From INRA Bordeaux, St. Julien GF 655/2 was selected, a rootstock commonly used in middle Europe, well adapted, but developing a lot of root and stem suckers. Damson is a group having oval to roundish fruits, with a dark blue to green skin and a bitter, spicy and sweet taste; they are aromatic and astringent. In France, it is called Damas and is used as rootstock, for example, Damas 1869, but now it is no longer recommended for use due to its tendency to sucker. In Germany, Switzerland and Austria, it is known as Krieche with black fruits and Zibarte with green-yellow fruits. Zibarte is cultivated for brandy production. It gives a high priced brandy having special taste and aroma. Bullace is yet another type of small, round fruits, which are usually dark blue and sweet. They are used for brandy production. It is called Haferpflaume in Germany.

P. spinosa (black thorn or sloe): The species is widespread in Europe, North Africa and Northern Turkey, but also found in North America imported from Europe. The sloe is tetraploid (2n = 4 = 32). One of the centres of origin is the Caucasian region where types of *P. spinosa* with chromosome set of 2n = 16, 24, 32, 40, 48, 64 and 96 were found. It is characterised by a broad adaptability and good viability. It survives on dry soils and often grows on eroded soils and stony slopes. It is a strongly branching shrub, but can also give trees reaching a height of 4–6 m. The young plants are very thorny, whereas old plants are usually free from thorns. Flowers are arranged as single flowers, and the blooming period is early in the spring before the leaves emerge. The fruits are small with a mass of 2–5 g, mostly clingstone with tart flesh and very astringent, sweet only after frost impact. There are some reports about sloes with sweet fruits, but these are bullaces and not sloes. Blackthorn is very drought resistant and often used for stabilising stony slopes. The sloe is used as dwarfing rootstock, but there are problems due to suckering. Because of its dwarfism and resistance traits, it is used as crossing partner in rootstock breeding programmes.

Japanese plum (*Prunus salicina* Lindl) and hybrids: It includes all the fresh market plums developed by hybridization of various diploid species with *P. salicina*. The species is primarily cultivated in warmer regions because of its lower chilling requirement and its sensitivity to winter and spring frost. Only few of these Japanese plums are pure; most of them are hybrids, primarily with different native American plum species like *P. americana, P. angustifolia* and others, starting with the breeding work of Luther Burbank. All cultivars of Japanese plums are diploid (2n = 2 = 16); most of them are self-sterile. Self-fertility was found to be an exceptional phenomenon among them. Some new varieties have now been released which are self-fertile or partially self-fertile. Self fertility is supposed to be determined by the so-called Se-allele, which seems to be identical with the S5-allele. The fruits of the present varieties are mostly big and round or heart shaped, but never oval or elongated. They are of attractive appearance and well suited for transport. For these reasons, there is an increasing demand in the market. The fruits are juicy, in most cases clingstone and have usually a much lower sugar and acid content than highly developed cultivars of the European plum, which are richer in aromatic compounds as well.

Apricot plum (*Prunus simonii*): It is cultivated in China, Japan and Central Asia since a long time and was described for the first time in 1872. The fruits are oblate and small with a diameter of 25–30 mm, dark to purple red. The flesh is firm, aromatic and clingstone. Its botanical position is unclear, and often it is considered to be an apricot–plum hybrid. Because of its firm flesh and strong flavour, it was used in Californian breeding programmes for the so-called Californian Japanese plum cultivars, for example, Shiro and Wickson.

Wild plums: Wild plums have been of special interest in variety and rootstock breeding as donors of resistance or ecological adaptability. The most interesting species are American wild plum species. There are many wild plum species present in North America. More than 20 species are known. Several native plum species have been characterised concerning their resistance traits.

P. americana: It is the most common wild plum with a small and usually thorny, spreading tree. Its native area is from Massachusetts to the Gulf of Mexico and to New Mexico. Mostly, fruits are red with a cling or free stone and an astringent skin. The flowering time is late. It is a donor of cold hardiness, but suckering is a big problem when used as rootstock. *P. americana* is the most common native species used in North American Japanese plum breeding programmes. Hybridizations with *P. domestica* are rarely successful.

P. angustifolia: It is the Chicksaw plum, native from Delaware to Florida and Texas and Southern Ohio. The small bushy tree is usually suckering. The fruit is small and cherry-like, bright to red coloured, sometimes also yellow. The species is abundant on sandy soil and has a low chilling requirement. It was successfully crossed with Japanese plum resulting in cultivars like Segunda, Robusta and Byrongold. *P. nigra* is the Canadian wild plum having red-orange to yellowish fruits with an astringent skin. Because of its cold hardiness, it was used in breeding for northern plum cultivars.

P. subcordata: It is the Sierra plum (also called Western or Pacific plum) and resembles more to the European and Asian species than other ones. The clingstone fruits are round to oblong with dark red to purplish skin and with sub-acid flesh. Some cultivars have been developed.

P. hortulana: It is the Hortulan plum and is a relatively tall upright tree (5–10 m) with small (25 mm in diameter) and red- to yellow-coloured fruits. The flesh is acid and clingstone with best fruit quality of all native North American *Prunus* species. The species is of some interest in rootstock breeding because it is dwarfing, no suckering and compatibility with plum and peach.

P. munsoniana: It is known as Wild Goose plum and is native in the regions of Kentucky, Texas, Tennessee and Kansas. Trees are 6–8 m high. The fruits are oval and bright, red and yellow coloured, with juicy flesh. Several varieties were cultivated but are not available now. The Marianna plum is a hybrid of *P. munsoniana* x *P. cerasifera.* It is used as a rootstock.

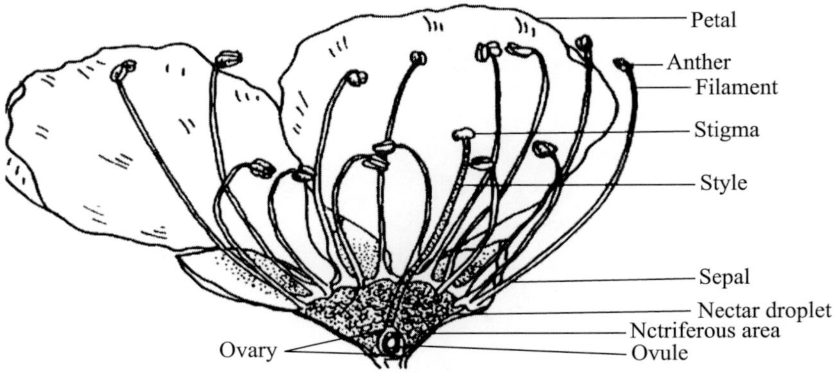

Fig. 20. Plum flower showing different parts.

(Source: McGregor, 1976)

Floral biology

The maximum anthesis takes place between 12 noon and 2 PM and the increase of temperature hastens the process. Anther dehiscence takes more than one day to complete. The anther dehiscence takes place on the day of flower opening and continuous up to next day. Dehiscence is more between 11 AM and 10 PM depending upon the temperature on the day. The stigma becomes receptive when it is quite plump and shiny. Brownish and dull colour indicates the non-functional stigma. The stigma is fully receptive from one day before anthesis and remains so up to four days. Natural pollination is 3-46.6%. Honey bees and insects are major pollinators. In Japanese plums pollinizers are necessary for commercial production for about half the cultivars. About half of the *domestica* type cultivars require pollinizers. Self-fruitful types exhibit an average of 30% fruit setting which is adequate for a commercial fruit production. The cultivars that are self-unfruitful set only 1—2% must inter-plant with pollinizing cultivars.

Breeding Objectives

(i) Development of varieties adapted to different climates with high productivity.

(ii) To breed cultivar with late blooming, self-fertility, short growing period, different ripening times.

(iii) To breed cultivar with subtropical climates with low chilling requirment and frost and winter resitance.

(iv) To breed cultivars for early and late maturing and completely sweet cultivars (unlike Santa Rosa).

(v) To breed cultivar for large sized fruits with excellent quality (good flavour), red to deep colour skin and firm flesh, free-stone (canning).

(vi) To breed cultivar for processing, drying types (prunes).

(vii) Varieties free from endocarp splitting and gummosis in the fruit flesh and resistant to insect and diseases (plum pox virus).

Limitations in Breeding

1. Long juvenile phase.
2. Sterility.
3. Self incompatibility.
4. Intersterility or cross incompatibility.
5. Lack of genetic resources.
6. Lack of information on inheritance pattern.
7. Restricted Adaptability.

Rootstock Breeding Objectives

1. Reduce vigour of the scion.
2. Adapted to cold and heat and resistance to diseases, insects and nematodes.
3. Compatible with most scion cultivars.

Germplasm Resources

The National Clonal Germplasm Repository in Davis, CA is a United States Department of Agriculture (USDA), Agricultural Research Service (ARS) genebank that houses national collections of various Mediterranean fruit and nut crops. The *Prunus* collection is the second largest in this genebank and is exceeded only by *Vitis* (grapes). There are more than 90 taxa and in excess of 1600 accessions of *Prunus* spp. in the collection that includes almonds, apricots, cherries, peaches and plums. The current database of plum contains 3530 maintained by 33 institutes from 16 countries of Europe.

In India, field gene banks have been maintained at NBPGR Station, Shimla, Dr YSPUHF, Solan, GBPUA&T, Pantnagar, SKUAST, Srinagar, CITH, Srinagar and NBPGR Station, Shimla (128 accessions). List of useful traits in plum breeding are given Table 19.

Table 19. Genetic sources (cultivars/species) of useful traits.

Dwarf or low vigour	Stanley, Marry Mather, d'Ente 698, Sugar, Ealta Dorata di Coe, Early Golden, Belle de September, Sticolase de Voinesti, Prun negru, Perje de toamna, Galbene de Aninosani, Grase de Becs and Grase de Pesteana.
Late blooming	*Prunus domestica.*
Self-fertility	Stanley, Anna Späth, Standard, Ontario, Ialomita, Diana, Romanta.
Early maturing	Beauty Burmosa, Early Golden, Methley, Premier, Red Beaut, Wison, Queensland Bellerose, Green Gage, Red Beauty.
Late ripening	*Prunus hortulana.*
Large fruit	Calita, Friar, Ozark Premier, Queen Rosa.
Fruit firmness	*Prunus simonii.*
Green colour	Green Gage.
Yellow colour	Golden Drop, Golden Transparent, Kanto.
Red flesh	Frontier, Kararl, Mariposa, Nishida, Satsuma.
High sugar content	*Prunus domestica.*
Drying quality (European plums)	D'Agen, Fellenberg, White Perdrigon.
Canning quality (European plums)	Early Laxton, Blaisdon Red, Eversham Wonder, Italian.

High yield	Italian Prune, French Prune, President, Giant Prune, Setter.
Stem canker	Burbank, Satsuma, Kelsey.
Leaf and fruit spots	Burbank, Mariposa, Santa Rosa.
Low chilling	Golden Lalri, Kala Amritsari, Titron.
Cold Hardiness	*Prunus nigra*, *P. ussuriensis*, *P. americana.*
Wet soil and drought	Myroblan and Marianna seedlings.
Adaptation under warmer climates	Au Amber, Au Rosa, Au Cherry, Gulf Ruby, Gulfgold, Florida 1-2.
Bacterial leaf spot	The cultivars derived from American species viz., Ember, Goff, Hazel, Kahinta, Monitor, Red Coat, Underwood, Wolf, Deep Purple, Hiawatha, Manor, Mansan, Sapa and Damson, Green Gage, Simon, Brice, Campass, Munson, *Prunus cerasifera.*
Sharka virus	Joja, Haroma.

Inheritance Pattern

Self fertility: It is supposed to be determined by the so-called Se-allele, which seems to be identical with the S5-allele.

Bloom time: Highly heritable and quantitively inheritable.

Full bloom: Hightly heritable.

Incompatiblity: Single genes.

Fruit size: Quantitatively inherited.

Fruit shape: Multiple genes.

Yellow skin colour: Recessive to black, purple or red colour.

Free stone character: Recessive to clingstone.

Sharka resistance: Multifactorial inheritance of the quantitative

Pre-selection criteria

The tree height and stem circumference correlate with precocity and fruit weight (Hartmann and Engelhorn, 1992). Pre-selection for resistance to PPV is mediated by hypersensitivity possibly using the double grafting method.

Breeding Methods

1. Introduction: The Japanese plum cultivars Alubukhara, Golden Zardalu, Lalri, Santa Rosa, Satsuma, Grand Duke, Mariposa, Kelsey, Methley, Burbank, Titron, Beauty, Red Heart, Au-cherry, Tarkol, Monarch, Au-Rosa, Frontier, Kanto-05, Red Beauty, Sweet Early, Duarret and Grand Duke, Howe, Gulfruby, Gulfbeauty, Gulfblaze, Gulfrose and Elephant Heart have been introduced from USA and other parts of world. Luther Burbank introduced cultivars of Japanese plum from China

The European plum cultivars namely President, Victoria, Prune d'Agen and German Prune Green Gage, Oullin and Graf Althans are introduced in India.

Clonal rootstocks:

***Prunus domestica*:** Black Damas, Brompton.

Myrobalan (*Prunus cersifera*): Myrobalan B, Myrobalan 29-C, Myrobalan GF 31, H29C, Myrobalan B, Myrobalan 29-C, Myrobalan GF 31.

Marianna (*P. cerasifera* and *P. munsoniana.***):** Marianna GF 8/1, and Marianna 2621.

Pixy and St. Julian A, St. Julian, GF 655-2: Selections of *Prunus insititia.*

2. Selection: In European plum, several selections viz., President (chance seedling), Plovdivska (Stanley – open pollination), Prune d Agen (selected at Bordeaux, France), Reine Claude Verte (selection in France), Kelsey (selected by J. Kelsy), Shiro (selected by Luther Burbank), Jubileaum (selected in Sweden), Avalon (selected from Reeves at LARS, Bristol, UK), Excalibur (selected from Cox's Emperor at LARS, Bristol, UK) and low chilling selections made in Japanese plum are Kala Amritsari, Allu-bukhara Amritsari and Satulej Purple are selections made in PAU, Ludhiana and selections viz., Pant Plum 1, Jamuni Meeruti, Aloo Bokhara Peshawari, Titron, Alucha Black were also made in different parts of India.

Bluebyrd: It was an open-pollinated seedling of NY H4, a selection of unknown parentage from the New York Agricultural Experiment Station, Geneva, N.Y. by H.W. Fogle at the U.S. Dept. of Agriculture–Agricultural Research Service (USDA–ARS), Agricultural Research Center, Beltsville, Md.

Kelsey: Selected by John Kelsey, Berkeley, California.

Red Beaut: Open pollinated, early-ripening, dark-skinned plumcot. Origin in Modesto, Calif., by Chris, Gary, and Grant Zaiger and Leith Gardner.

Flavor Gold: Open-pollinated seedling of 336LC131 (=36ED128 (=Friar plum x Autumn Giant plum) x 7HC160{=(Friar op x 4G1180 plumcot [=Red Beaut op]) x 4G1180}).Origin in Modesto, Calif., by Chris, Gary, and Grant Zaiger and Leith Gardner.

Flavorglo: Open-pollinated seedling of 35EA376=(Red Beaut OP. An early ripening yellow plumcot. Origin in Modesto, Calif., by Chris, Gary, and Grant Zaiger and Leith Gardner.

Clonal selection: The variety Bu¨ hler Fru¨ hzwetsche, clones were found showing differences of three weeks in ripening time. There are also some clonal selections of the cultivar Italian Prune. Most important ones are with a higher specific yield. Some clones were selected within the variety Mirabelle de Nancy as well. The most interesting one is clone No. 1725 with large, pink spotted fruits. No. P 2778 is a clone of Mirabelle de Mete with high sugar content and a typical aroma.

Clonal rootstocks:

***Prunus domestica*:** Black Damas, Brompton.

Myrobalan (*Prunus cersifera*): Myrobalan B, Myrobalan 29-C, Myrobalan GF 31, H29C, Myrobalan B, Myrobalan 29-C, Myrobalan GF 31.

Marianna (*P. cerasifera* and *P. munsoniana*)**:** Marianna GF 8/1, and Marianna 2621.

Pixy and St. Julian A, St. Julian, GF 655-2: Selections of *Prunus insititia.*

3. Hybridization: European plum varieties released in USA/Canada are:

Victory: Vision x Valor originated at Ontario, Canada.

Plum varieties released in Europe:

Felsina Hohenheim: Italian Prune x Ersinger.

Elena Hohenheim: Italian Prune x Stanley

Czar: Prince Engelbert x Rivers Early Pacific, developed by Thomas Rivers in England.

Sonora: La Crescent × Jefferson.

Pozna Plava: Cacanska Najbolja × C. Najbolja.

Zlatka: Large Sugar Prune × Zolta Butilkovidna.

Splendor: French' ('Agen') pollinated by 'Pond' ('Hungarian'); it originated in 1886.

Giant' prune: It is a cross of French (Agen) x Pond (Hungarian).

Table 20. Important varieties in Germany with their origin.

Variety	Origin
German Prune	Unknown
Bu¨ hler	Germany, 1854
Cacˇanska lepotica	Yugoslavia (Cˇ acˇak), 1961
Auerbacher	Germany, 1875
Ortenauer	Germany, 17th century
Hanita Germany	German (Hohenheim), 1980
Cacˇanska rodna	Yugoslavia (Cˇ acˇak), 1961
Ersinger	Germany, 1896
Top Germany	Geisenheim, 1985
President	England, 1894
Elena Germany	Hohenheim, 1980
Presenta Germany	Hohenheim, 1981
Herman Sweden	Balsgaard, 1952
Katinka Germany	Hohenheim, 1982

Japanese plum hybrids

Luther Burbank (1875) did commendable work in development of cultivars in Japanese plum. Most of the cultivars were developed by crossing *Prunus salicina* with *P. simonii*, *Prunus americana*, *P. besseyi, P. subcordata, P. munsoniana and P. hortulana* e.g., Beauty, Burbank Durate, Formosa, Santa Rosa and Wickson.

Table 21. Origin/breeder of plum cultivars.

Cultivar	Origin/breeder
Friar	Fresno
Angeleno	Garabadien
Blackamber	Fresno
Santa Rosa	Burbank
Simka	Kazarian
Casselman	Casselman
Royal Diamond	Kitahara
Black Beaut	Anderson
Fortune	USDA Fresno
Red Beaut	Anderson
Kelsey	Japan
Howard Sun	Chamberlin
Grand Rosa	Anderson
Rosemary	Anderson
Roysum	Sumruld
Catalina	Krause
Black Diamond	Superior
Laroda	UCD
July Santa	Rosa Friesen
Black Flame	Superior
Wickson	Burbank
Autumn Beaut	Zaiger Prima
Prima Rosa	Gerawan
Freedom	USDA–Fresno
Autumn Giant	Zaiger
Queen Rosa	USDA–Fresno
Eldorado	Burbank
Show Time	Wuhl
First Beaut	Neufeld

In USA, several private companies, viz., Zaiger Genetics (Modesto), Bradford Farms (Le Grand) and SunWorld International (Bakersfield) started breeding programs for good quality plum.

The cultivars released by Zaiger Genetics viz.: Hiromi Red, Joanna Red, Golden Globe® (syn. Zairobe), Earliqueen and CrimsonGlo.

SunWorld International also created new Japanese plums. Important cultivars are Black Diamond®, Red Giant™, Applum™, Black Giant®, Red Diamond®, PL1347RB and PL454YY selections, respectively. In California, some private breeders and growers have selected new cultivars such as the late-ripening Angeleno® (syn.: Suplumsix), Green Sun, October Sun, Tracy Sun and TC Sun, selected by J. Garabedian.

Angeleno: Cross between Queen Ann x Unknown, originated by Garbadien developed in California.

Amber: Cross between Methley x Unknown, Al. USA.

Black Amber: Cross between Friar x Queen Rosa. Developed at Fresno, CA.

Bruce: Cross between Abundance x *P. angustifolia*, TX, USA.

Burmosa: Cross between Burbank x Formosa.

Byrongold: Cross between (Gaviota F2) x (Ozark Premier x *P. angustifolia*), GA, USA.

Catalina: Cross between Angeleno x unknown, CA, USA.

Climax: Cross of Simon (*P. simonii*) x Abundance.

Duarte: Originated as a cross of seed parent (*P. munsoniana* x *P. salicina*) and pollen parent Climax.

Eldorad or **El Dorado:** Originated as a cross *of P. salicina* x *P.simonii*, introduced 1904, developed by Luther Burbank.

Formosa: Developed from mixture of *salicina* species and several others.

Fortune: Crossing betwen Laroda x [Queen Anne x Late Santa Rosa], made at the US Horticultural Field Station, Fresno.

Friar: Selected from a Gaviota x Nubiana. It was developed and released by John Weinberger at the USDA-ARS research station in Fresno, CA.

Gaviota: It is said to have originated in 1900 as a cross of *P. salicina* x *P. americana*.

Ozark Premier: Cross beween Burbank x Methely.

Premier: Cross beween Burbank x Methley, MO, USA.

Queen Rosa: Cross beween Queen x Santa Rosa.

Roadside: Cross beween Ozark Premier x Unknown, AL,USA.

Robusto: Cross beween (Queen Ann x Barstow) x (Ozark Premier x *P. angustifolia*).

Rubysweet: Cross beween Mariposa x Morris, GA, USA.

Santa Rosa: Complex hybrid containing a mixture of *Prunus salicina, P. simonii, and P. americana,* with the *salicina* characters predominating. It was introduced by George C. Roeding of the Fancher Creek Nurseries, Fresno, CA, in 1906.

Segundo: (Queen Ann x Santa Rosa) x (Ozark Premier x *P. angustifolia*), GA, USA.

Wilson's Early: Possibly cross between Burbank and Cherry plum.

Wickson: Resulted from a cross *P. salicina* x *P. simonii* made by Burbank.

At the University of Florida, from a low-chill breeding programme varieties **Gulfbeauty, Gulfblazze, Gulfruby, Gulfgold** and **Gulfrose** have been released.

Interspecfic Hybridization

Interspecific hybridisation has been done between different species of the genus *Prunus*. The rootstock Marianna is an interspecific hybrid between *P. cerasifera* and *P. munsoniana*.

Private companies like Zaiger Genetics, Bradford Farms and Sun World International have developed several hybrids by interspecific crossings.

Pluots, **apriums**, **apriplums**, or **plumcots** are some of the hybrids between different *Prunus* species that are also called interspecific plums. Whereas plumcots and apriplums are first-generation hybrids between a plum parent (*P. salicina* or *P. cerasifera* or their hybrids) and an apricot (*P. armeniaca*), pluots and apriums are later generations. Both names "plumcot" and "apriplum" have been used for trees derived from a plum seed parent, and are therefore equivalent.

Apriums

Apriums [(*P. armeniaca* x *P. domestica*) x *P. armeniaca*)] are complex plum-apricot hybrids that show primarily apricot traits and flavor. Apriums resemble apricots on the outside. The flesh is usually dense and notable for its sweet taste due to a high content of fructose and other sugars. Apriums are usually only available early in the fruit season, like apricots and unlike pluots, which include some very late ripening varieties. Aprium trees grow quickly and are smaller compared to other common home grown apricots. The fruit is gold, with red coloration. Semi-mature fruit is hard and does not ripen if picked before completely mature.

Aprium varieties:

- **Honey Rich Aprium®:** Bright orange skin, freestone orange flesh, juicy, rich, intensely sweet flavor. Developed by Zaiger Genetics.
- **Cot-N-Candy**: harvests in early to mid June, flesh is extra sweet and juicy, with a plummy aftertaste, size is 2.0 to 2.5 inches in radius on average, self-fruitful.
- **Flavor Delight**: Resembles an apricot, but with a distinctive flavor and texture all its own, pleasant, lingering aftertaste, bigger crops if pollenized by any apricot.
- **Tasty Rich Aprium**: Very early harvests, in May. Light yellow-orange, firm, freestone. Flavor predominantly apricot, with pleasant aftertaste.

Rose aprium

(Source: Floyd Zaiger)

Plumcots (*Prunus armeniaca* x *Prunus domestica*) **and Apriplums**

Natural plumcots (also called apriplums) have been known for hundreds of years from regions of the world that grow both plums and apricots from seed. The name plumcot was created by Luther Burbank. The plumcot tree can reproduce asexually by budding whereas the apriplum tree resulted from hybridized seedlings and cannot reproduce.

Cultivars of plumcot

Flavor Ann: Hybrid of 87EB90= (apricot 15W297 x plumcot [=Red Beaut op]) x Royal Zee plum, originated by Chris, Gary, and Grant Zaiger and Leith Gardner at Modesto, Calif.

Flavorfall: Hybrid of 65EC752 (=unknown late plumop) x 4G1180 plumcot (=Redbeaut op). Originated: by Chris, Gary, and Grant Zaiger and Leith Gardner at Modesto, Calif.

Flavor Gem: Hybrid of 21GB794 (=plumcot of unknown parentage x King David plum) x Royal Zee plum. Originated in Modesto, Calif., by Chris, Gary, and Grant Zaiger and Leith Gardner.

Flavor Heart: Hybrid of 24EB412 plum (=Red Beaut op) x 4G1180 plumcot (=Red Beaut op). Originated by Chris, Gary and Grant Zaiger and Leith Gardner at Modesto, Calif.. An interspecific plum x plumcot with dark red skin.

Flavor King: Hybrid of 29EB30 =(Mariposa x plumcot =Red Beaut OP) x 42GA580=(plumcot= Red Beaut OP), invented by Chris, Gary, and Grant Zaiger and Leith Gardner at Modesto, Calif.

Flavorosa: A Hybrid of 65EC664=(Royal Zee OP) and 42GA580 plumcot= (irradiated).

Flavorite: Hybrid of 16GG159 (=25GB924 (=unknown plum) x 4G1180 plumcot [=Red Beaut op]) x 26GF10 (=21GB794 [=unknown plumcot x King David] x Royal Zee, originated by Chris, Gary, and Grant Zaiger and Leith Gardner.

Harmony: Originating from a cross between Soldam plum and Harcot apricot. The cross was made in 1999 at the National Institute of Horticultural & Herbal Science (NIHHS) of the Rural Development Administration (RDA) in Korea.

Honey Rich: Apricot open pollinated (op) x plumcot (=Red Beaut op)] x Royal Zee plum} x Tom Cot apricot. A complex interspecific plumcot with apricot appearance. originated by Chris, Gary, and Grant Zaiger and Leith Gardner.

Geo Pride: Hybrid of 29EB30=(Mariposa x plumcot from Red Beaut OP) and 42GA580=(plumcot from irradiated Red Beaut OP). Originated**:** in Modesto, Calif., by Chris, Gary and Grant Zaiger and Leith Gardne.

Miwang: Hybrid of an open-pollinated apricot (*Prunus armeniaca*) x native plum (*P. domestica*). Origin: near JungPyong, KeoSan, ChungBuk, Korea, by Yun-Won Lee.

Spring Satin: It is a plum x apricot hybrid introduced by Dr. Dick Okie, USDA-ARS, Bryron, GA.

Tiffany: Originated from a cross between 'Soldam' Japanese and 'Harcot and released from the National Institute of Horticultural & Herbal Science (NIHHS) of the Rural Development Administration (RDA) in the Republic of Korea. Redfleshed Plumcot.

Pluots= plumcot x plum (*P. armeniaca* x *P. domestica*) x *P. domestica*)

Pluots are later generations of complex hybrid between the Japanese plum *Prunus salicina* (providing the greater amount of parentage) and the apricot, Prunus armeniaca. The fruits exterior has smooth skin closely resembling that of a plum. Pluots were developed in the late 20th century by Floyd Zaiger.

Pluot varieties

***Flavorosa* pluot**

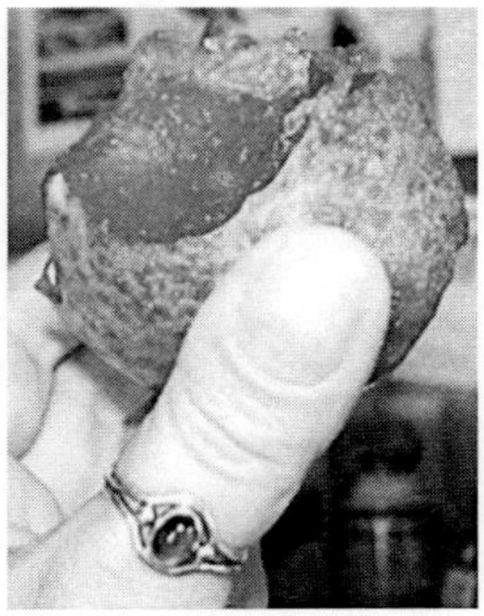

***Dapple Dandy* pluot**

***Splash* fruit tree**

***Splash* pluot**

(Source: Floyd Zaiger)

- **Dapple Dandy**: large size with mottled pale green to yellow, red-spotted skin, red or pink juicy flesh, firm flesh, moderately late ripening. Dinosaur egg is a trademarked name for Dapple Dandy variety.
- **Early Dapple**: good flavour, medium-sized, mottled green over red skin with pink flesh, early ripening.

- **Emerald Drop**: medium to large size, green skin and yellow-orange flesh, moderately late ripening.
- **Flavor Delight**: medium-sized, fuchsia-honey colored skin with pink flesh, early ripening.
- **Flavor Fall**: large size, average flavour, red skin with yellow flesh, late ripening.
- **Flavor Finale**: medium to large size, purple-red skin with amber-red flesh, exceptional complex flavour, late ripening.
- **Flavor Grenade**: large size, oblong shape with red blush on green background, yellow juicy flesh, moderately late ripening.
- **Flavor Heart**: very large, black with a heart shape, and yellow flesh.
- **Flavor Jewel**: sweet flavour, heart shaped, red over yellow skin with yellow flesh.
- **Flavor King**: Fruit punch flavour, medium size, with burgundy skin and red super sweet juicy flesh, moderately late ripening, flesh is hard until fully ripe.
- **Flavor Prince**: large round and purple, with red flesh.
- **Flavor Penguin:** medium size, early ripening grown in a humid climate.
- **Flavor Queen**: medium to large size, very juicy flesh, very sweet, golden yellow when fully ripe, midseason.
- **Flavor Rich**: medium-sweet, large black round fruit with orange flesh.
- **Flavor Royal**: very sweet, medium-sized, dark purple with crimson flesh, very early ripening.
- **Flavor Supreme**: medium or large, greenish purple skin, juicy red flesh.
- **Flavorich**: large size, dark purple skin and firm, sweet, yellow-orange flesh, moderately late ripening.
- **Flavorosa**: very sweet, medium-sized, flat, round, dark-purple fruit with red flesh, very early ripening.
- **Geo Pride**: medium size, red-skin and yellow flesh, balanced acid-sugar, predominately sweet with unique plum/apricot Flavor, moderately late ripening.
- **Raspberry Jewel**: medium, dark red skin, brilliant red, honey-sweet flesh.
- **Red Ray:** medium bright red with dense, sweet orange flesh.
- **Splash:** small to medium red-orange fruit, with very sweet orange flesh, midseason.
- **Pluerry™:** Cross between plum and sweet cherry.

Rootstock Breeding

For the improvement of rootstocks, interspecific hybridisation between different species of *Prunus* is commonly done.

Marianna: Interspecific hybrid between *P. cerasifera* and *P. munsoniana.*

St. Julien GF 655/2: Selected from INRA Bordeaux.

INRA GF-677: The almond x peach selection from INRA (drought resistant)

GF-557: Open pollinated selection of peach x almond (nematode resistant).

4. Mutation: There are some reports on the induction of mutagenesis in European plum using x-rays in order to obtain spur types of some plum cultivars. Now the induction of point mutations, nucleotide insertions or deletions plays no role in plum breeding as it is not expected that important breeding aims can be achieved by single small scale mutations of existing cultivars. The mutation rate can be increased using x-rays but heat therapy is also used to produce virus free material. For example, in Germany a virus-free clone of German Prune, called clone Rheinland, resulted from heat therapy and was propagated for some years in the nurseries., The trees of this clone had good flowering every year but never set fruits. On an individual tree of German Prune, clone Rheinland, one branch was found with high annual crop load. This showed the mutation character of the clone Rheinland. Usually, mutations affect only one gene.

Among the many cultivars developed as sports or mutations from Santa Rosa, or using Santa Rosa as a parent or grandparent, were Bella-Rosa, Black Amber, Black Ruby, Casselman, Campagne, Explorer, Fire Queen, Fortune, Fresno Rosa, Gar-Rosa, July Santa Rosa, Laroda, Midrosas, Padre, Premier, Roysum, Queen Rosa, Queensland Bellarosa, Rosa Grande, Rosanna, Salsa-Pride, Santa Rosa-Two, Segundo, Showtime, Sierra, Star Rosa, and Weeping Santa Rosa.

5. Biotechnology:

Micropropagation: Micropropagation methods have been developed for plums by using stem node, shoot tip, axillary bud, leaf, cotyledon and seed explants. Propagation of rootstocks by micropropagation *in vitro* has been widely used in commercial nurseries, because it allows the production of a large number of plants from one mother tree.

Molecular markers: Random amplified polymorphic DNA (RAPD) markers has been used to distinguish different cultivars, which focus on the utilization of DNA fingerprints in identifying cultivars and also to develop a new strategy to properly utilize DNA markers in the separation of plum cultivars. RFLP markers are used to estimate polymorphism and genetic diversity of plum cultivars, whereas SSR markers are available for molecular genotyping

Tranformation: A major limitation to cultivars improvement through genetic engineering and transformation techniques is poor understanding of the genetic control of important traits and the present scarcity of single, dominant genes of significant value for plant transformation. Despite large efforts, genetic transformation and regeneration in plum has only been successful in few cases.

The PPV coat protein (CP) gene isolated, sequenced and cloned was used for *Agrobacterium*-mediated transformation of plum. A part of the coat protein gene of the *Plum pox virus* was transferred to the genome of seedlings of the *Prunus domestica* genotype B 69 158. HoneySweet is a plum variety developed through genetic engineering to be highly resistant to plum pox potyvirus (PPV) the causal agent of sharka disease that threatens stone fruit industry world wide, most specifically in Europe.

APRICOT

Origin: Three important regions have been described as centres of origin for cultivated apricots: the Chinese centre (China and Tibet), the Central Asian centre (from Tien Shan to Kashmir), and the Near-Eastern centre (Iran, the Caucasus and Turkey) (Vavilov, 1951).

Taxonomy status:

Order: Rosales

Family: Rosaceae

Subfamily: Amygdaloideae

Genus: *Prunus*

Section: Armeniaca

Botanical name: *Prunus armeniaca* L.

Chromsomes number n=8; 2n=16

Genome Size: 240 Mb.

Closely related species are *P. brigantiaca* (Briancon apricot from the French Alps); *P. ansu*; *P. mume* (Japanese apricot); *P. sibirica*; *P. mandshurica* and *P. dasycarpa* (black apricot). *P. dasycarpa* is a natural hybrid between *P. armeniaca* and *P. cerasifera.*

Kostina divided the cultivated apricot according to their adaptability into four major ecogeographical groups: (1) the Central Asian group, (2) the Iran-Caucasian group, (3) the European group, and (4) the Dzhungar-Zailij group.

Botany

The apricot is a small tree, 8–12 m tall, with a trunk up to 40 cm in diameter and a dense, spreading canop with reddish bark. The leaves are ovate or round-ovate, 5–9 cm long and 4–8 cm wide with a rounded base, a pointed tip and a finely serrated margin. The flowers are solitary, 2–4.5 cm in diameter with five white to pinkish petals produced singly or in pairs in early spring before the leaves. Flowers are perfect, white or pinkish with 5 sepals, 5 petals, 20 stamens, perigynous ovary and single pistil. The fruit is a drupe, 1.5–2.5 cm in diameter (larger in some cultivars), from yellow to orange, often tinged red on the side most exposed to the sun; its surface can be smooth or velvety with very short hairs. The flesh is usually firm and not very juicy. Its taste can range from sweet to tart. The single seed is enclosed in a hard, stony shell, often called a "stone", with a grainy, smooth texture except for three ridges running down one side.

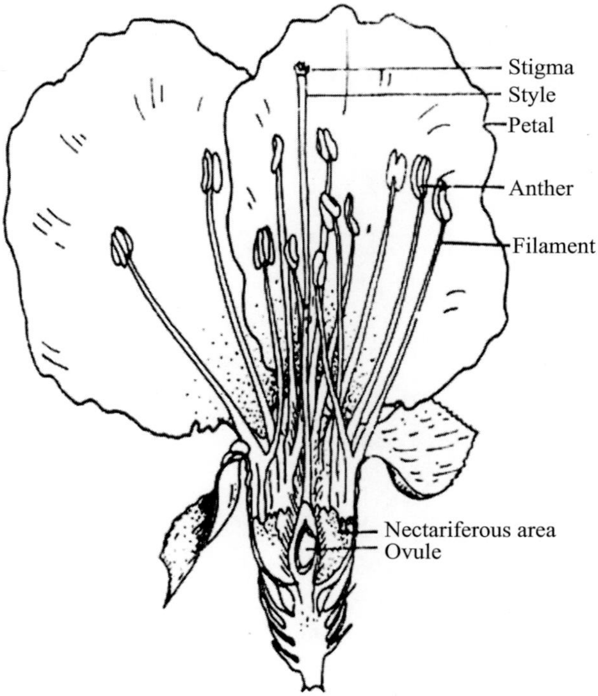

Fig. 21. L.S. of apricot flower.

(Source: McGregor, 1976)

Floral Biology

Maximum anthesis occurs 2-3 hrs. after sunrise. Dehiscence of anthers starts with the opening of flowers. Stigma receptivity occurs immediately after anthesis and continues up to 3 days.

Scion Breeding Objectives:

(i) To breed for climatic adaptation, low chilling, early and late ripening habit, regular bearing, high yielding and self-fruitfulness.

(ii) To breed cultivars considering fruit characteristics like large fruit, firmness, shape, reddish colour and transportability.

(iii) To breed cultivars for fresh and dry market fruits, canned fruits, other post harvest uses.

(iv) Resistance to fruit cracking.

(v) To breed cultivars resistant to brown rot (*Monilinia spp.*), leaf roll, sharka virus (PPV), chlorotic leaf roll and bacterial canker.

(vi) To breed cultivars tolerant to frost.

Limitations in Breeding

1. Long juvenile phase.
2. Self incompatibility.
3. Heterozygous.
4. Narrow genetic base.
5. Pollen sterility.
6. Self-incompatiblity.
7. Adapted to specific climate only.
8. Inheritance pattern is unknown.
9. Irregular yields.

Rootstock breeding objectives:

1. Reduction in tree size.
2. Increased scion precocity.
3. Scion graft compatibility and case of propagation.
4. Resistance to resistance to diseases, such as viruses and *Phytophthora* and *Armillaria* root rots, insects and nematodes.
5. Tolerance to cold, heat, drought and salinity.

Germplasm Resources

The current database of apricot contains 1982 accessions physically represented by 1947 individuals maintained by 18 institutes from 13 countries. The Davis, California National Clonal Germplasm Repository (NCGR) houses most Mediterranean-adapted fruit and nut crop collections in the US, including apricot. In India, field gene banks have been maintained at Regional Horticultural Research and Training Insitute, Dr YSPUHF, Mashobra accessions), Temperate Horticulturral Research Station Dr YSPUHF, Kotkhai, Shimla (37), SKUAST, Srinagar, Hill Fruit Research Centre, Chaubattia NBPGR Research Station, Phagli, IARI RRS, Amartara, Shimla (23) and CITH, Srinagar.

Table 22. Sources of useful traits in apricot.

Trait	Culivars/species
Dwarfness	*Prunus besseyi.*
Low-chilling (200-400 hrs)	Adedi, Amal, Arengt, Bedri, Currot, Earligold, Hamtdi, Jazil, Louzi, Precoce, Zaraqosa.
Large fruit size	Harcot, Goldrich, Hargrand, King, Nugget, Perfection, Sundrop, Venus.
Firm flesh	Goldrich, Hargrand, Harogem, NJA 32, Sundrop, Abate, Roksana, Fracasso.

Drying types	Khurmani, Supkhant, Isfarak, Ere Satent, Hasanbey Blenril, Moorpark, Hunqerian Best, Babal, Royal, Blenheim.
White flesh	Chuanzhihong.
Canning varieties	Boccuciza Spinosa, Pelummella, San Castrese, Murciana, Viragold Budapest, Hargrand, Harlayne, Goidest, Harglow.
Suitability for freezing	Cafona, Don Gactano, San Francesco.
Sweet kernels	Moorpark, Harcot.
Early maturing	Abid, Bedri, NJA 43, NJA 53, NJA 54, NJA 87, Zaragozano, Jerseycot, Jubilee, Large Early, Ranna Samarkand.
Late maturing	Bergeron, Budapest, Comandor, Litorol, Olimp, Sulmona, Olimp, Dacia, Tirzii de Bucuresti and Venus.
Cold tolerant cultivars	Zard, Suphony, Tokshion.
Long ripening cultivars	Harcot, Harogem, Harlyne, Hargrand, Hargiow.
Fruit cracking	Boreale, Fournes, Goldrich, Moniqui, and San Castrese.
Root knot nematode	*P. armeniaca.*
Bacterial gummosis	Mariana GF 8/1, Mariana 2624.
Plum pox virus or sharka (PPV)	*Prunus mandshurica.*

Inheritance Pattern

Flowering date, maturity date, fruit size, fruit weight and yield: Polygenic inheritance.

Skin colour, red blush, flesh colour and red blush: Independently inheritance.

Self incompatbilty, male sterility, self incompatibility, seed bitterness and PPV: Oligogenic inheritance.

Late flowering season, long dormancy, small fruit, large pit,

Endo PG gene (Endopolygalacturonase), high TSS: Dominant.

Breeding Methods

1. Introduction: The cultivars like St. Ambriose, Kaisha, New Castle, Blenheim (Royal), Early Shipley, Communis Holy, Harcot, Saffaida, Charmagaz, Shakarpara, Frogmore Early, Gillgati Sweet, Amba, Moorpark, Halman, Rakchey Karpu, Narmo, Afgani selection, Tokpapa, Koban, Tilton Royal, Farmingdale, Alfred, Coninos, Nugget, Turkey, Nugget, Baiti, Beladi, and Nari have been introduced in India.

Promising cultivars to be introduced in India are Tomcot, Goldcot, Lehrman, Patterson, Robada, Earlicot, Jordane, Gold Bar, Gold Strike, Alfred, Arzami, Earliorange, Sundrop and Gold.

The clonal rootstocks that need to be introduced are Marianna GF 9/1, Marianna-2624, Pollizo and Haggit.

2. Selection: Several selections such as CITH Apricot-1, CITH Apricot-2, CITH Apricot-5 and CITH Apricot-6 have been done at Central Temperate Research Institute, Srinagar.

3. Hybridization: The hybridization work was intiated at Chaubattia, India and following hybrids were released:

Chaubattia Alankar: Kaisha x Charmagz. Low chilling and early bearer.

Chaubattia Madhu: Turkey x Charmagz. Early and regular bearer.

Chaubattia Kesri: St. Ambroise x Charmagz.

Hybrids released by research stations of the world:

Dipartimento di Produzione Vegetale at Milano, Bologna Universities and Dipartimento di Coltivazione e Difesa delle Specie Legnose at Pisa University: This was a publicly funded apricot breeding program.

Gheriana: Cross between Portici and Harcot.

Silvana: Cross between Bergeron and Canino Tardivo.

USDA-ARS San Joaquin Valley Agricultural Sciences Center, Parlier, CA:

Twocot: Cross between Apricot No. 73 × P301-110, originated by C. Ledbetter.

***Austria*:**

Danubu: Cross between Sam No 1 and Mari de Cenad.

Fortuna: Cross between Sam no. 1 and Mori de Cenad.

Auras: Cross between Sam no. 1 and Stark Early Crange.

Cristal: Cross between Sam no. 1 and Maride Cenad.

Promising NJAES apricot cultivars and selections:

Early Blush®: Cross between NJA53: RR17-62 and NJA13.

SunGem® (NJA54): Cross between NJA2 and RR17-62.

NJA97: Cross between RR17-62 and NJA13.

Orangered® (NJA32): Cross between Lasgerdi Mashhad and NJA2.

Jerseycot: Cross between RR18-64 and CR3-135.

NJA113: Cross between Jerseycot and NJA93.

NJA114: Cross between Jerseycot and NJA93.

NJA82: Cross between I13-36 and NJA11.

Danubu: Sam no. 1 x Mari de Cenad.

Fortuna: Sam no. 1 x Mori de Cenad.

Auras: Sam no. 1 x Stark Early Crange.

Cristal: Sam no. 1 x Maride Cenad.

Interspecific hybridization: Intra-specific crossing is the most widely spread method for apricot scion breeding. Inter-specific crossing between *Prunus* species is common for rootstock breeding and for trait improvement in scion cultivar. Among the most interesting apricot-like species employed in breeding programs are *P. ansu, P. dasycarpa, P. mandshurica*, *P. mume, P. salicina* and *P. sibirica*.

Cherry Kose: Cross between *P. armeniaca* x *P. besseyi.*

Crosses between plums and apricots (Plumcot) and in other *Prunus* species for regional adaptation by incorporating characteristics such as late blooming, disease resistance, cold hardiness and modified tree types into apricots would be of value.

Honey Sun®: (apricot × peach hybrid) developd by Zaiger Genetics, California, USA.

P. dasycarpa: Natural hybrid between *P. armeniaca* and *P. cerasifera.*

4. Mutation: Early Blenheim was selected following thermal neutron irradiation. This is the only report on mutation breeding in apricot.

5. Biotechnology: Biotechnology has obvious implications as a new tool that can be employed in apricot breeding. Potential areas for its implication are regeneration and micropropagation, virus elimination and genetic improvement which could include somaclonal variation, protoplast culture and fusion, embryo rescue and haploid induction. The high degree of polymorphism of molecular markers such as microsatellite markers (Single Sequence Repeats (SSR)) and AFLPs provide efficient tools for identification and revealing genetic diversity among apricot germplasm. Recombinant DNA technology might also be employed to carry out genetic transformation and gene characterization. With the exception of micropropagation, embryo rescue and virus elimination, biotechnology has not yet been fully employed as a tool in apricot breeding.

CHERRY

Origin: Asia minor (situated with Black sea and Caspian sea).

Taxonomy status:

Order: Rosales

Family: Rosaceae

Subfamily: Amygdaloideae

Genus: *Prunus*

Subgenus: *Cerasus*

Tribe: Amygdaleae

Section: Eurocerasus

Botanical name: Sweet cherry, *Prunus avium* L. 2n=2x= 16; Sour cherry *Prunus cerasus* L. n= 8; 2n=4x=32.

Chromsomes Number: n= 8; 2n=2x=16; 4x=2n=32.

Genome size: 352.9 Mb (Sweet cherry)

Cherry consists of three main groups: the sweet cherry (*Prunus avium* L., 2n=16); the tart (sour) cherry (*P. cerasus* L., 2n=32); and the ground cherry (*P. fruiticosa* L., 2n=32). Duke cherries are natural hybrids between non-reduced gametes of sweet and tart cherries (2n=32). The origins of the three species overlap and include Central Europe and areas surrounding the Black Sea, the sweet cherry as far east as central Russia.

Botany

The cherry is a deciduous, upright tree growing to 15–32 m height, with a trunk up to 1.5 m in diameter. Young trees show strong apical dominance with a straight trunk and symmetrical conical crown, becoming rounded to irregular in old trees. The bark is smooth purplish brown with prominent horizontal grey brown lenticels on young trees, becoming thick dark blackish brown and fissured in old trees. The leaves are alternate, simple, ovoid acute, 7–14 cm long and 4–7 cm broad, glabrous matt or sub-shiny green above, variably finely downy beneath, with a serrated margin and an acuminate tip, with a green or reddish petiole 2–3.5 cm long bearing two to five small red glands. The tip of each serrated edge of the leaves also bear small red glands. Flowers are white or rose colored, usually with pedicel, solitary or fascicled or raceme; calyx tube cup-shaped or tubular and flowers sessile; 5 sepals, 5 petals, 20 stamens and 1 carpel. The fruit is a drupe 1–2 cm in diameter (larger in some cultivated selections), bright red to dark purple when mature in mid summer, edible, variably sweet to somewhat astringent and bitter to eat fresh. Each fruit contains a single hard-shelled stone 8–12 mm long, 7–10 mm wide and 6– 8 mm thick, grooved along the flattest edge; the seed (kernel) inside the stone is 6–8 mm long. Stone is smooth or furrowed and pitted.

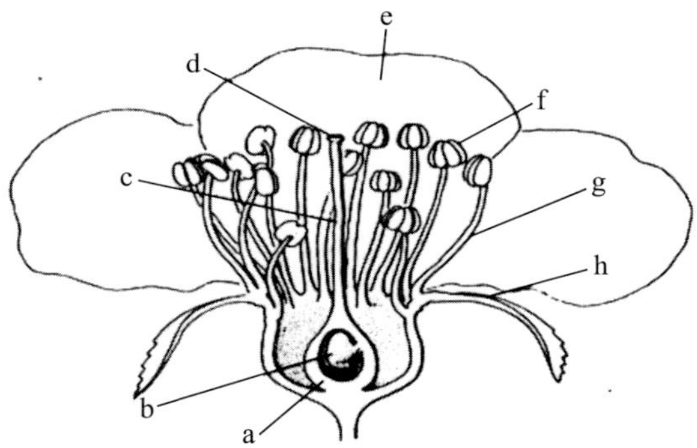

Fig. 22. L.S. of cherry flower a. ovary, b. ovules, c. style, d. stigma, e. petal, f. anther, g. filament, h. sepal

Floral Biology

The anthesis occurs in the morning. Dehiscence of anthers takes place with the opening of flowers. Maximum stigma receptivity occurs immediately after anthesis.

Scion Breeding Objectives

(i) Compact growth habit, precocious, extension harvest season, improved fruit quality (large fruit, bright red colour, sweet, firm, flavour and round pit and higher yield).

(ii) Late flowering.

(iii) Resistance to rain cracking.
(iv) Self fruitfulness.
(v) Suitable cultivars for warmer areas.
(vi) Extension of season.
(vii) Mechanical harvesting.
(viii) Tolerant to biotic and abiotic stress.

Limitations in Breeding

1. Heterozygous.
2. Long juvenile phase.
3. Self incompatibility.
4. Adapted to specific climate.
5. Inheritance pattern is not known.

Rootstock Breeding Objective

1. Reduction in tree size.
2. Increased scion precocity.
3. Scion graft compatibility.
4. Resistance to resistance to diseases such as viruses and *Phytophthora* and *Armillaria* root rots, insects and nematodes.
4. Adapted to cold and heat.

Germplasm Resources

The current database of cherry contains 5585 accessions physically represented by 5136 individuals maintained by 42 Institutes from 17 countries of the world. The National Clonal Germplasm Repository (NCGR) at Davis is one of the 32 federally funded repositories in the National Plant Germplasm System (NPGS) that is dedicated to the *Prunus* collection and preservation of plant germplasm from around the world. The Plant Genetic Resources Unit in Geneva, New York also houses tetraploid cherry *Prunus* species.

In India, large number of collections was made through introductions and indigenous collections and has been conserved *ex-situ* at national field gene banks at CITH and at NBPGR stations. Field gene banks are maintained at Regional Horticultural Research and Training Insitute, Dr YSPUHF, Mashobra (47 accessions), Temperate Horticultural Research Station Dr YSPUHF, Kotkhai, Shimla (37), NBPGR Research Station, Phagli, IARI, RRS, Shimla (15), SKUAST, Srinagar, Hill Fruit Research Centre, Chaubattia and CITH, Srinagar. Genetic resources used for different characters in cherry breeding are listed in Table 23.

Table 23. Sources of useful traits in cherry.

Trait	Cultivars/species
Compact growth habit	Compact Stella, Compact Lambert, Garden Bing, Early Black, *Prunus tomentosa.*
Fruit cracking	Viscount, Brooks, Summit, Regina, Adriana and Flemings Srim.
Self fruitfulness	Kordia, *Regina*, Stella, Sunburst, Lapins and Sweet Heart.
Fruit size	Celeste and Sunburst, Big Lory, Regina and Nanyo.
Firmness	Duroni series of cultivars and Corniola and Kordia.
Early ripening	Chelan, Burlat, Rita, Empress and Primulat.
Late ripening	Regina and Staccato.
Yield efficency	Meteror.
Easy fruit stem abscission	Vittoria and Linda.
Drought	*Prunus tomentosa.*
Cold	Van, *Prunus tomentosa.*
Rain damage	Adriana and Flemings Srim.
Bacterial canker	Merpet, *P. mugus.*
Powder mildew	*Prunus cerasoides*, *P.nepaulensis.*
Root rot	*Prunus cerasoides.*

Inheritance pattern

Precocity: Quantitatively.

Stem length and thickness characters: Closely linked.

Fruit size: Controlled bymultiple factors.

Heart shaped: Dominant over oblate shape.

Red fleshed: dominant over yellow.

Ripening time: Quantitatively.

Pre-selection criteria:

A significant correlation exists between the time of bloom, full bloom and fruit picking (de Vries, 1967).

Breeding Methods

1. Introduction: The cultivars Bing, Black Heart, Black Tartarian, Napoleon Bigarreau (Royal Ann or Royal Anne or also called Napoleon), Sue Sam, Lambert, Bedford Prolific, Black Heart (Siyah Gole), Guigne Noir Grossa Lacenta (Jontal), Governor's Wood, Guigne Pour ova Precece (Awal number), Bigarreau Napolean (Double), Bigarreau Noir Grossa (Mishri), Guigne Noir Hative (Makhmali) heve been introduced from different parts of the world. Recent introductions are Van, Rainier, Regina, Sweet Heart, Bing, Stella and Lapins.

Clonal rootstocks:

Colt: Semi dwarfing.

Mahaleb, Mazzard and **F-12/1**: Vigorous.

Gisela-10: Dwarfing, high producing and winter hardiness.

2. Selection: Several superior types have been identified at CITH, Srinagar.

CITH Cherry–01: Trees are semi-spreading, precocious, regular and prolific bearing. The cultivar has been selected from Bigarreau Napolean (Double Gilass) cherry orchard.

CITH Cherry-02: Trees are upright, precocious, prolific and regular bearer selected from (Bigarreau Noir Grossa) Local Mishri. The other clonal selections are CITH Cherry–5, CITH Cherry–19 and CITH Cherry–21.

Van: It originated from open pollination of Empress Eugenie at the Summerland Research Station, Summerland, British Columbia.

Lambert: It was selected from seedling at Oregon, USA.

Staccato™: Originated as an open pollinated seedling of Sweetheart from seeds collected by R. MacDonald in 1982.

3. Hybridization:

As the cherry flowers are self incompatible, there is no need of emasculation. However, the branches will need to be enclosed using bags or cages that exclude insects to prevent pollination. The protective covering should be removed following fruit set.

The strength of the sweet cherry industry throughout Europe and Asia is due to the breeding efforts in France, Germany, Hungary, Italy, Denmark, the UK, USA, Canada, the Czech Republic, Romania, Estonia, Ukraine, Turkey, Japan and China. Several varieties have been developed and released for the growers. These are:

Sylvia: Comact Lambert x Van.

Summit: Van x Sam.

Sweetheart: Van x New Star developed at Summerland, British Columbia, Canada.

Sunburst: Van x Stella.

Merton Bigarreau: Knight's Earl Black x Napolean.

Stella: Lambert x self fertile John Innes.

Starkrimson: Stella x Garden Bing.

Penny: Colney x Inga. Developd at EMRS, East Malling.

Benton: Stella × Beaulieu.

Black Star: Lapins × Burlat.

Brooks: Rainier × Early Burlat.

Celeste: Van × Newstar.

Chelan: Stella x Beaulieu.

Giant Red: Large Red × Ruby.

Kiona: Stella × Early Burlat PC7144-3.

Chelan: Hybrid of Stella and Beaulieu, first developed in Prosser, WA.

Australian varieties are being bred by the National Cherry Breeding Program based in Lenswood, South Australia. The program has released 6 varieties: Sir Tom and Sir Don (1998), **Dame Roma** (2001) and **Sir Douglas, Sir Hans** and **Dame Nancy** (2002).

***International Fruit Genetics, Bakersfield, CA 9331*:**

Newly developed cultivars are Cherry Burst™, Cherry Crunch™, Cherry Blush™, Cherry Treat™, and Cherry Moon™.

Zaiger Genetics:

Royal Crimson: Fruit large, tasty, crimson red with chilling requirement 200-300 hrs.

Cherry rootstocks

Colt: Cross of *P. avium* and *P. pseudocerasus*, released in 1977.

Gisela 6: Cross between *P. cerasus* and *P. canescens.*

Inmil: Cross between *P. incisa* and *P. serrula.*

F12/1: was released in 1924; **Cob** was released in 1980 and **Charger** was released in 1986.

Interspecific Hybridization

The following interspecific hybrids have been identified to have potential in cherry rootstock breeding: *P. avium* x *P. pseudocerasus*; *P. incisa* x *P. serrula*; *P. cerasus* x *P. maackii*; *P. cerasus* x *P. avium*; *P. cerasus* x *P. canescens*; *P. cerasus* x *P. fruticosa*; *P. fruticosa* x *P. avium*; *P. subhirtella* x *P. yedoensis*; *P. mahaleb* x *P. avium*; *P. avium* x *P. kurilensis*; *P. avium* x *P. incisa*; *P. canescens* x *P. incisa*; *P. canescens* x *P. tomentosa*; and *P. cerasus* x *P. pensylvanica*. The others are: Rosa (*P. subhirtella* autumno), Gisela 6 (*P. cerasus* × *P. canescens*), Colt (*P. avium* × *P. pseudocerasus*), Inmil (*P. incisa* × *P. serrula*) and Damil (*P. dawyckensis*).

4. Mutation: The cultivars developed by exposing gamma rays, X-rays: Burlat C1, Ferrovia spur, Compact Stella 35B11 showing Dwarfness and Compact growth.

5. Biotechnology:

Micropropagation: Plant regeneration has been reported for 19 sweet cherry (*Prunus avium*) cultivars, 12 sour cherry cultivars and 24 for other cherry genotypes species such as cherry hybrids, black cherries, and wild cherries.

***In vitro* selection via somaclonal variation**

Exploitation of stable somatic variants from cell and *in vitro* tissue culture has proved feasible but practical results have not yet been achieved. Protoplast cultures have not proved encouraging. Research successes include tissue culture for somaclonal variability in timber cherry; whole plant regeneration from leaves and cotyledons; organogenesis, i.e. direct regeneration from disks of *in vitro* grown leaves, to breed clones of Hedelfinger, which appear to exhibit reduced apical dominance; and somatic embryogenesis in Summit from root explants of *in vitro* grown plantlets.

Molecular markers: Several molecular markers have been assessed in sweet cherry, especially AFLPs and SSRs and used for genetic fingerprinting, estimation of genetic diversity and mapping and in order to improve breeding efficiency. Several linkage maps have been developed for *Prunus* sp., including sweet cherry. Boritzki *et al.* (2000) used AFLP and SSR primer combinations to separate almost 200 cherry cultivars and important qualitative trait loci (QTL) have been investigated in stone fruits including cherry.

S-allele variability of self-incompatible genotypes: A list of 17 incompatiblity groups is now available and specific PCR S-allele from S1 to S16 have been detected for cultivars including self-compatible ones. The most advanced studies of the S-locus have demonstrated multiple codominant alleles such as *S4I* making it possible to recognize compatible cultivars for practical use (Sonneveld *et al.*, 2001; Tassinari, 2005). A pollen expressed gene protein linked to S-RNase has been found by Yamane *et al.* (2003).

Genetic Transformation: The standard *Agrobacterium* mediated and less frequently the biolistic gun as vectors have so far produced clones of the Colt GM rootstock carrying the rice phytochrome A to induce more compact internodes. Plants of the rootstocks Damil and Inmil have been transformed in Belgium using GUS and rol genes via *Agrobacterium* (Druart *et al.*, 1988). Stable transgenic plants have been reported for sour cherry 'Montmorency' cherry rootstocks 'Gisela 6' and 'Colt', and *Prunus subhirtella* Miq. 'Rosa', but only reporter genes were used in these successful transformations.

PERSIAN OR ENGLISH WALNUT

Origin: Mountain regions of Central Asia

Taxonomy status:

Order: Fagales
Family: Juglandaceae
Sub family: Juglandoideae
Section: *Juglans*
Subgenus: *Juglans*

Botanical name: *Juglans regia* L.

Chromsomes number: n = 16; 2n = 32.

Genome size: 606 Mbp.

Related species are *J. nigra*, *J. cinerea*, *J. hindsii, J. neotropica*, *J. cathayensis* , *J. ailantifolia* , *J. mandshurica* and *J. sieboldiana.*

Botany

Walnuts are deciduous, tall, broad headed, dichogamous and monoecious trees with scaly furrowed bark branches (with lamellate pith); buds with few scales, rarely short-stalked; leaves alternate, pinnate, large, aromatic, extipulate, opposite, serrate or entire; flowers monoecious, the staminate flowers on last years branchlets lateral pendulous catkins, each consisting of bract bearing 8-40 stamens, 2 bractlets and 4 calyx lobes; pistillate flowers borne on current seasons wood in few to many flowered terminal racemes, with 4 calyx lobe and a 3-lobed involucre consisting of a bract and 2 bractlets; style divided into 2 plumose stigmas; fruit a large indehiscent drupe; nut thick walled, incompletely 2—4 celled, in dehiscent or finally separating into 2 valves; seed 2-4 lobed, remaining within the shell in germination.

Floral Biology

Flowering occurs in the spring from mid April for early cultivars to mid May for late blooming cultivars. The female is receptive when the two stigmatic surfaces separate to form a V-shape and secrete an exudate that makes them appear moist. Once the lobes have achieved a 45° angle, the surface begins to dry, and the female flower is no longer receptive. Pollen of walnut is relatively short lived wind borne and can be carried great distances by wind. The flower will expand for the next 3 weeks to the size of a marble and abscise if it has not been fertilized. The stigma surfaces stay moist if the flower has not been fertilized. The cause of pistillate flower abscission (PFA) is caused by excess pollination. During this, the flowers shrivel and abscise shortly after they lose receptivity. PFA occurs to some extent in all cultivars but is most serious in the cultivar Serr.

Scion Breeding Objectives

(i) To develop precocious, dwarf and better adaptated cultivar.

(ii) Develop cultivars with late leafing and tolerant to pistillate flower abscission.

(iii) High number of fruitful lateral, early harvesting, self-fertile and homogamy are desired.

(iv) Develop cultivars with reduced winter chilling requirement.

(v) Develop cultivars with early harvesting, higher yield and regular bearing.

(vi) Develop cultivars with good shell quality (smooth shell texture, light colour, round to oblong shape, large nut, paper shell, strong shell seal).

(vii) Develop cultivars with high shell out percentage.

(viii) Develop cultivars with good kernel qualty: red kernels (Rouge de la Donan), kernel of light coloured, high kernel-shell ratio, plump, bold, easy to remove kernel halves, light in colour, sweet in taste, well-filled, high oil content.

(ix) Develop a commercial cultivar with resistance (hypersensitivity) to the Cherry Leaf Roll Virus (CLRV), which causes blackline disease.

(x) Resistant to insect (Moths, Aphids, Walnut Husk Fly) and diseases bacteriosis (*Xanthomonas juglandis*), anthracnose

(xi) Tolerant to heat and winter hardiness.

Rootstock Breeding Objectives

1. Development of dwarf rootstock.
2. Compatibility between stock and scion.
3. Easy to propagate through vegetative means.

4. Resistance to nematode, crown gall, *Phytophthora* and blackline problem caused by the CLRV, and abiotic stresses.

Limitations in Breeding

1. Long juvenile phase.
2. Heterozygosity.
3. Heterodichogamy (protogyny and protogynous).
4. Lack of good pollinizers.
5. Polygenic control of most horticultural traits.

Germplasm Resources

In USA, germplasm collections are maintained by the Walnut Improvement Program at the University of California, Davis, CA, and the USDA National Clonal Germplasm Repository, Winters, CA.

In India, large number of collections was made in nuts (359 accessions) at CITH, Sringar. Germplasm has also been maintained at NBPGR Station, Shimla, Dr YSPUHF, Solan, GBPUA&T, Pantnagar, and SKUAST, Srinagar.

Genetic resources used for different characters in walnut breeding are given in Table 24.

Table 24. Sources of useful traits in walnut.

Trait	Cultivars/species
Frost	*Juglans cathayensis*, *J. mandshurica*
Precocity and lateral bearing	Payne, Marchetti, Chandler, Chico, Ideal, Lara, Rego, Precocious, Amigo, Sunland, Vina, Cisco.
Late Leafing	Come, Marbot, Grande Jean, Parisienne, Mayetter, Hartley, Franquette, Howard, Pedro, Tehama, Cisco.
Pistillate flower abscission (PFA)	Serr, Algaida and Franquette.
Quality of kernel	Franquette, Chandler.
Walnut blight	*Juglans hindsii.*

Inheritance Pattern

Hypersensity to blackline disease: Single dominant gene.

Table 25. Estimated heritability of traits in walnut.

Traits	h^2
Leafing date	0.96
Percent fruitful laterals	0.39
Harvest date	0.85
Shell seal	0.38
Shell thickness	0.9

Nut weight	0.86
Kernel weight	0.87
Ease of kernel removal, color (extra light, light, light amber, amber)	0.52

Pre-selection indices

Wood/bark and the rate of wood/transverse area of walnut seedlings have been used as index to predict tree vigour (Zhang *et al.*, 2007).

Breeding Method

1. Introduction: The cultivars Lake English, Colby Blackmore, Franquette, Lake Franquette, Opex Caulchery, Mayette, Lozeronne, Laciniata 93-5, Meylan, Chinese 85-8, Manregian, Sharkey, Turtle 16, Marbot, Grande Jean, Howard, Waterloo, Midland, Pioneer, Carmello, Adams, Hansen, Mahan, Payne, Sunland, Vina, Ashley, Serr, Lompoc, Chico, Tehama, Eureka, Gustine, Placentia, Marchetti, Cisco, Chandler, Amigo, Hartley, Turtle-31, Nelson, Tulare, Early Ehrhardt, Conway mayette, Pedro, Trinta, Poe, Weeper Drainovsky and Cascade etc., have been introduced from University of California, Davis, USA; University of Oregon; University of Idaho, University of Washington and INRA, France.

2. Selection: The seedlings with high shelling percentage, better kernel quality and soft shell have been identified and released for cultivation, e.g. Gobind, Eureka, KS-1, Sarawari, K-2, SH-23 and SR-11, Roopa, Ratna, Pratap in HP; Chakrata No-2, 4, 6, 13 and 14 in Uttarakhand; and Sulaiman, Hamdan in Kashmir. The selections made at Central Institute of Temperate Horticulture, Srinagar, J & K are CITH Walnut 1, CITH Walnut 2, CITH Walnut 3,4, CITH Walnut 5, CITH Walnut 6, CITH Walnut 7, CITH Walnut 8, CITH Walnut 9 and CITH Walnut 10.

Pusa Khor: Developed by IARI Regional station, Shimla. Lateral bearer and suited for high density plantation.

Franquette: It is a chance seedling selected by a grower in Merced, USA.

Ashley: It is a seedling or bud sport first noticed by Pallas Neal Ashley about 1945 in the orchard of Wendall Payne located in southern Colusa County.

Ehrhardt: It is a chance seedling, or possibly a bud sport of Ehrhardt found by George Rutherford in Refugio Canyon near Goleta, California.

Westside: It is a chance seedling from the San Joaquin Valley.

Trinta: It is bud sport or chance seedling of Waterloo.

Payne: It is a chance seedling, with fruitful lateral buds, found by George Payne, a farmer from Santa Clara County.

Marchetti: It is a chance seedling which occurred in the orchard of Guilio Marchetti near Stockton, California.

Idaho: It is a protogynous selection from among the large bijou-type nuts.

Howe: It is a chance seedling discovered by Cecil B. Howe of Brownsville, Oreg.

3. Hybridization: The UC breeding program used two distinct procedures for crossing parent material. In the first method, wind blown pollen is excluded from female flowers of interest by covering them with tightly secured bags that have small plastic windows. Pollen is collected from the other parent of interest and stored frozen over saturated magnesium chloride until use. When bagged female flowers open and are receptive, pollen is applied through the bags with a hypodermic needle. Bags are later removed and nuts marked for collection in the fall. The male parent is known with this method, but the costs are high and seedling production is low. The second method is to locate geographically isolated young trees of the desired female parent. Using young trees is important, because as a cultivar matures, the female flowers are usually present 2–3 years before the male flowers. This often requires the cooperation of a grower with a recently planted orchard. Any male flowers on these trees are removed manually before bloom to prevent selfing. Once the female flowers begin to bloom, pollen of the desired male parent or parents is applied by airbrush several times during the bloom period. At harvest, the cooperating grower either donates or is compensated for the nuts. This method produces many more seed at lower cost but with low certainty of the male parent. Male parents of selections can be determined later by DNA analysis. Some selfing also occurs, which results in stunted, twisty trees with russetted hulls and small kernels.

Prior to the Serr-Forde breeding program (1948–1978) in California, most cultivars grown in northern California, where the industry now resides, were cultivars brought from France by Felix Gillet in the late 1800s or chance seedlings. Gene Serr and Harold Forde made remarkable progress in breeding new cultivars that revolutionized the industry. Their primary breeding objectives were to combine the late leafing and quality of the French types with the lateral fruitfulness and precocity of Payne. They made 196 crosses, evaluated about 6000 progeny, and released 13 cultivars, 10 in 1968 and 3 in 1978. The most important of these are Vina, Serr, Howard and Chandler. In 1993, Tulare was released from a cross made 27 years earlier by Serr and Forde. Four new cultivars have now been released. Robert Livermore is a red skinned walnut. Sexton, Gillet and Forde are all precocious, laterally fruitful, high yielding, midseason with low blight scores and high quality kernels.

Newly developed hybrids:

Gustine (52-61): It is originated from a cross between Waterloo and Payne made by Serr and Forde in 1951 and released by the University in 1968.

Tulare: It came from a Tehama x Serr cross in 1966 by Forde.

Tehama: It is from a cross between Waterloo and Payne made by Serr and Forde.

Midland: It is a cross between Franquette and Payne.

Pioneer: Cross between Franquette and Payne.

Howard: Cross between Pedro and 56- 224.

Sunland: It is a cross between Lompoc and PI 159568.

Vina: Cross between Franquette and Payne.

Serr: It is a cross between Payne and PI 159568.

Lompoc: It is a cross between Waterloo and Payne.

Chico: Cross between Sharkey and Marchetti.

Tehama: It is a cross between Waterloo and Payne.

Gustine: Cross between Waterloo and Payne.

Cisco: It is a cross between Meylan and Pedro.

Chandler: Cross between Pedro and 56-224.

Amigo: Cross between Sharkey and Marchetti.

Hartley: It is a cross between Franquette and Mayette.

Pedro: Conway Mayette and Payne.

Interspecific hybrids:

Paradox: *J. nigra* x *J.regia.*

Rx1: *J. microcarpa* x *J. regia.*

Vx 211: *J. hindsii* x *J.regia.*

Paradox: is a hybrid between a California black walnut (either *J. hindsii* or *J. californica*) and the English or Persian walnut (*J. regia*).

Royal hybrids: *J.hindsii* x *J. nigra.*

Butterjaps or butternuts: *J. cinerea* x *J. ailantifolia*; *J. regia* x *J. mandshurica.*

J. nigra* x *J. ailantifolia: Cross named Leslie Burt, which exhibits anthracnose resistance.

4. Backcrossing: Marker-assisted backcross breeding is being used to develop a commercial quality, *J. regia*-like cultivar with resistance (hypersensitivity) to the CLRV, which causes blackline disease (Woeste *et al*., 1996). Studies showed that a single dominant gene from *J. hindsii* confers hypersensitivity and that progeny from backcrosses (*J. hindsii* x *J. regia*) x *J. regia* segregates 1:1 hypersensitive: tolerant.

5. Biotechnology: Many of the biotechnology tools have been applied in micropropagation and somatic embryogenesis in *Juglans regia* and *J. hindsii* was developed in 1984. Marker-assisted selection is being used successfully in a backcross breeding program designed to transfer hypersensitivity to CLRV from *J. hindsii* into a commercially acceptable Persian walnut cultivar. This greatly reduces the time required to screen progeny in each generation. DNA finger printing is becoming routine for cultivar identification and DNA sequence markers are used to identify the species involved in Paradox seedlings. Gene cloning from walnuts is under way, and genes of interest include those for tannin, naphthaquinone, unsaturated fatty acid, and flavonoid biosynthesis. In spite of activities in walnut biotechnology, genome mapping should take place only in the distant future when the tools are readily available and become less expensive.

Genetic engineering has been used successfully in *J. regia* since methods for somatic embryogenesis from immature cotyledons and plant regeneration were developed. McGranahan and C. Leslie have been in use for walnuts and field trials of mature transgenic trees are under way. Genes of interest include Bt from *Bacillus thuringiensis* for insect resistance and crown gall silencing for resistance to crown gall. Tree architecture has been modified by the rolABC genes from *A. rhizogenes*, but the goal of increasing rootability could not be achieved. The reticence of the public to accept genetically engineered organisms has prevented commercialization of transgenic walnut trees, but it is expected that transgenic rootstocks will prove more acceptable.

STRAWBERRY

Origin: The cultivated strawberry is an accidental hybrid between the North American species or Virginia strawberry (*F. virginiana* Duch.) and the pistillate South American species *F. chiloensis* (L.) Duch. in France.

Taxonomy status:

Order: Rosales

Family: Rosaceae

Subfamily: Rosoideae

Tribe: Potentillae

Genus: *Fragaria*

Botanical name: *Fragaria* x *ananassa* Duch.

Chromosomes number: n = 7; 2n = 2x = 14 = 4x = 28 = 6x = 42: 8x = 56(cultivated strawberry).

There are about 20 recognized species of strawberries in five chromosome groups (x = 7): ten diploids, four tetraploids, one pentaploid, one hexaploid and four octoploids. The cultivated strawberry is an octoploid (2n = 8x = 56).

Genome size: 813.4 Mb; *Fragaria vesca* ssp. *vesca:* 240 Mb

A genome: *Fragaria vesca*

B genome: *F. iinumae*

The most recently proposed genome composition model for the octoploid *Fragaria* species, AAAABBBB (Bringhurst, 1990), implies the presence of up to four distinct subgenomes (A, A, B and B). The genomic complexity of the octoploid species has prompted attention to diploid relatives such as *Fragaria vesca* (2n=2x=14), as model systems for strawberry genetics and genomics (Davis and Yu, 1997; Sargent *et al.* 2004).

The closest relatives of strawberry are *Duchesnea* Smith and *Potentilla* L. can be used in improvement programme

Botany

Strawberry is acaulescent, more or less hairy perennial herb with basal leaves and long filiform runners from the axils, which root and form new plants; petioles mostly long and channeled above; stipules adnate at the base of the petiole, large, mostly scarious and brown, persistent and covering the stock. Leaves are 3- foliate, or sometimes unequally imparipinnate (i.e., with pair of much smaller lateral leaflets below the nodal leaflets sharply dentate, but entire at the more or less wedge-shaped base, the lateral ones oblique, the inner half usually smaller); scape mostly about as long as the petioles. The inflorescence is cyme terminating with primary blossoms. The flower is polygamodioecious, rarely hemaphrodite, the male flowers larger and showier, all 5-parted, central flowers open firsts often 6 to 8-parted and larger than the later one; calyx has ten sepals and calyx lobes form a flat hypanthium, augmented by as many short, mostly narrower outer calyx lobe bractlets. Flowers have five or more white obovate petals that surround up to 30 stamens, which in turn surround a raised, conical receptacle covered in up to 500 pistils, each atop an individual carpel stamens about 20 or less, or abortive/filaments mostly shorter than the receptacle, anthers oblong; receptacle roundish or conical bearing numerous pistils with lateral styles, at maturity the receptacle becoming enlarged and juicy, an aggregate fruit. The strawberry fruit is not actually a berry but instead a receptacle for potentially hundreds of tiny fruits, known as achenes.

Floral Biology

The time of flowering and its duration is recorded between February to June and flowers open in the morning. The maximum rate of anthesis and dehiscence in strawberry plants is found between 10 AM to 12 PM. Pollen matures before the anthers open, but it is not dehisced until the floewer opens. It remains viable for 2 -3 days. Stigma remains receptive to pollen for 2-3 days. Fertilization occurs 24-48 hr after fertilzation. Rise in temperature and fall in humidity are found to hasten anthesis as well as dehiscence. The pollination is cross pollination.

Breeding Objectives

1. To develop varieties for vigour, erect type, early flowering and higher yielder.
2. To develop varieties for blossom frost hardiness.
3. To develop varieties to extend the harvest season.
4. To develop varieties having good fruit characteristics (large size, flavor, symmetry, shape, skin firmness and high flesh firmness) and improved fruit quality (increased anthocyanin content, and soluble sugar content).
5. To develop varieties with low susceptibility to pathogens.
6. To develop varieties for resistance to red stele root rots, *Verticillium* wilt, leaf spot, leaf scorch, leaf blight, powdery mildew, gray mold and anthracnose and red mite.
7. To develop varieties having winter hardiness.

Limitations in Breeding

1. Heterozygosity.
2. Narrow genetic base.
3. Inbreeding depression.
4. Blossom sterility.
5. Polyploidy.

6. Polygenic control of most of horticultural traits.
7. Inheritance pattern is not known.

Germplasm Resources

The US Department of Agriculture, Agricultural Research Service (USDA-ARS), National Clonal Germplasm Repository at Corvallis, Oregon houses the *Fragaria* gene bank for the US National Plant Germplasm System (NPGS). This collection includes 411 unique *F.* x *ananassa* Duch. cultivars and selections in addition to 1079 representatives of 33 taxa from 29 countries.

In India, the germplasm is maintained at Regional Station IARI, Shimla (96), NBPGR Station, Bhowali, Dr YSPUHF, Solan, GBPUA&T, Pantnagar, SKUAST, Srinagar and CITH, Srinagar, India. Large number of collections was made through introductions and indigenous collections and has been conserved *ex-situ* at national field gene bank CITH and at NBPGR stations.

Sources of useful taits:

Early ripening: Sumas.

Large fruit size: *F. chiloensis*, Vesper, Belrubi.

Fruit flavour: *F. vesca*, *F. moschata* and *F. viridis.*

Day neutral character: *F. virgiana* ssp. *glauca,* Seascape, Albion.

Warm climate/Heat tolerance: *F. virginiana*, *F. vesca, F. orientalis.*

Cold: *F. moschata,* Joliette, Redcoat and Bounty.

Drought: *F. vesca*, Genotypes of *F. virginiana* and *F.chiloensis.*

Salt: *F.chiloensis* and Lassen, Festival naya, Fresno.

Heavy Soils: Joliette.

Alkaline soils: *F. viridis.*

Winter hardiness: *F. virginiana* spp. *glauca,* Kent, Joliette.

Two spotted spider mite: *F.chiloensis.*

Aphids: *F. nilgerrensis.*

Botrytis rot: Darselect, Sumas.

Red Stele root rot: Allstar, Arking, Sumas.

Leaf spot: Arking, Joliette.

Anthracnose: Sweet Charlie.

Powdery mildew: *F. moschata, F. vesca,* Arking, Vantage, Joliette.

Crown rot: *F. vesca.*

Verticillium wilt: Tristar, Vantage.

Virus tolerance: Toteum.

Genetics of Strawberry

Out of 20 recognized species in five chromosome groups (x=7): 10 diploids, 4 tetraploids, one pentaploid, one hexaploid and 4 octoploids, the cultivated strawberry is an octoploid (2n = 8x = 56). The centres of diversity of strawberry species are spread over Eurasia and North and South America.

Table 26. Polyploidy in *Fragaria* species from different habitats.

Ploidy	Chromosomes	Species/ subspecies	Habitat
Diploid	14	*Fragaria daltoniana*	Himalayas
Diploid	14	*Fragaria iinumae*	Japan, Eastern Russia
Diploid	14	*Fragaria nilgerrensis*	Southern Asia, Southeast Asia
Diploid	14	*Fragaria nipponica*	Western side of the Japanese island of Honshū, Japanese island of Yakushima
Diploid	14	*Fragaria nubicola*	Himalayas
Diploid	14	*Fragaria vesca*	Northern Hemisphere
Diploid	14	*Fragaria viridis*	Central Asia, Europe
Diploid	14	*Fragaria yezoensis*	Northeast Asia
Tetraploid	28	*Fragaria moupinensis*	China
Tetraploid	28	*Fragaria orientalis*	China, Mongolia, Russia
Pentaploid	35	*Fragaria × bringhurstii*	West Coast of the United States
Hexaploid	42	*Fragaria moschata*	Europe
Heptaploid	49	*Fragaria* x *comarum* hybrids	
Octoploid	56	*Fragaria × ananassa*	A hybrid species that is cultivated worldwide
Octoploid	56	*Fragaria chiloensis*	Pacific Ocean coasts of North and South America and Hawaii
Octoploid	56	*Fragaria chiloensis* subsp. *chiloensis* forma *chiloensis*	
Octoploid	56	*Fragaria chiloensis* subsp. *chiloensis* forma *patagonica*	Argentina, Chile
Octoploid	56	*Fragaria chiloensis* subsp. *lucida*	Coastal regions of British Columbia, Washington, Oregon, California
Octoploid	56	*Fragaria chiloensis* subsp. *pacifica*	Coastal regions of Alaska, British Columbia, Washington, Oregon, California

Octoploid	56	*Fragaria chiloensis* subsp. *sandwicensis*	Hawaii
Octoploid	56	*Fragaria iturupensis*	Iturup, Kuril Islands
Octoploid	56	*Fragaria virginiana*	North America
Decaploid	70	*Fragaria* x *vescana* *F. iturupensins*	Unnatural hybridization between *Fragaria* × *ananassa* and *Fragaria vesca*

Genetics of economically important traits (inheritance pattern)

Adaptation:

Chilling requirement: Quantitatively inherited.

Frost tolerance: Quantitatively inherited.

Flowering time: Quantitatively inherited.

Photoperiod sensitivity: Single dominant gene to quantitative inheritance.

Productivity:

Flower number, fruit size/weight, runner number, total yield: Quantitatively inherited.

Concentrated ripening: Quantitatively inherited; negatively correlated with yield.

Fruit quality:

Firmness: Quantitatively inherited; flesh firmness and skin toughness often correlated positively; QTL identified.

Ease of calyx removal: Quantitatively inherited; much additive variation.

Fruit colour: Skin and flesh colour quantitatively inherited; largely additive with a few major genes.

Pedicle length: Quantitatively inherited.

Sugar content: Controlled with varying levels of additive and dominance control depending on the population.

Acidity: Controlled with varying levels of additive and dominance control depending on the population; QTL identified for pH and titratable acidity.

Vitamin C: Quantitatively inherited with partial dominance for high levels.

Pest and diseases:

Aphid: Resistant genotypes were identified; regulated by more than one locus, with partial dominance and additive action.

Two-spotted spider mite: Resistant genotypes were identified; resistance may be biotype specific; strong additive and dominance effects involved in resistance.

Black vine weevil: Tolerant genotypes were identified; probable quantitative inheritance.

Alternaria leaf, spot, black leaf spot: Susceptibility is inherited as a single locus expressing dominance.

Verticillium wilt: Additive fashion with partial dominance.

Breeding Methods

1. Introduction: There is need to introduce cultivars of three categories: June-bearing, Everbearing and Day-neutral varieties based on photoperiodic requirement. These are Elsanta, Douglas, Vista, Bolero, Everest, Addie, Bangalore, Belruby, Blackmore, Brighton, Camarosa, Chandler, Dana, Douglas, Elasta, Etna, Fenny, Gorella, Majestic, Missionary, Ofra, Oso Grande, Pajaro, Robinson, Selva, Shasta, Steel Master, Sweet Charlie, Pajaro, Etna, Confictura, Dana, Belrubi, Gorella, Majestic, Phenomenal, Brighton and Fern. Some of these have been introduced and others need to be introduced from Europe and North America.

2. Selection: The earliest cultivars resulted from selection of superior phenotypes from chance crossings. Tioga was selected at the Wolfskill Experimental Orchards, University of California, Davis, in 1955 from a population of about 300 plants. Cambridge Favourite was selected from Blakemore in England. Pusa Early Dwarf was selected at IARI Research Station, Shimla. The cultivar Jutogh Special and Shimla Delicious strawberry varieties developed for early fruiting at IARI, Regional station, Shimla.

3. Hybridization: Thomas Knight was first to initiate breeding in UK and developed several hybrids like Downton, Elton and Keens Seedling. Thomas Laxton released a number of important cultivars including Noble and Royal Sovereign. In England, D. Boyle produced a large series of cultivars with the prefix 'Cambridge', e.g. Cambridge Favourite. The public and private breeding programmes across the world are in progress to develop new strawberry cultivars.

***East Malling Reseach Station, East Malling*:** Strawberry breeding has been successful at East Malling (NIAB-EMR), Kent, UK. East Malling Strawberry Breeding Club (EMSBC), a public-private partnership, was formed in 2008 to fund the development of new cultivars. As a result new cultivars Malling™ Star, Strawbino, Serenity, Malling™ Glow and Malling™ Sunrise were released.

Pandora: (Van Humboldt x Redstar) x Merton.

Sweetheart: It is a cross between Rosie and Onebor and was first selected in 1999.

Flamenco: It was selected in 1995 from progeny of the cross Evita x EMR77. The male parent is from crosses involving Selva, Tioga, Gorella and Gento.

Buddy: Originated by cross Diamante × DNBL205, developed by David William Simpson, Adam Barrie Whitehouse, Abigail Wendy Johnson at EMRS, East Malling.

Florence: Cross between [Tioga x (Redgauntlet × (Wiltguard × Gorella)] × (Providence × self).

Elegance: It is a cross between EM834 and EM1033.

Malling Centenary: It was selected at East Malling Research Station in 2006 from a cross between two East Malling breeding lines (SDBL102 and EM1315).

***Plant Research International B.V., Wageningen Netherlands*:**

Elsanta: Cross between Gorella and Holiday.

***Germany*:**

Malwina: Cross of 'Sophie' x a clone from Schimmelpfeng, weihenstefan, bred by Peter Stoppel in Germany.

***USDA-ARS Beltsville, Maryland, MD*:**

Allstar: Cross between US 4419 and MDVS 3184.

Earliglow: It resulted from a cross made in 1964 between Fairland and Midland seedling with a Redglow x Surecrop seedling by D.H.Scott and A.D. Draper.

Lateglow: Cross between Tamella and MDUS 3184 (NCUS 1768 × Surecrop) made in 1974 and selected at Wye Inst., near Queenstown, Md., in 1976 by A.D. Draper and P. Rosati.

Lester: Result of cross made in 1969 between Raritan and MDUS 3413 ([(Tennessee Shipper × Maytime) × Stelemaster] × Earlidawn); selected at Wye Institute and originated in Beltsville, Md., by D.H. Scott, USDA.

Surecrop: Fairland x Mdus 1972 AAFC, St. Jean-sur-Richelieu, Que, Canada.

St. Pierre: Chandler x Jewel.

***University of Florida, Florida*:**

Sweet Charlie: Cross between A.L. 80-456 and Pajaro and released in 1992 by Craig K. Chandler, University of Florida.

Rosa Linda: Originated by crossing FL87-418 with FL 87-200.

Strawberry Festival: Cross between Rosa Linda and Oso Grande.

Earlibrite: Cross between Rosa Linda and FL 90-38.

Florida Brilliance: Cross between unreleased selection FL 11.31-14 and FL 10-153.

Carmine: Cross beteween Rosa Linda and FL.93-54.

Florida Beauty: Cross of selection 2010-119 × Florida Radiance at University of Florida, Wimauma, FL, by V. Whitaker, originated from Queensland, Australia. It is a day neutral cultivar resistant to root rot disease.

Florida Radiance: Cross between Winter Dawn and FL 99-35.

Florida Elyana: Cross between FL 96-114 and FL 95-200.

Florida127: Originated from a cross between Winterstar™ FL 05-107 and unreleased selection FL 02-58 (male parent).

Winterstar (FL 05-107): Cross between Florida Radiance and Earlibrite.

***University of California, Davis*:**

Albion: Cross between Diamante and Cal 94.16.-1.

Aromas: Developed from the cross Cal 87.112-6 × Cal 88.270-1.

Camarosa: Originated by crossing Douglas × Cal 85.218-605 and was released in 1993 by V. Voth, R. Bringhurst and D. Shaw of the University of California.

Albion: Originated by crossing Diamante and Cal 94.16-1.

Chandler: Developed from Douglas x 3 Cal 72.361-105 (C55) cross. It was introduced by R.S. Bringhurst and V. Voth of the University of California.

***New York State Agr. Expt. Sta. NY*:**

Jewel: It is a cross of (Tennessee Shipper × Fairfax) × Geneva] × Senga Sengana) × Holiday, developed by J.C. Sanford, D.K. Ourecky, and J.E. Reich.

Canoga: It is a cross of NY1123 'Senga Sengana' × 'Midland') × Holiday.

***University of Arkansas, Fayetteville* :**

Cardinal: Cross of Earlibelle and Ark. 5063, made in 1967 by J.N. Moore, H.L. Bowden and W.A. Sistrunk.

***Agr. Canada Res. Station, Nova Scotia*:**

Cavendish: Cross between Glooscap and Annapolis made in 1981, selected in 1983, originated in Kentville, Nova Scotia, by D.L. Craig, of Agr. Canada Res. Sta.

Sumas: Cross between Cheam and Tioga; made in 1969, originated in Vancouver, B.C., by H.A. Daubeny, Agriculture Canada Research Station.

Niigata S3: It was selected from a cross between Kei812 and Asuka-Ruby at Tohoku Agricultural Research Center. The cultivar is early flowering and the fruit possesses high soluble solid content and good coloration.

***Insititute for Horticultural Development, Knoxfield, Victoria, Australia*:**

Tallara: Result of a cross between Parker and Pajaro.

Interploid crosses are often quite difficult but species with the same ploidy level can often be successfully crossed.

Pineberry: Hybrid crof *F. Chiloensis* and *F. Virginians*. Fruit is white. Disease resistant.

Intergeneric hybridization

- *F. ananassa* x *Potentilla patustris*
- *F. ananassa* x *Potentilla erecta*
- *F. ananassa* x *Potentilla reptans*
- *F. chiloensis* x *Potentilla glandulosa*
- *F. virginiana* x *Potentilla anserine*
- *F. virginiana* x *Potentilla fruticosa*

Fragaria and *Potentilla* have been hybridized but most of the hybrids were sterile and died at an early age.

Rubus ideaus subsp. *strigosus* Michx. x *Fragaria* x *ananassa* Duch., originated in South America.

Interspecific hybridization

Fragaria chiloensis x *Fragaria virginiana.*

4. Biotechnology:

***Micropropagation*:** Micropropagation of strawberry is widely used in commercial propagation as well as in breeding programs. This technique has been adopted by most nurseries producing millions of disease free plants. Amongst a vast number of protocols developed for *in vitro* regeneration, recently developed protocols enable strawberry micropropagation in a single step where shoot multiplication and rooting takes place in the same culture medium.

***Molecular markers*:** Molecular marker techniques are useful to verify the clonal fidelity of micropropagated strawberries. There is still need to utilize the full potential *in vitro* selection, somaclonal variation, haploid recovery, somatic hybridization and in cryopreservation of strawberries.

Genetic transfomation: Genetic transformation has been developed using tissue culture systems with varying degree of success. *Agrobacterium*-mediated transformation has been a technique of choice for transformation even though a few direct gene transfer protocols have been reported. However, so far no commercial transgenic varieties have been released. The first transformation of strawberry was attempted in 1990. Nehra *et al.* (1990) inoculated strawberry leaf disks and calli with *A. tumefaciens* strain MP90 which contains nptII and gus genes to yield a transformation frequency of 6.5% for leaf disks and 3% for calli where all of the transgenic calli and shoots expressed varying levels of nptII and GUS enzyme activity. Integration of both marker genes into the strawberry genome was further confirmed by Southern blot analysis. The most important aspect of transformation efficiency is the choice of species and cultivars. Genotype is an important factor that significantly influences the efficiency of transformation. High variability of efficiency was observed among strawberry cultivars with different genetic backgrounds, even within the same cultivar.

6

Epilogue

Modern fruit breeding is both an art and science. It has been practised for thousands of years as humans continued to use various methods to change the traits of plants in order to add desired characteristics. It has been in a state of flux because the traditional support by government experiment stations declined throughout the world as a reaction to the high costs of long term programmes that are necessary. However, the globalization of the fruit industry has underscored the economic value of improved germplasm indicating that support for fruit breeding needs to be increased. Returns from patents offer one solution to impede the decline in funding and to encourage breeding efforts. Patents are essentially a direct tax on the industry to improve breeding and also offer incentive to breeders when they share the patent royalties. In some countries the costs of foreign patents has encouraged investment in local breeding programmes. The objectives of breeding efforts in the future must be twofold. The most important is to increase quality in order to increase consumption. This is important to reverse the decline in per capita fruit consumption in developed countries. The second effort must be to increase the efficiency with special emphasis on productivity, annual bearing, and pest resistance. However, the future of fruit breeding is crop specific. For example the tremendous advances in seedless table grapes have intensified breeding efforts. In contrast, the breeding of wine grapes is impeded by the association of traditional cultivar name with the name of the wine. As a result the wine industry has not encouraged cultivar change except for clonal selection. Practically no advances have been made in the genetic improvement of banana but there are major projects via international efforts of Bioversity International. The lack of diversity in export of banana makes breeding essential due to disease pressure (such as Panama wilt, Sigatoka leaf spot) but seedless banana is notoriously difficult to hybridize. Similarly, citrus industry has increased efforts to produce new seedless, easy-to-peel clones while the threat of diseases such as hanglongbing (citrus greening) is an upcoming problem that may require a genetic solution. The success of breeding in *Prunus* has encouraged a number of private breeding organizations especially in peach and plums. Partially stoneless plum described by Luther Burbank suggests that the hard stone might be eliminated in *Prunus.* Similarly the success of breeding efforts to produce late ripening cherries in British Columbia, Canada has indicated the high economic returns made possible by sustained breeding efforts. Apple breeding has produced many new cultivars and the success of Fuji and Honeycrisp has underscored the returns from improved quality. There are large efforts in breeding loquat in China including triploid seedless clones. The success in blackberry with the introduction of thornlessness, primocane fruiting, reduced seediness and large, tasty fruits have created a new global industry. In the case of underutilized or new fruit crops, knowledge of domestication should be used to predict future changes. Thus one might anticipate that in kiwifruit, hermaphroditic mutations would eliminate the need for staminate clones as pollinizers, and that fruit skin mutations or breeding could lead to non-pubescent clones with more attractive and more edible fruit surface. The interspecific hybridization should be encouraged in *Citrus, Prunus* and *Rubus* to develp new fruits. Similarly, several underexploited crops hold lot of potential in view of the climate change scenario. One of the likely candidates for domestication is pitaya (species of *Selenicereus*, *Hylocereus* and *Ceresus*), an extremely attractive fruit of columnar cactus, which requires breeding efforts since many selections have been insipid. In the future, the greatest progress in fruit breeding will be the improvement of underutilized fruits such as jaboticaba in Brazil. A number of temperate fruits such as kiwi fruit, *Vaccinium*, cranberry, blueberry, lingonberry are candidates for commercialization provided postharvest technology can be improved. The fruit species such as North American pawpaw, blue honeysuckle (*Lonicera caerulea*), ōhelo (*Vaccinium reticulatum*), Indian ber (*Ziziphus manuritiana*) and aonla (*Phyllanthus emblica*) that have commercial potential are currently in a status of low economic importance. The development of these fruits depends on factors including the cultivation limitations, yields, uses of the fruit and marketing potential.

Many of our fruit cultivars are ancient and based on grower-selected seedlings and somatic mutations. There is search for specific mutations (parthenocarpy and seedlessness, breakdown of dioecy, loss of compatibility, high sugar, loss of astringency) in these fruits.

In conventional breeding systems, elite clones are selected followed by fixation by vegetative propagation, to provide raw material for subsequent cycles of hybridization and selection. The development of fruit culture is based on an interaction between improved clones and cultural practices. Deficiencies are compensated by such techniques as artificial pollination, use of disease resistant and size controlling rootstocks, extensive methods of disease control involving complex schedules of pesticide application, the control of fruit size and annual bearing by manual and chemical fruit and flower thinning, control of fruit abscission by growth regulators and extensive pruning and training systems. The dilemma is to determine if breeding emphasis should concentrate either on quality for consumers or to overcome production problems for growers by genetic solution. Obviously, both must be addressed in modern breeding program. Since the fruit breeding depends on genetic diversity, the development of germplasm resources must be strengthened by collections *in situ*. In some cases these collections suggest that a number of fruits that have been ignored, such as quince, are candidates for increased breeding efforts. Finally advances in biotechnology must be attached to breeding efforts. Marker assisted selection could offer significant help in disease and pest resistance breeding but much genetic work must still be pursed to locate these quality traits. The US Department of Agriculture funded programme in 2009 called RosBREED is underway to translate progress in marker-assisted breeding in rosaceous fruit crops including apple, peach, sweet and tart cherries and strawberry. Similarly transgene technology could find a place in fruit breeding. The example of virus resistant papaya has demonstrated that success can be achieved. However, the rationale for transgene technology assumes that the present germplasm is optimum.

This is a conservative approach somewhat similar to backcross breeding but it is true that many high quality fruits are difficult to replace.

Two important international research programmes have been initiated to improve fruit breeding (Laurens *et al.*, 2012); the USA based international research program RosBREED (www.rosbreed.org) and the European research project FruitBreedomics (www.fruitbreedomics.com). The project FruitBreedomics comprises 28 Institutes and companies representing 14 countries and 367 stake holders in Europe. The project was to focus primarily on two major fruits in Europe, i.e. apple and peach. However, the knowledge gained on the latest methodologies, technologies should also benefit to improve the other species of the Rosaceae family in view of the strong genomic and functional ancestral relatedness by adding efficiency in current fruit breeding approaches. The project has already made significant breakthroughs in developing tools, methodologies and the latest knowledge in fruit breeding.

Recent advances on molecular genetics, genomics and biotechnology allow developing more efficient breeding programmes aimed to improve sustainability by obtaining new varieties with high fruit quality, enriched with health promoting compounds, seedless and biotic and abiotic stress resistance. The efficiency of breeding programmes will be continuously improved as a result of acquired new knowledge on gene-trait associations and gene regulations, especially from new genome editing experiments. New gene functions discovery and directed mutation to improve elite mandarin cultivars should be acheived through CRISPR/Cas system which has positioned as the most promising strategy for genome editing. Future advances should be aimed to transfer of genetic resistance to biotic and abiotic stresses through increased use of biotechnology including marker assisted selection of new early flowering, productive genotypes having quality traits with emphasis on consumer satisfaction.

Suggested Future Reading

Al-Khayri, J., Jain, S. M. and Johnson, D. V. (eds). (2018). Advances in Plant Breeding Strategies: Fruits. Springer International Publishing, Berlin, Germany.

Badenes, M.L. and Byrne, D.H. (eds.) (2012). Fruit Breeding, Handbook of Plant Breeding 8, Springer Science + Business Media.

Hancock, J.F. (ed.) (2008). Temperate Fruit Crop Breeding: Germplasm to Genomics. Springer Science + Business Media B.V, New York.

Hanke, M.V. and Flachowsky, H. (2010). Fruit Crops. In Kempken, F. and Jung, C. (eds.), Genetic Modification of Plants, Biotechnology in Agriculture and Forestry, Vol. 64, Springer, Berlin Heidelberg, Germany.

Jaganathan, Deepa., Ramasamy, Karthikeyan., Sellamuthu, Gothandapani., Jayabalan, Shilpha and Venkataraman, Gayatri (20018). CRISPR for Crop Improvement: An Update Review. Front Plant Sci. 9: 985.

Janick, J. and Moore, J.N. (eds.). (1996). Fruit Breeding vol.1-3, John Wiley & Sons, Inc. New York.

Kole, C. (ed.) (2011). Wild Crop Relatives: Genomic and Breeding Resources, Tropical and Subtropical Fruits. Springer-Verlag, Berlin Heidelberg.

Kumar, N. (2006). Breeding of Horticultural Crops: Principles and practices. New India Publishing Agency, India.

Ray, P.K. (2002). Breeding Tropical and Subtropical Fruits. Narosa Publishing House, New Delhi.

Shunell, R.J. and Priyadarshan, P.M. (eds.). (2012). Genomics of Tree Crops. Springer, New York, pp. 209–239.

Song, G., Jia, M., Chen, K., Kong, X., Khattak, B., Xie, C., Li, A. and Mao, L. (2016). CRISPR/Cas9: a powerful tool for crop genome editing. Crop J. 4(2):75–82.

Song, Guo-qing., Prieto, Humberto and Orbovic, Vladimir. (2019). *Agrobacterium*-mediated transformation of tree fruit crops: methods, progress, and challenges. Front. Plant Sci. 10: 226.

Thormann, E., Dulloo, M.E. and Engels, J.M. M. (2006).Techniques for *ex situ* plant conservation. In : Plant Conservation Genetics, Chapter: 2, Publisher: Haworth Press, Editors: Henry, R.J., pp.7-36.

References

Abadie, Catherine., Chilin-Charles, Yolande and Huat, Joël. (2009). New approaches to select cultivars of banana with durable resistance to *Mycosphaerella* leaf spot diseases. Acta Horticulturae 828: 171-178.

Acquaah, G. (2012). Principles of plant genetics and breeding. 2nd edn. Wiley-Blackwell, Oxford.

Adato, A., Sharon, D. and Lavi, U. (1995). Application of DNA fingerprints for identification and genetic analyses of mango (*Mangifera indica*) genotypes. Journal of the American Society for Horticultural Science 120: 259–264.

Ahloowalia, B. S., Maluszynski, M. and Nichterlein, K. (2004). Global impact of mutation-derived varieties. Euphytica 135: 187–204.

Aiyappa, K.M. and Nanjappa, P.P. (1959). Coorg Honey, a new find in papaya. Indian Hort. 3: 3.

Albrigo, L.G. Stelinski, L.S. and Timmer, L.W. (2019). Citrus, 2nd ed. CABI, UK. p.38.

Aleza, P., Juárez, J., Cuenca., J., Ollitrault., P. and Navarro, L. (2012a). Extensive citrus triploid hybrid production by 2x×4x sexual hybridizations and parent-effect on the length of the juvenile phase. Plant Cell Rep 31:1723–1735.

Aleza, P., Juárez, J., Hernández, M., Ollitrault., P. and Navarro, L. (2012b). Implementation of extensive citrus triploid breeding programs based on 4x×2x sexual hybridisations. Tree Genet. Genomes 8:1293–13 06.

Allard, R.W. (1960). Principles of plant breeding. John Wiley and Sons, New York.

Alston, F.H. and Briggs, J.B. (1970) Inheritance of hypersensitivity to rosy apple aphid *Dysaphis plantaginea* in apple. Can J Genet Cytol 12:257-258.

Alston, F.H., Phillips, K. and Evans, K. (2000). A *Malus* gene list. Acta Horticulturae 538: 561-570.

Ammal, Janaki and Raghvan, R. S. (1957). Polyploidy and vitamin c in *Emblica officinalis* Gaertn. Proc. Ind, Acad. Sci. 47 (5): 312-314.

Angiosperm Phylogeny Group. (2003). An update of the Angiosperm Phylogeny Group classification for the orders and families of flowering plants: APG II. *Bot.* J. Linn. Soc., 141: 399-436.

Aquilizan, F.A. (1983). Development of papaw hybrid varieties. In: Driscoll, C.J.J. (ed.) Proceedings of Australian Plant Breeding Conference, 14–18 February 1983, Adelaide. Plant Breeding Conference Organizing Committee, Adelaide, Australia, pp. 267–277.

Aquilizan, F.A. (1987). Breeding systems for fixing stable papaya inbred lines with breeding potential for hybrid variety production. In: The breeding of horticultural crops. Proc. of Intern. Symp. held at Taichung, Taiwan, 1-5 December 1986, pp. 101-106. FFTC Book Series No. 35. Taipei, Taiwan.

Aron, Y., Czosnek, H., Gazit, S. & Degani, C. (1998). Polyembryony in mango (*Mangifera indica* L.) is controlled by a single dominant gene. HortScience 33 (7): 1241-1242.

Aslam, Junaid., Khan, Saeed Ahmed and Azad, M.A.A. (2015). *Agrobacterium* mediated genetic transformation of date palm (*Phoenix dactylifera* L.) cultivar "Khalasah" via somatic embryogenesis. Plant Science Today 2 (3): 93-101.

Azad, M. A. K.., Rabhani, G. and Amin, I. (2012). Plant regeneration and somatic embryogenesis from immature embryos derived through interspecific hybridization among different *Carica* species. International Journal of Molecular Science 13: 17065–17076.

Azhar, M. and Heslop-Harrison, J.S. (2008). Genomes, diversity and resistance gene analogues in *Musa* species. Cytogenetic and Genome Research 121(1): 59-66.

Babo, A. von and Mach, E. (1909). Handbuch des Weinbaues und. der Keller wirtschaft, Berlin.

Badenes, M.L. & Parfitt, D.E. (1995). Phylogenetic relationships of cultivated *Prunus* species from an analysis of chloroplast DNA variation. Theor. Appl. Genet. 90: 1035-1041.

Bakry, F. and Horry, J.P. (1992). Evidence for androgenesis in bananas. In Book of Poster Abstract, XIIIth EUCARPIA Congress, July 06–11th 1992, Angers, France, pp. 133–134.

Banno, K., Hirano, Y., Ishikawa, H. & Kakegawa, M. (2003). Breeding and characteristics of symmetric intergeneric hybrids between apple and pear. Acta Horticulturae 622: 265–276.

Beakbane, A.B. (1941). Anatomical studies of stems and roots of hardy fruit trees, In: the anatomical structure of some clonal and seedling apple rootstocks stem- and root-grafted with a scion variety. J. Pomol. Hort. Sci. 18:344-367.

Beakbane, A.B. and Thompson, E. C. (1939). Anatomical studies of stems and roots of hardy fruit trees, 11: the internal structure of the roots of some vigorous and some dwarfing rootstocks and the correlations of structure with vigor. J. Pom. 17:141-149.

Beakbane, A.B. and Thompson, E. C. (1947). Anatomical studies of stems and roots of hardy fruit trees, 4: the root structure of some new clonal apple rootstocks budded with Cox's Orange Pippin. J. Pom. 23: 206-211.

Beakbane, A.B. and Majumder, P. K. (1975). A relationship between stomatal density and growth potential in apple rootstocks. J. Hort. Sci. 50: 285-289.

Bentham, G. and Hooker, W. J. (1862-1883). Genera Plantarum. L. Reeve & Co., London.

Bettencourt, E., Hazekamp, T. and Perry, M.C. (1992). Directory of Germplasm Collections. 6.1. Tropical and Subtropical Fruits and Tree Nuts. International Board for Plant Genetic Resources, Rome.

Bhattacharya S.C. and Dutta, S. (1956). Classifications of Citrus fruits of Assam. ICAR Sci. Monograph 20: 1–110

Bhujanga Rao, C. and Rangacharlu, V.S. (1958). Breeding new mango varieties in South India. Indian J. Hort. 15:173-83.

Biswajit, Das., Ahmed, Nazeer., Ranjan, J.K., Attri, B. L., Sofi, A.A., Verma, R.K., Verma, M.K. and Tripathi, P. C. (2012). New apple Variety Released. Indian J. Hort. 69(2): i-iv.

Boritzki, M., Plieske, J. and Struss, D. (2000). Cultivar identification in sweet cherry (*Prunus avium* L.) using AFLP and micro-satellite markers. Acta Horticulturae 538: 505-510.

Boskovic, R., Tobutt, K.R., Batlle, I., Duval, H., Martínez-Gómez, P. and Gradziel, T.M. (2003). Stylar ribonucleases in almond: correlation with, and prediction of incompatibility genotypes. Plant Breed. 122: 70-76.

Boss, P.K. and Thomas, M.R. (2002). Association of dwarfism and floral induction with a grape 'green revolution' mutation. Nature 416: 847–850.

Bramel, P.J. and Volk, G. (2019). A global strategy for the conservation and use of apple genetic resources. Global Crop Diversity Trust. Bonn, Germany.

Bringhurst, R.S. (1990). Cytogenetics and evolution in American Fragaria. HortScience 25(8): 879–881.

Brown, J. and Caligari, Peter. (2008). An Introduction to Plant Breeding. Blackwell Publishing , Oxford.

Burns, W. and Prayag, S.H. (1920).The Book of Mango. Bulletin Department of Agriculture, Bombay, 103 p.

Cain, D.W., Emershad, R.L. and Tarailo, R.E. (1983). *In-ovulo* embryo culture and seedling development of seeded and seedless grapes (*Vitis vinifera* L.). Vitis 22: 9–14.

Campalans, Anna., Pagès, Montserrat. and Messeguer, Ramon. (2001). Identification of differentially expressed genes by the cDNA-AFLP technique during dehydration of almond (*Prunus amygdalus*). Tree Physiology 21: 633–643.

Campos, M.K.F., Carvalho, K., Souza, F.S., Marur, C.J., Pereira, L.F.P., Bespalhok, J & Vieira, L.G. E. (2011). Drought tolerance and antioxidant enzymatic activity in transgenic 'Swingle' citrumelo plants over-accumulating proline. Environmental and Experimental Botany 72: 242-250.

Candolle, A. de. (1886). Origin of cultivated plants. Haffner, New York.

Carbonneau, A. (1985). The early selection of grapevine rootstocks for resistance to drought conditions. Am. J. Enol. Vitic. 36 (3): 195-198.

Chan, Y.K. (1987). Backcross method in improvement of papaya (*Carica papaya* L.). Malaysian Journal of Applied Biology 16: 95–100.

Chan, Y.K. (1992). Progress in breeding of F_1 papaya hybrids in Malaysia. Acta Horticulturae 292: 41–49.

Chan, Y.K. (2001). Heterosis in Eksotika × Sekaki papaya hybrids. Journal of Tropical Agriculture and Food Science 29 (2): 139–144.

Chan, Y.K. (2009). Breeding papaya (*Carica papaya* L.) In: Jain, S.M. and Priyadarshan, P.M. (eds) Breeding Plantation Tree Crops: Tropical Species. Springer Science and Business Media, New York.

Chan, Y.K. and Lam, P.F. (2003). Irradiation induced mutation in papaya with special emphasis on papaya ringspot resistance and delayed fruit ripening. In: Improvement of Tropical and Subtropical Fruit Trees through Induced Mutations and Biotechnology, Second Research Coordination Meeting, 2–6 September 2002, Vienna, Austria. Food and Agriculture Organization of the United Nations (FAO)/International Atomic Energy Agency (IAEA), Rome, p. 3545.

Chan, Y.K., Lee, H.K. and Rusana, I. (2007). Irradiation induced variations in M1 populations of Eksotika papaya. Journal of Tropical Agriculture and Food Science 35(1): 49–57.

Chan, Y.K. and Mak, C. (1996). Simultaneous selection for yield and stability of papaya (*Carica papaya* L.) MARDI Res. J. 24: 1-11.

Chang, C.C. and Wu, C.C. (1974). Trials of Hawaiian papaya varieties in Taiwan and performance of F_1 hybrids. Journal of Taiwan Agricultural Research 23: 273–283.

Chang, L.S. and Guo, T.T. (2002). Heritability of tolerance to papaya ringspot virus among papaya populations by parent–offspring regression. In: XXVIth International Horticultural Congress and Exhibition (IHC 2002*),* 11–17 August 2002, Toronto, Canada. International Society for Horticultural Science (ISHS), Toronto, Canada, Abstract S20-P-20, p. 507.

Chan-Tai, C., Yen, C.R., Chang, L.S., Hsiao, C.H. and Ko, T.S. (2003) All hermaphrodite progeny are derived by self-pollinating the sunrise papaya mutant. Plant Breeding 122: 431–434.

Chao, Chih Cheng T. and Krueger, Robert R. (2007). The Date Palm (*Phoenix dactylifera* L.): Overview of Biology, Uses, and Cultivation**.** HortScience 47: 1077-1082.

Cheema, G.S. and Deshmukh, G.B. (1927). Culture of guava and its improvement of selection in western India. Bombay Dept. Agril. Bull., 148.

Chen, G., Ye, C.M., Huang, J.C., Yu, M. and Li, B.J. (2001). Cloning of the Papaya ringspot virs (PRSV) replicase gene and generation of PRSV-resistant papayas through the introduction of the PRSV replicase gene. Plant Cell Reports 20: 272-277

Collins, J.L. (1933). Morphological and cytological characteristics of triploid pineapples. Cytologia 4: 248–256.

Collins, J. L. (1960). The Pineapple: Botany, Cultivation and Utilization. Interscience Publishers, New York.

Collins, J.L. and Kerns, K. (1938) Mutations in pineapples: a study of thirty inherited abnormalities in the Cayenne variety. Journal of Heredity 29: 169–173.

Campos, M. K. F. *et al.* (2011). Drought tolerance and antioxidant enzymatic activity in transgenic 'Swingle'citrumelo plants over-accumulating proline. Environ. Exp. Bot. 72, 242–250.

Conover, R. A., Litz , R. E. and Malo, S. E. (1986). 'Cariflora'- a Papaya ringspot virus-tolerant papaya for South Florida and the Caribbean. HortScience 21: 1072.

Corner, E. J. H. (1952). Wayside Trees of Malaya, 1. Second edition. Singapore: Government Printing Office.

Cruz-Hernández, A. and Gómez-Lim, M.A. (1995). Alternative oxidase from mango (*Mangifera indica*, L.) is differentially regulated during fruit ripening. Planta 197 (4): 569-76.

Dandekar, A.M., Jacobson, A., Ibáñez, Ana M., Gouran, Hossein, Dolan, David L., Agüero, Cecilia B., Uratsu, Sandie L., Just, R. and Paulo, A. Zaini. (2019). Trans-Graft protection against Pierce's disease mediated by transgenic grapevine rootstocks. Front. Plant Sci. 10: 1-10.

Darlington, C. D. (1956). Chromosome Botany and the Origins of Cultivated Plants. Revised 2nd Edn. Crows Nest, NSW: G. Allen & Unwin Ltd.

Dash, D.K., Mishra, R.S. and Singhsamant, P.K. (1988). Selection parameters for improvement in papaya (*Carica papaya* L). Orissa Journal of Horticulture 26 (1): 47–49.

Davis , T. M. and Yu , H. (1997). A Linkage Map of the Diploid Strawberry, *Fragaria vesca*. Journal of Heredity 88(3): 215 -221.

De Candolle, A. (1886) Plants Cultivated for Their Seeds. In: Origin of Cultivated Plants, 2nd Edition, International Scientific Series Vol. XLIX, Kegan Paul Trench & Co., London, 376-384.

de Vries, D.P. (1967). Phenological stages in sweet cherry with regard to pre-selection. Euphytica 16: 177–182.

Dinesh, M.R., Iyer, C.P.A. and Subramanyam, M.D. (1991). Combining ability studies in *Carica papaya* L. with respect to yield and quality characters. Gartenbauwissenschaft 56 (2): 81–83.

Dinesh, M.R., Rekha, A., Ravishankar, K. V., Praveen, K. S. and Santosh, L. C. (2007). Breaking the intergeneric crossing barrier in papaya using sucrose treatment. Sci. Hort. 114: 33-36.

Dinesh, M.R. and Yadav, I.S. (1998). Half-sib analysis in guava (*Psidium guajava*). Ind. J. Hort., 55: 20-22.

Drew, R. (2014). The Use of Non-Transgenic Technologies for the Development of Papaya Ringspot Virus Resistance in *Carica papaya.* Acta Horticulturae 1022: 23-29.

Drew, R.A., O'Brien, C.M. and Magdalita, P.M. (1998). Development of *Carica* interspecific hybrids. Acta Horticulturae 461: 285–291.

Drew, R.A., Siar, M., O'Brien, C.O. and Sajise, A. (2005). Introgression of genes for PRSV-P resistance to papaya from *Vasconcellea quercifolia*. In: 1st International Symposium on Papaya, 22–24 November 2005, Genting Highlands City, Malaysia. International Society for Horticultural Science (ISHS), Leuven, Belgium, Abstract S2-9, p. 21.

Druart, Ph., Delporte, F., Brazda, M., Ugarte-Ballon, C., da Câmara Machado, A., Lamier da Câmara Machado, M., Jacquemin., J. and Watillon, B. (1998). Genetic transformation of Cherry trees. Acta Horticulturae 468: 71-76.

Duarte, I.A., Ferreira, J.M. and Nuss, C.N. (2003). Potencial discriminatorio de tres testadores em top crosses de milho. Pesquisa Agropecuária Brasileira 38: 365–372.

Dwivedi, A.K. (1998). Coheritability and predicted response to selection in papaya (*Carica papaya* L.). South Indian Horticulture 46 (1&2): 1–4.

Eastwood, A; Lazkov, G. and Newton, A.(2009). The Red list of Trees of Central Asia. Fauna and Flora Internation, Cambridge, UK. p.13.

Emershad, R.L. and Ramming, D.W. (1984). *In-ovulo* embryo culture of *Vitis vinifera* L. cv. Thompson Seedless. Amer. J. Bot. 71: 873–877.

Endo, T., Shimada, T., Fujii, H., Kobayshi, Y., Araki, T. and Omura, M. (2005). Ectopic expression of an FT homolog from citrus confers an early flowering phenotype on trifoliate orange (*Poncirus trifoliata* (L.) Raf.). Transgenic Research 14: 703-712.

Engelmann, F. (2000). Importance of cryopreservation for the conservation of plant genetic resources. In: Engelmann, F. and Takagi, H. (eds.) Cryopreservation of tropical plant germplasm—current research progress and applications. JIRCAS/IPGRI, Tsukuba/Rome, pp 8–20.

Engelmann, F. (2004) Plant cryopreservation: Progress and prospects. In vitro Cell.& Dev. Biol. Plant 40: 427-433.

Fellman, Shanna and Walter-Peterson, Hans (2015). VitisGen: Mapping the Way to the Next Generation of Grapes. Research Focus 2015-1: Cornell Viticulture and Enology, 1-5

Firoozbady, E. & Young, T. R. (2015). Pineapple plant named "Rose". US Patent PP25,763 P3.

Fitch, M.M.M., Leong, T., Akashi, L., Yeh, A., White, S. *et al.* (2002). Papaya ringspot virus resistance genes as stimulus for developing new cvs and new production system. Acta Horticulturae 575: 85–91.

Fitch, M.M.M. and Manshardt, R. M. (1990). Somatic embryogenesis and plant regeneration from immature zygotic embryos of papaya (*Carica papaya* L.). Plant Cell Reports 9 (6) : 320–324.

Fitch M.M.M., Manshardt, R. M., Gonsalves, D., Slightom, J.L. and Sanford, J.C. (1990). Stable transformation of papaya via microprojectile bombardment. Plant Cell Reports 9(4):189–194.

Fitch, M.M.M., Manshardt, R.M., Gonsalves, D., Slightom, J.L. and Sanford, J.C. (1992).

Virus resistant papaya plants derived from tissues bombarded with the coat protein gene of papaya ringspot virus. Biotechnology 10: 1466–1472.

Flachowsky, H., Le Roux, Pierre-Marie., Peil, Andreas., Patocchi, Andrea., Richter, Klaus and Magda-Viola, H. (2011). Application of a high-speed breeding technology to apple (*Malus* x *domestica*) based on transgenic early flowering plants and marker-assisted selection, New Phytologist 192: 364-377.

Flachowsky, H., Peil, A., Sopanen, T., Elo, A and Hanke, M.V. (2007). Over-expression of BpMADS4 from silver birch (*Betula pendula* Roth.) induces flowering in apple (*Malus x domestica* Borkh.). Plant Breeding 126: 137-145.

Floyd W.F. (1920). Peach growing in Florida. Fla Coop Ext Bul. 27:5–8

Franks, T., Botta, R. and Thomas, M.R. (2002). Chimerism in grapevines : implications for cultivar identity, ancestry and genetic improvement. Theor. Appl. Genet. 104: 192–199.

Freiman , A., Shlizerman, L., Golubowicz, S., Yabloviz, Z., Korchinsky, R., Cohen, Y., *et al*. (2012). Development of a transgenic early flowering pear (*Pyrus communis* L.) genotype by RNAi silencing of PcTFL1-1 and PcTFL1-2. Planta 235: 1239–1251.

Frost, H.B. and Soost, R.K. (1968). Seed reproduction: Development of gametes and embryos, p. 290–324. In: W. Reuther, L.D. Batchelor, and H.J. Webber (eds.). The citrus industry. vol. 2 (revised ed.). Univ. of California Press, Berkeley.

Gan,Y.Y., Zaini, S. and Idris, A. (1981). Genetic variation in the grafted vegetatively propagated mango (*Mangifera indica* L.). Pertanika 4:53-62.

Gayral, P., Noa-Carrazana, J.C., Lescot, M., Lheureux, F., Lockhart, B.E., Matsumoto, T., Piffanelli, P. and Iskra-Caruana, M.L. (2008). A single Banana streak virus integration event in the banana genome as the origin of infectious endogenous pararetrovirus. J. Virol. 82: 6697–6710.

Geng, J., Wei, T., Wang, Y., Huang, X. & Liu, J. H. (2019). Overexpression of PtrbHLH, a basic helix-loop-helix transcription factor from *Poncirus trifoliata*, confers enhanced cold tolerance in pummelo (*Citrus grandis*) by modulation of H2O2 level via regulating a CAT gene. Tree Physiol. 39 : 2045–2054 .

Gonsalves, D. (1998). Control of papaya ringspot virus in papaya: a case study. Annual Review of Phytopathology 36: 415–437.

Grosser, J.W. (2002). Somatic Hybridization in Citrus — A Relevant Technique for Variety Improvement in the 21st Century and beyond.. Plant Biotechnology, Springer verlag pp. 537-540.

Grosser, J.W. and Gmitter Jr., F.G. (1990). Protoplast fusion and citrus improvement. Plant Breeding Rev. 8: 339–374.

Grosser, J.W., Graham, J. H., McCoy, C.M., Hoyte, A., Rubio, H. M., Bright, D. B. and Chandler, J.L.(2003). Development of "Tetrazyg" rootstocks tolerant of the

Diaprepes/Phytophthora complex under greenhouse conditions. Proc. Fla. State Hort. Soc. 116: 262–267.

Hamilton, R.A. (1954). A quantitative study of growth and fruiting in inbred and crossbred progenies from two Solo papaya strains. Hawaii Agric. Expt. Sta. Bull. No. 38: 16.

Hang, N.T. and Chau, N.M. (2010). Radiation induced mutation for improving papaya variety in Vietnam. Acta Horticulturae 851: 77–80.

Hartmann, W. and Engelhorn, E. (1992). Some characteristics of young seedlings for pre-selection for precocity and fruit size in plum breeding. Acta Horticulturae 317: 125-132.

Hawkes, J.G., Maxted, Nigel and Ford-Lloyd, B.V. (2000). The *Ex Situ* Conservation of Plant Genetic Resources. Kluwer Academic Publishers, Dordrecht.

Hedrick, U. P. (1925). Systematic Pomology. Macmillan, New York.

Helliot, B., Panis., B., Poumay, Y., Swennen, R., Lepoivre, P. and Frison, E. (2002). Cryopreservation for the elimination of cucumber mosaic and banana streak viruses from banana (*Musa* spp.). Plant Cell Rep. 20 (12): 1117-1122.

Hofmeyr, J.D.J. (1938). Genetical studies of *Carica papaya* L. South African Department Agriculture Forest Science Bulletin 187: 64.

Hofmeyr, J.D.J. (1945). Further studies of tetraploidy in *C. papaya* L. South African Journal of Science 41: 225–230.

Horovitz, S. and Jiménez, H. (1967). Cruzamientos interespecificos e intergenericos en Caricaceas y susimplicaciones fitotecnias. Agronomy Tropical 17: 353–359.

Huang, S., Ding, J., Deng, D., Tang, W., Sun., Liu., Dongyuan., Zhang, Lei., Niu, X., Zhang, X., *et al.* (2013). Draft genome of the kiwifruit *Actinidia chinensis*. Nat. Commun. 4: 2640.

Hume, H. H. (1957). Citrus Fruits. Macmillan Publishing Co., New York.

Husaini, A.M. and Abdin, M.Z. (2008). Overexpression of tobacco osmotin gene leads to salt stress tolerance in strawberry (*Fragaria* × *ananassa* Duch.) plants. Indian J. Biotech. 7: 465–471.

Husselman, J.H., Daneel, M.S., Sippel, A.D. and Severn-Ellis, A.A. (2016). Mutation breeding as an effective tool for papaya breeding in South Africa. Acta Horticulturae 1111: 71–78.

Hutchinson, J. (1969). Evolution and phylogeny of flowering plants : dicotyledons: facts and theory. Academic Press, London.

Huxley, A. (1978). An Illustrated History of Gardening. Paddington Press, London.

Hwang, S.C. and Ko, W.H. (1987). Somaclonal variation of banana and screening for resistance of Fusarium wilt. p.157-160. In: G.J. Persley and E.A. De Langhe (eds.), Banana and Plantain Breeding Strategies. ACIAR Proceedings No. 21, ACIAR, Canberra, Australia.

IARI. (1996). Promising Grape Hybrids. IARI, News. 12(4): 2.

Ibrahim, R., Hamzah, A., Jan Jam, Z., Bahagia, M. & Joyo, M. (2009). Gamma Irradiation-Induced Mutation for the Improvement of Josapine Pineapple Against Bacterial Heart Rot Disease and Improved Fruit Quality. In. Q.Y. Shu (ed.), Induced Plant Mutations in the Genomics Era. Food and Agriculture Organization of the United Nations, Rome, 2009, 276-278.

Iyer, C. P. A. (1991) Recent advances in varietal improvement in mango. Acta Horticulturae 291: 107–111.

Iyer, C.P.A. and Randhawa, G. S. (1965). Chemical induction of male sterility in grapes. Curr Sci (Bangalore) 34:411-412.

Iyer, C.P.A. and Subramanyam, M.D. (1984). Using of bridging species in interspecific hybridization in genus *Carica*. Current Science 53 (24): 1300–1301.

Iyer, C.P.A., Subramanyam, M.D. and Dinesh, M.R. (1987). Interspecific hybridization in genus *Carica*. In: IIHR Annual Report. Indian Institute of Horticultural Research, Bangalore, India, pp. 32–33.

Jaillon, O., Aury, J.M., Noel, B., Policriti, A., Clepet, C., Casagrande, A. *et al.* (2007). The grapevine genome sequence suggests ancestral hexaploidization in major angiosperm phyla. Nature 449: 463–467.

James, D. J., Passey, A. J., Barbara, D.J. and Bevan, M. (1989). Genetic transformation of apple (*Malus pumila* Mill.) using a disarmed Ti-binary vector. Plant Cell Rep 7: 658-661.

Janick, Jules. (2012). Fruit breeding: past, present and future. Papers presented at the XXII Congress Brasileiro de Fruticoltura, Bento Goncalves-RS, and October 22-26.

Janick, Jules and Moore, James N. (1975). Advances in Fruit Breeding. Purdue University Press, West Lafayette IN, USA.

Jayavalli, R., Balamohan, T.N., Manivannan, N. and Govindaraj, M. (2011). Breaking the intergeneric hybridization barrier in *Carica papaya* and *Vasconcellea cauliflora*. Scientia Horticulturae 130: 787–794.

Jia, Hongge. and Wang, Nian. (2014).Targeted Genome Editing of Sweet Orange Using Cas9/ sgRNA. PLoS One. 9 (4): 1-6.

Jin, W. *et al.* (2009). Improved cold-resistant performance in transgenic grape (*Vitis vinifera* L.) overexpressing cold-inducible transcription factors AtDREB1b. HortScience 44: 35–39.

Kaneyoshi, J., Kobayashi, S., Nakamura, Y., Shigemoto, N. and Doi, Y. (1994). A simple and efficient gene transfer system of trifoliate orange. Plant Cell Rep 13:541–545

Karale, A. R. and Desai, U. T. (2000). Study of heterosis for fruit characters in inter cultivar crosses of pomegranate (*Punica granatum* L.). Indian Journal of Genetics and Plant Breeding 60 (2): 191-196.

Karunakaran, G. and Singh, Pritee. (2019). Jackfruit varieties serves niche for growers. Indian Horticulture, March–April 2019, pp. 18-19.

Khanvilkar, S.V., Salvi, M. J. and Joshi, G.D. (1986). Evaluation of seedling types in Kokum I. Studies on Certain Strains of Kokum Seedlings for Yield and Fruit Characters. J. Maharashtra Agric. Univ.11(2): 174-175.

Kosterman, A.J.G.H. and Bompard, J.M. (1993). The mangoes: Their botany, nomenclature, and utilization. International Board for Plant Genetic Resources, Rome and the Linnean Society Publishers, London, 232 p.

Kulkarni, V.J. (1986). Graft induced off-season flowering and fruiting in the mango (*Mangifera indica* L.). J. Hort. Sci. 61(1):141-45.

Kumar, G. B. S., Ganapathi, T. R., Revathi, C. J., Srinivas, L., & Bapat, V. A. (2005). Expression of hepatitis B surface antigen in transgenic banana plants. Planta 222(3): 484–493.

Kumar, M., Prakash, S., Rao, S., Prasad,Y., Chand, P. and Singh, M.K. (2017a) Effect of seed treatment with gamma rays on fruit quality of papaya (*Carica papaya* L.). Research in Environment and Life Science 10(2): 182–184.

Kumar, M., Prasad, M. and Choudhary, V. (2017b) Effect of seed treatment by ethyl methane sulphonate (EMS) on fruit quality of papaya. (*Carica papaya* L.) cv. Pusa Dwarf. International Journal of Applied Chemistry 13 (1): 145–150.

Ku¨ ppers, H. (1976) Die St.-Julien-Pflaumen-Unterlage–was ist sie? Deutsche Baumschule 28: 194–199.

Kurian, R. M.. and Iyer , C. P. A. (1992). Stem anatomical characters in relation to tree vigour in mango. Scientia Horticulturae 50: 245-48.

Lancaster, P.S. and Bose, T.K. (1965). Studies on hybrids of *Syzygium*. Indian J. Hort. 22: 87-88.

Laurens, Francois., Aranzana, Maria José., Pere Arus, Bassi., Daniele, Marco Bink., Bonany, Joan Andrea Caprera., Luca Corelli-Grappadelli, Costes., Evelyne Charles-Eric Durel, Jehan-Baptiste Mauroux, Hélène Muranty, Nelson Nazzicari., Thierry Pascal, Andrea Patocchi, Andreas Peil, Quilot-Turion, Bénédicte., Rossini, Laura., Stella, Alessandra., Troggio, Michela., Velasco, Riccardo & van de Weg, Eric. (2018). An integrated approach for increasing breeding efficiency in apple and peach in Europe. Horticulture Research 5: 11

Laurens, F., Durel, C.E., Patocchi, A., Peil, A., Salvi, S., Tartarini, S., Velasco, R. and van de Weg, W.E. (2012). Review of fruit genetics and breeding programmes and a new European initiative to increase fruit breeding efficiency. Acta Horticulturae 929: 95-102.

Li, Jun., Wang, Xianhang., Wang, Xiping., Wang, Yuejin and Wang, Yuejin. (2015). Embryo rescue technique and its applications for seedless breeding in grape. Plant Cell Tiss. Organ Cult. 120:861–880.

Li, Ming-fang., Zheng, Xue-qing., Zhu, Yongqiang., Wang, Xiangshe., Liang, Suyu., LI, Lei. and Wu, Xingrong. (2006). Development and characterization of SSR markers in lychee (*Litchi chinensis*). Molecular Ecology Resources 6 (4): 1205-7.

Liebhard, R., Koller, B., Gianfranceschi, L. and Gessler, C. (2003). Creating a saturated reference map for the apple (*Malus* × *domestica* Borkh.) genome. Theor. Appl. Genet. 106: 1497–1508.

Liou, P.C., Gmitter Jr. F.G. and Moore, G.A. (1966). Characterization of the Citrus genome through analysis of restriction fragment length polymorphism. Theor. Appl. Genet (1996) 92:425-435.

Litz, R.E., Knight, R.K. and Gazit, S. (1982). Somatic embryos from cultured ovules of polyembryonic *Mangifera indica* L. Plant Cell Rep 1: 264–6.

Litz, R.E., Mathews, H., Bharathan, N. and Narayan K.R.1990. Transformation of somatic embryos of mango. Proceedings of the VII International Congress on Plant Tissue and Cell Culture. Amsterdam; p. 67. Abstrac.

Liu, Xiaomei and Ya, G. (2011). Clonal propagation of guava (*Psidium guajava* L.) on nodal explants of mature elite cultivar. International Journal of Plant Biology 2(1): 7-10.

Loredana, Abbate., Nicasio, Tusa., Del Bosco, SergioFatta., Strano, T., Renda, A. and Ruberto, G. (2012). Genetic improvement of Citrus fruits: New somatic hybrids from *Citrus sinensis* (L.) Osb. and *Citrus limon* (L.) Burm. F. Food Research International 48 (1): 284-290.

Ma, H., Moore, P.H., Liu, Z., Kim, M.S., Yu, Q., Fitch, M.M.M. *et al*. (2004). High-density linkage mapping revealed suppression of recombination at the sex determination locus in papaya. Genetics 166(1): 419–436.

Mabberley, D.J. (1998). Australian Citreae with notes on other Aurantioideae (Rutaceae). Telopea 7(4): 334-344.

McGregor, S. E. (1976). Insect pollination of cultivated crop plants. USDA, Tucson, Arizona.

Magdalita, P.M., Adkins, S.W., Godwin, I.D. and Drew, R.A. (1996). An improved embryo rescue protocol for a *Carica* interspecific hybrid. Australian Journal of Botany 44 (3): 343–353.

Magdalita, P.M., Drew, R.A., Godwin, I.D. and Adkins, S.W. (1998). An efficient interspecific hybridization protocol for *Carica papaya* L. × *Carica cauliflora* Jacq. Australian Journal of Experimental Agriculture 38(5): 523–530.

Majumder, P.K., Singh, R.N., Sharma, D.K. and Mukherjee, S.K. (1972). Preliminary studies on inheritance in *Mangifera indica* L. Acta Horticulturae 24:101-106.

Mak, C., Ho, Y.W., Tan, Y.P. and Ibrahim, R. (1996). Novaria—a new banana mutant induced by gamma irradiation. InfoMusa 5:35–36.

Malnoy, M., Jin, Q., Borejsza-Wysocka, E. E., He, S. Y. & Aldwinckle, H. S. (2007). Overexpression of the apple MpNPR1 gene confers increased disease resistance in *Malus* × *domestica*. Mol. Plant-Microbe Interact. 20: 1568–1580.

Manshardt, R.M. (1992). Papaya. In: Hammerschlag, F.A. and Litz, R.E. (eds). Biotechnology of Perennial Fruit Crops. CAB International, Wallingford, UK, pp. 489–511.

Manshardt, R.M. and Drew, R.A. (1998). Biotechnology of papaya. Acta Horticulturae 461, 65–73.

Manshardt, R.M. and Wenslaff, T.F. (1989). Interspecific hybridization of papaya with other *Carica* species. Journal of the American Society for Horticultural Science 114: 689–694.

Martinelli, Federico., Uratsu., Sandra. L., Albrecht, Ute, , Reagan, Russell. L. , Phu, My. L., Britton, Monica ., Buffalo, Vincent. , Fass , Joseph., Leicht, Elizabeth ., Zhao, Weixiang., Lin, Dawei., D'Souza, Raissa., Davis, Cristina .E., Bowman, Kim. D. and Dandekar, Abhaya. M.(2012). Transcriptome profiling of citrus fruit response to huanglongbing disease. PLoS One 7(5):1-16.

Mathews, H., Wilde, H.D., Litz, R.E. and Wetzstein, H.Y. (1993). Genetic transformation of mango. Acta Horticulturae 341: 93-97.

Matsuda, N., Ikeda, K., Kurosaka, M., Takashina, T., Isuzugawa, K., Endo, T. and Omura, M. (2009). Early flowering phenotype in transgenic pears (*Pyrus communis* L.) expressing the CiFT gene. J. Jpn. Soc. Hort. Sci.78: 410-416.

Matsumoto, K., Vilarinhos, A. and Oka, S. (2002). Somatic hybridization by electrofusion of banana protoplasts. Euphytica 125:317–324.

Mekako, H.U. and Nakasone, H.Y. (1975). Interspecific hybridization among six *Carica* species. Journal of the American Society for Horticultural Science 100: 145–148.

Mengesha, M. H. (1984). International germplasm collection, conservation, and exchange at ICRISAT. In: Conservation of crop germplasm-international perspective. Crop Science Society of America, Madison, pp 47–54.

Menzel, Christopher. (2002). The Lychee Crop in Asia and the Pacific. RAP Publication: 2002/16. Food and Agriculture Organization of the United Nations Regional Office for Asia and the Pacific, Bangkok, Thailand.

Mezzetti, B., Landi, L., Scortichini, L., Rebori, A., Spena, A. and Pandolfini, T. (2002). Genetic engineering of parthenocarpic fruit development in strawberry. Acta Horticulturae 567:101-104.

Mishra, M., Jalil, Syed. Uzma., Sharma, Nimisha and Hudedamani, Umesh. (2014). An Agrobacterium mediated transformation system of guava (*Psidium guajava* L.) with endochitinase gene. Crop Breeding and Applied Biotechnology 14 (4): 232-237.

Mitra, Sisir and Sharma, Suneel. (2020). Varieties and Crop Improvement . In The Papaya: Botany, Production and Uses. (ed. S. Mitra), pp. 68-101. CAB International, U.K.

Moore, G. and Tymowski, W. (2005). Explanatory guide to the international treaty on plant genetic resources for food and agriculture. IUCN, Gland/Cambridge.

Moura˜o Filho, F.A.A., Gmitter, F.G. and Grosser, J.W. (1996). New tetraploid breeding parents for triploid seedless citrus cultivar development. Fruit Varieties J. 50, 76–8.

Mourão Filho, Francisco de Assis Alves., Pio, R., Mendes, Beatriz., Madalena, Januzzi., Azevedo., Fernando Alves de *et al.* (2008). Evaluation of citrus somatic hybrids for tolerance to *Phytophthora nicotianae* and citrus tristeza virus. Scientia Horticulturae 115 (3): 301-308.

Mukherjee, S.K. (1963). Cytology and breeding of mango. Punjab Hortic. J 3:107–115.

Mullins, M.G. and Srinivasan, C. (1976). Somatic embryos and plantlets from an ancient clone of the grapevine (cv. Cabernet Sauvignon) by apomixis *in vitro*. J. Exp. Bot. 27 : 1022–1030.

Mullins, M.G., Tang, F.C.A. and Facciotti, D. (1990). *Agrobacterium*-mediated genetic transformation of grapevines: transgenic plants of *Vitis rupestris* Scheele and buds of *Vitis vinifera* L. Nature Biotech 8: 1041-1045.

Naik, K.C. (1948). Improvement of the mango (*Mangifera indica* L.) by selection and hybridization. Indian J. Ag. Sci., 18: 35-41.

Nehra N.S., Chibbar, R.N., Kartha, K.K., Datla, R.S.S., Crosby, W.L. and Stushnoff, C. (1990). Genetic transformation of strawberry by *Agrobacterium tumefaciens* using a leaf disk regeneration system. Plant Cell Rep 9: 293–8.

Nicolosi, E., Deng, Z.N., Gentile, A., Malfa, S. La., Continella G.& Tribulato , E. (2000). Citrus phylogeny and genetic origin of important species as investigated by molecular markers. Theoretical and Applied Genetics 100:1155–1166.

Nixon, R. W. (1928). The immediate influence of pollen on the size and time of ripening of the fruit of the date palm. J. Hered. 19: 240-254.

O'Brien, C.M. and Drew, R.A. (2009). Potential for using *Vasconcellea parviflora* as a bridging species in intergeneric hybridization between *Vasconcellea pubescens* and *Carica papaya*. Australian Journal of Botany 57: 592–601.

Oliveira, E.J. de, dos Santos Silva, A., de Carvalho, F.M., dos Santos, L.F., Costa, J.L. *et al.* (2010). Polymorphic microsatellite marker set for *Carica papaya* L. and its use in molecular assisted selection. Euphytica 173: 279–287.

Oliveira, E.J., Fraife Filho, G. de A., de Freitas, J.P.X., Dantas, J.L.L. and de Resende, M.D.V. (2012). Plant selection in F_2 segregating populations of papaya from commercial hybrids. Crop Breeding and Applied Biotechnology 12 (3): 191–198.

OECD (2019). Consensus Document on the Biology of Apple. Environment Directorate, OECD , Paris.

Pareek, O. P. and Sharma, Suneel (2017). Systematic Pomology; Vol. 1: 1-496 & Vol. 2:499-909. Scientific Publishers (India), Jodhpur & New Delhi, India.

Painter, R.H. (1951). Insect Resistance in Crop Plants. Macmillan, New York.

Pasquali, G., Biricolti, S., Locatelli, F., Baldoni, E. and Mattana, M. (2008). Osmyb4 expression improves adaptive responses to drought and cold stress in transgenic apples. Plant Cell Rep 27:1677–1686.

Pathak, R.K. (2003). Status Report on Genetic Resources of Indian Gooseberry-Aonla (*Emblica officinalis* Gaertn.) in South and Southeast Asia. IPGRI Office for

South Asia, National Agriculture Science Centre (NASC), DPS Marg, Pusa Campus, New Delhi, India, pp 1–96.

Pati, R. and Muthukumar, M. (20013). Genetic Transformation of Bael (*Aegle marmelos* Corr.). In book: Biotechnology of Neglected and Underutilized Crops, pp.343-365.

Peace, Cameron. P. (2017). DNA-informed breeding of rosaceous crops: promises, progress and prospects. Horticulture Research 4: 1-13.

Pereira, F.M., Carvalho, C.A., and Nachtigal, J.C. (2003). Seculo XXI: Nova cultivar de goiabeira ′ de dupla finalidade. Revista Brasileira de Fruticultura 25: 498–500.

Perry, B. A. (1943). Chromosome number and phylogenetic relationships in the Euphorbiacae. Amer. J. Bot. 30: 527- 543.

Planchon, J. (1887). Monographie des Ampélidées vrais. In *Monographiae Phanaerogamarum*; De Candolle, A.F.P.P., De Candolle, C., Eds.; Sumptibus, G. Masson: Paris, France; Volume 5.

Potter, D., Eriksson, Torsten., Evans, Rodger. C. and Oh, Sang-Hun. (2007). Phylogeny and classification of Rosaceae. Plant Systematics and Evolution 266 : 5-43.

Priyanka, P., Kumar, D., Yadav, A., Yadav, K. and Dwivedi, U.N. (2017). Analysis of simple sequence repeats information from floral expressed sequence tags resources of papaya (*Carica papaya* L.). American Journal of Plant Sciences 8: 2315–2331.

Pujar, D.U., Vasugi, C., Adiga, D., Honnabyraiah, M.K., Vageeshbabu, H.S. *et al.* (2019). Study on cross compatibility of intergeneric hybridization between *Carica* with *Vasconcellea* species. International Journal of Current Microbiology and applied Science 8 (4): 1514–1537.

Puhringer, H., Moll, D., Hoffman-Sommergruber, K., Watillon, B., Katinger, H. and Laimer da Camara Machado, M. (2000). The promoter of an apple Ypr gene, encoding the major allergen Mal d1, is stress- and pathogen-inducible. Plant Science 152: 35–50.

Purseglove, J.W. (1968) Tropical Crops: Dicotyledons Vol 1 and 2, Longman Scientific and Technical., John Wiley and Sons Inc., New York.

Ram, M. (1982). Papaya improvement through selection and breeding technique. Punjab Horticulture Journal 22: 8–14.

Ram, M. (1993). Improvement of papaya. In: Chadha, K.L. and Pareek, O.P. (eds) Advances in Horticulture, Vol. 1. Malhotra Publishing House, New Delhi, pp. 383–397.

Ram, M. and Majumder, P.K. (1981). Dwarf mutant of papaya (*Carica papaya* L.) induced by gamma rays. Journal of Nuclear Agriculture Biology 10: 72–74.

Ram, M., Majumder, P.K., Singh, B.N. and Akhtar, A. (1999). Study on hybrid vigour in papaya (*Carica papaya* L.) varieties. Indian Journal of Horticulture 56(4): 295–298.

Ramesh, A.N., Hanur, V.S., Manmohan, M., Rekha, A., Santosh, D.B. and Santosh, G.M. (2019). Effect of gamma radiation for improving quality in papaya (*Carica papaya* L.) cv. Arka Prabhat in M1 generation. International Journal of Pure and Applied BioScience 7(3): 515–524.

Ramming, D.W., Emershad, R.L. and Tarailo, R. (2000). A stenospermocarpic seedless *Vitis vinifera* x *Vitis rotundifolia* hybrid developed by embryo rescue. HortScience 35:732–734.

Ramos, H.C.C., Pereira, M.C., Viana, A.P., Luz, L.N., Cardoso, D.L. and Ferreguetti, G.A. (2014). Combined selection in backcross population of papaya (*Carica papaya* L) by mixed model methodology. American Journal of Plant Science 5(20), 2973–2983.

Ravishankar, H. and Purohi, A.G. (1991). Note an extra-large fruited and leaf spot tolerant sapota selection. Paper presented in National Seminar on optimization of productivity and utilization of sapota, at GAU, Navsari, 8th October, 19

Reustle, G., Harst, M. and Alleweldt, G. (1994). Regeneration of grapevine (*Vitis* sp.) protoplasts. Vitis 33: 173–174.

Reynolds, A.G., Fuleki, T. and Evans, W.D. (1982). Inheritance of methyl anthranilate and total volatile esters in *Vitis* spp. Am. J. Enol. Vitic. 33: 14–19.

Roberts, E.H. (1973). Predicting the storage life of seeds. Seed Sci. Technol. 1: 499-514.

Robertson, K. R., Phipps, J. B., Rohrer, J. R. and Smith, P. G. (1991). A synopsis of genera of the Maloideae (Rosaceae). Syst. Bot. 16: 376–394.

Roose, M.L. and Williams, T.E. (2007). Mandarin tree named 'Tango'. U.S. Patent No. PP17,863.

S′agi, L., Panis, B., Remy, S., Schoofs, H., de Smet, K., Swennen, R. and Cammue, B. (1995). Genetic transformation of banana and plantain (*Musa* spp.) via particle bombardment. Bio/Technology 13: 481–485.

Sakai, A., Matsumoto, Toshikazu., Hira, D. and Niino, Takao. (2000). Newly developed encapsulation-dehydration protocol for plant cryopreservation. Cryo Letters 21(1):53-62

Sambamurthy, K. and Ramalingam, V. (1954). A note on hybridization in the sapota (*Achras zapot*a L.). Indian J. Hort. 11: 57-62.

Saran, P.L., Solanki, I.S. and Choudhary, R. (2016). Papaya – Biology, Cultivation, Production and Uses. CRC Press, Boca Raton, Florida.

Saraswathi, M.S., Siva, S.A., Uma, S., Sathiamoorthy, S. and Valdivel, E. (2007). RAPD markers in flesh colour determination of papayas (*Carica papaya* L.). Acta Horticulturae 740: 127–132.

Sargent, D. J, Clarke, J., Simpson, D. W., Tobutt, K. R., Arús, P., Monfort., A., Vilanova, S. B., Rousseau, M., Folta, K. M., Bassil, N. V. & Battey, N. H. (2004). An enhanced microsatellite map of diploid *Fragaria.* Theoretical and Applied Genetics_112: 1349–1359.

Sen, P.K., Mallik, P.C. and Ganguly, B.D. (1946). Hybridization of mango. Indian Hort. 4:4

Schneider, S., Reustle, G. and Zyprian, E. (1996) . Detection of somaclonal variation in grapevine regenerants from protoplasts by RAPD-PCR. Vitis 35: 99–100.

Schnell, R.J. and Knight, R. J. (1993). Genetic relationships among *Mangifera* spp. based on RAPD markers. Acta Horticulturae 341: 86-92.

Schnell. R.J., Ronning, C.M. and Knight, R.J. (1995). Identification of cultivars and validation of genetic relationships in *Mangifera indica* L. using RAPD markers. Theor. AppI. Genet. 90: 269-271.

Scorza, R. and Okie, W.R. (1990). Peaches (*Prunus*). pp 177-231 In: Moore J.N, Ballington J.R (eds.) Genetic resources of temperate fruit and nut crops. ISHS, Wageningen.

Scorza, R. and Okie, W.R. (1991). Peaches (*Prunus*). Acta Horticulturae 290: 175–231.

Scorza, Ralph., Ravelonandro , Michel., Callahan, A., Zagrai , I., Polak, J., Malinowski, T., Cambra, M., Laurene, Levy., Damsteegt, Vern., Krska, Boris., Cordts, J., Gonsalves, D. and Dardick, Chris. (2016). HoneySweet (C5), the First Genetically Engineered Plum pox virus–resistant Plum (*Prunus domestica* L.) Cultivar. HortScience 51 (5): 601-603.

Sharma, D.K. and Majumder, P.K. (1988). Further studies on inheritance in mango. Acta Horticulturae 231: 106–11.

Sharma, D.K. and Singh, R.N. (1970). Self-incompatibility in mango (*Mangifera indica* L.). Hort. Res.10:108-15.

Sharpe, R.H., Webb, T.E. and Lundy, H.W. (1954). Peach cultivar tests. Proceedings of the Florida State Horticultural Society 67: 245–246.

Shen, Z., Confolent, C., Lambert, P., Poëssel, Jean-Luc., Quilot-Turion, B., Yu, M., Ma, Ru. and Pascal, T.(2013) Characterization and genetic mapping of a new blood-flesh trait controlled by the single dominant locus DBF in peach. Tree Genetics & Genomes 9 : 1435–1446.

Sherman, W.B., Lyrene, P.M. and Sharpe, R.H. (1996). Low-chill peach and nectarine breeding at the University of Florida. Florida State Horticultural Society 109: 222–223.

Shao, J., Chen, C. and Deng, X. (2003). *In vitro* induction of tetraploid in pomegranate (*Punica granatum*). Plant Cell Tissue and Organ Culture. 75(3): 241-46

Simmonds, N.W. and Shepherd, K. (1955). The taxonomy and origins of the cultivated bananas. Journal of the Linnean Society of. Botany 55:302-312.

Simpson, B.B. and Conner-Ogorzaly, M. (1986). Economic botany: plants in our world. MacGraw-Hill, New York.

Singh, H. and Chadha, K.L. (1981). Improvement of Dashehari by clonal selection. National Symp. on Trop. and Sub-trop. Fruit crops. Hort. Soc. India, Bangalore, p. 5 (Abstr).

Singh, L.B. (1960). The Mango-Botany, Cultivation and Utilization. Leonard Hill (Books) Ltd., London. World Crop Book Series, 438 p.

Singh, R. and Murthy, B.N.S. (1993). Improvement of grape. In: Advances in Horticulture. Vol. 1. (Eds K.L. Chadha and O.P. Pareek) Malhotra Publishing House, New Delhi. pp. 349 – 381.

Singh, R.N. (1955). Further studies in colchicine induced polyploidy in papaya *(Carica papaya* L.). Indian Journal of Horticulture 12: 63–71.

Singh, R.N. (1990). Mango. ICAR low-priced books series no.3, ICAR, New Delhi.

Singh, R.N., Gorakh, S., Rao, O.P. and Mishra, J.S. (1985). Improvement of Banarsi Langra through Clonal Selection. Prog. Hort.17: 273–7.

Singh, R.N., Majumder, P.K. and Sharma, D.K. (1962). Self-incompatibility in mango var. Dashehari. Current Science 4: 209.

Singh, R.N., Sharma, D.K. and Majumdar, P.K. (1980) An efficient technique of mango hybridization. Scientia Horticulturae 12 (3): 299-301.

Smigocki, A. C. & Hammerschlag, F. A. (1991). Regeneration of plants from peach embryo cells infected with a shooty mutant strain of *Agrobacterium*. J. Am. Soc. Hort. Sci. 116: 1092–1097.

Sofi, A. A., Verma, M.K., Ahmed, N., Verma, R.K., Kumar, D., Mir, J.I., Singh, D.B., Choudhary, H., Lal, S., Singh, S.R., Srivastava, K.K., Sheikh, M.A., Pushkar, B.A. and Dar, G.A. (2009). New apricot varieties released. Indian J. Hort. 66 (4): i-iii.

Sofi, A.A., Verma, M.K., Ahmed, N., Verma, R.K., Mir, J.L., Singh, D.B., Kumar, Dinesh., Lal, S., Singh, S.R., Srivastava, K.K., Bhat, H.A., Pandit, A.H., Sheikh, M.A. and Shah, Shazia. (2012). New walnut varieties released. Indian J. Hort. 67(1) : i-vii.

Sofi, A.A., Ahmed, N., Srivastava, K.K., Verma, M.K., Verma, R.K., Kumar, Dinesh., Mir, J.I., Mahindiran, G., Rather, J.A. and Shaf, Wajida (2012). New Sweet Cherry Varieties Released . Indian J. Hort. 69(4): i-ii.

Somsri, S., Fletcher, R.J., Jobin, M., Drew, R., Lawson, W. and Graham , M.W. (1998). Developing molecular markers for sex prediction in papaya (*Carica papaya* L.). Acta Horticulturae 461: 141-148.

Sonneveld, T., Robbins, T.P., Bośković, R. and Tobutt, K.R. (2001). Cloning of six cherry self-incompatibility alleles and development of allele-specific PCR detection. Theor. Appl. Genet. 102:1046–1055.

Soost, R. K. (1987). Breeding citrus-genetics and nucellar embryony. In: Abbott , A.J. and Atkin, R.K, eds. Improving vegetatively propagated crops. London: Academic Press, pp. 83–110.

Soost, R. K. and Cameron, J. W. (1980). Oroblanco: A new grapefruit hybrid. California Agriculture, November-December 1980: 16-17.

Soost, R.K. and Cameron, J.W. (1985). 'Melogold', a triploid pummelo-grapefruit hybrid. Hort Science 20: 1134-1135.

Spiegel-Roy, P., Sahar, N., Baron, J. and lavi, U. (1985). *In vitro* culture and plant formation from grape cultivars with abortive ovules and seeds. J. Amer. Soc. Hort. Sci. 110: 109-112.

Srinivasan, C. & Mullins, M.G. (1978). Control of flowering in grapevine (*Vilis vinifera* L.) : Formation of inflorescences *in vitro* by isolated tendrils. Pl. Physiol. 61: 127- 130.

Srinivasan, K. (1970). "Muttom Varikka"- a promising jackfruit variety. Agricultural Research Journal of Kerala 8 (1): 51-52

Storey, W.B. (1941). The botany and sex relationships of the papaya. In: Papaya production in the Hawaiian Islands. University of Hawaii Agriculture Experiment Station Bulletin 87: 5–22.

Storey, W.B. (1958). Modification of sex expression in papaya. Hort. Adv. 2: 49-60.

Storey, W.B. (1969). Papaya (*Carica papaya* L.). In: Outline of Perennial Crop Breeding in the Tropics, pp. 389- 407 (Ferwerda, P. and Wit, F., ed.). H. Veenman and Zonen, Wageningen.

Subhadrabandhu, S. and Nontaswatsri, C. (1997). Combining ability analysis of some character of introduced and local papaya cultivars. Scientia Horticulturae 71: 203–212.

Subramanyam, M.D. and Iyer, C.P.A. (1981). Crossability and interspecific relationship in the genus *Carica*. Paper presented at the National Symposium on Tropical and Subtropical Fruit Crops, 21–24 January 1981, Bangalore, India.

Subramanyam, M.D. and Iyer, C.P.A. (1984). Exploitation of heterosis in papaya. Indian J. Hort. 41: 40-46.

Subramanyam, M.D. and Iyer, C.P.A. (1998). Report, Fruit Research Workshop on Tropical and Subtropical Fruits. Rajendra Agril. Univ., Pusa, India, pp. 81-84.

Talón, M. and Gmitter Jr., F.G. (2008). *Citrus* Genomics. Intern. J. Plant Genomics. 2008, article ID528361, 17 pages, doi:10.1155/2008/528361.

Tamaki, M., Urasaki, N., Nakamura, I., Motomura, K. and Adaniya, S. (2011). Shortening the breeding cycle of papaya (*Carica papaya* L.) by culturing embryos treated with ethrel. Plant Cell Tissue Organ Culture 106: 225–233.

Tanaka, T. (1961). Citrologia semi centennial commemoration papers on citrus studies. Osaka: Citrologia Supporting Foundation. 114 pp.

Tassinari, P. (2005). Biologia fiorale e autoincompatibilità nel ciliegio. Riv. Frutticoltura3: 27–29.

Tripathi, Jaindra. N., Ntui, Valentine. O., Ron, Mily., Muiruri , Samwel K., Britt, Anne and Tripathi, Leena (2019). CRISPR/Cas9 editing of endogenous banana streak virus in the B genome of Musa spp. overcomes a major challenge in banana breeding. Commun Biol. 2: 46. doi: 1 0.1038/s42003-019-0288-7.

Tripathi, L., Ntui, V. O., and Tripathi, J. N. (2020). CRISPR/Cas9 based genome editing of banana for disease resistance. Curr. Opin. Plant Biol. 56: 118–126.

Ulubelde, M. & Tan, A. (1986). A numerical taxonomic study of citrus species based on leaf phenolic compounds. Acta Horticulturae 182: 349–358.

Ushijima, K., Sassa, H., Dandekar, A.M., Gradziel, T.M., Tao, R. and Hirano, H. (2003). Structural and transcriptional analysis of the self-incompatibility locus of almond: Identification of a pollen-expressed F-box gene with haplotype-specific polymorphism. Plant Cell 15:771–781.

ValizadehKaji, Babak. and Ahmadreza, Abbasifar. (2017). Transformation of pomegranate (*Punica granatum* L.) a difficult-to-transform tree. Biocatalysis and Agricultural Biotechnology 10: 46-52

Vardi, A., Levin, I. and Carmi, N. (2008). Induction of seedlessness in citrus: from classical techniques to emerging biotechnological approaches. J. Am. Soc. Hort. Sci. 133: 117-126.

Vardi, A., Spiegel-Roy, P., Elchanati, A., Frydman-Shani, Ahuva and Neumann, Hana. (2003). Citrus tree Named Orri. United States Plant Patent No. US PP13,616 P2 Date of Patent: Mar. 4, 2003.

Vasconcelos, H.S.V., Sampaio, L.S.V. and Luna, J.V.U. (1982). Compartamento de linhas endogamas de mamao (*Carica papaya* L.) e seus hibridos, em solo infestado com *Phytophthora* sp. Proceedings of the Tropical Region 25: 301–304.

Vavilov, N.I. (1949/1950). The origin, variation, immunity and breeding of cultivated crops. Chron. Bot. 13, 364 p.

Vavilov, N.I. (1951). The Origin, Variation, Immunity and Breeding of Cultivated Plants (Translated by K Starr Chester.). Chronica Botanica 13:1-366.

Vishnevetsky, J., White Jr., Thomas L., Palmateer., Flaishman, Moshe. *et al.* (2011). Improved tolerance towards fungal diseases in transgenic Cavendish banana (*Musa* spp. AAA group) cv. Grand Nain. Transgenic Research 20: 61–72.

Viruel , M. A. and Hormaza, J. I. (2004) . Development, characterization and variability analysis of microsatellites in lychee (*Litchi chinensis* Sonn., Sapindaceae). Theoretical Applied Genetics 108: 896–902.

Waugh, F. A. (1912). Fruit culture. Orange Judd Company, New York.

Weigl, K., Wenzel, S., Flachowsky, H., Peil, A. and Hanke, M.V. (2015). Integration of *BpMADS4* on various linkage groups improves the utilization of the rapid cycle breeding system in apple. Plant Biotechnology Journal 13: 246-258.

Wisniewski, M., Fuchigami, L, Wang, Y., Srinivasan, C. and Norilli, J. (2002). Overexpression of a cytosolic ascorbate peroxidase gene in apple improves resistance to heat stress. In: XXXVI International Horticultural Congress and Exhibition, p 147 (Abstr.).

Woeste, K., McGranahan, G.H. and Bernatzky, R. (1996). Randomly amplified polymorphic DNA loci from a walnut backcross [(*Juglans hindsii* × *J. regia*) × *J. regia*]. J. Am. Soc. Hort. Sci.121: 358 –361.

Wong, W.S., Li, G.G., Ning, W. and Xu, Zeng-Fu. (2001). Repression of chilling-induced ACC accumulation in transgenic citrus by over-production of antisense 1-aminocyclopropane-1-carboxylate synthase RNA. Plant Science 161(5): 969-977.

Wongs-Aree, C. and Noichinda, S. (2011). Sugar apple (*Annona squamosa* L.) and atemoya (*A. cherimola* Mill. × *A. squamosa* L.). In Postharvest Biology and

Technology of Tropical and Subtropical Fruits Mangosteen to White Sapote. Woodhead Publishing Series in Food Science, Technology and Nutrition. pp. 399-426.

Wu, Guohong Albert., Terol, Javier., Ibanez, Victoria., López-García, Antonio., Pérez-Román, Estela., Borredá, Carles., Domingo, Concha., Tadeo, Francisco R., Carbonell-Caballero, Jose., Alonso, Roberto., Curk, Franck ., Du, Dongliang ., Ollitrault, Patrick., Roose, Mikeal L., Dopazo, Joaquin., Gmitter Jr, Frederick, G., Rokhsar, Daniel S. & Talon, Manuel. (2018). Genomics of the origin and evolution of *Citrus*. Nature 554: 311–316.

Wu, J., Wang, Z., Sh, Z., Zhang, S, Ming, R., Zhu, S., Khan, M.A., Tao, S., Korban, S.S., Wang, H., Chen, N.J., Nishio, T., Xu, X., Cong, L., Qi , K., Huang, X., Wang, Y., Zhao, X., Wu, J., Deng, C., Gou, C., Zhou, W., Yin, H., Qin, G., Sha, Y., Tao, Y., Chen, H., Yang, Y., Song, Y., Zhan, D., Wang, J., Li, L., Dai, M., Gu, C., Wang, Y., Shi, D., Wang, X., Zhang H., Zeng, L., Zheng , D., Wang, C., Chen, M., Wang, G., Xie, L., Sovero, V., Sha, S., Huang,W., Zhang, S., Zhang, M., Sun, J., Xu, L., Li, Y., Liu, X., Li, Q., Shen J., Wang, J., Paull, R. E., Bennetzen, J.L., Wang, J. and Zhang, S. (2013). The genome of the pear (*Pyrus bretschneideri* Rehd.). Genome Res 23(2):396–408.

Xie, Xiaoqing., Agüero, Cecilia. B. and Walker, Andrew, M. (2016). Genetic transformation of grape varieties and rootstocks via organogenesis. Plant Cell Tissue and Organ Culture 126: 541–552.

Xu, K. (2013). An overview of Artic apples: basic facts and characteristics. New York State Horticultural Society. 21 (3): 8-10.

Yamamoto, T., Kimura, T., Sawamura, Y., Manabe, T., Kotobuki, K., Hayashi, T., Ban, Y. and Matsuta, N. (2002). Simple sequence repeats for genetic analysis in pear. Euphytica 124: 129–137

Yamane, H., Ikeda, K., Ushijima, K., Sassa, H. and Tao, R. (2003). A pollen-expressed gene for a novel protein with an F-box motif that is very tightly linked to a gene for S-RNase in two species of cherry, *Prunus cerasus* and *P. avium*. Plant Cell Physiol. 44: 764–76.

Zeng, Huaping Zeng., Kan Bao, Lin Lin., Ya, Min, Hua Zhang., Min Miao., Xiaofeng Tang., Yunye Zhu., Yuan Sui, Guangwei, Li, Hanju Sun., Junyang Yue., Jiaqi Sun., Fangfang Liu., Liangqiang Zhou., Lin Lei, Xiaoqin Zheng, Ming Liu, Long Huang., Jun Song., Chunhua Xu., Jiewei Li., Kaiyu Ye, Silin Zhong, Bao-Rong Lu, Guanghua, He., Fangming, Xiao., Hui-Li Wang, Hongkun Zheng., Zhangjun, Fei, and Yongsheng, Liu. (2013). Draft genome of the kiwifruit *Actinidia chinensis*. Nat Commun. 18:1-9.

Zerpa, D.M. (1957). Triploides de *Carica* papaya L. Agronomica Tropical 7: 83–86.

Zerpa, D.M. (1959). Cytology of inter-specific hybrids in *Carica*. Agronomica Tropical 8: 135–144.

Zeven, A.C. and de Wet, J.M.J. (1982). Dictionary of cultivated plants and their regions of diversity. PUDOC, Wageningen. 259 p.

Zeven, A.C. and Zhukovsky, P.M. (1975). Dictionary of cultivated plants and their centres of diversity. PUDOC, Wageningen. 219 p.

Zhang, Zhai-hau., Xin-cai, Liu., Yan-hong, Liu., Hong-xia, Wang., Fa-chao., Chu., Mo-lin, Gong. and Xin-ying, Shen. (2007). Study on the relationship between the anatomical structure of shoots and growth vigor of *Juglans regia* L. Scientia Agricultura Sinica 4 (6):1303-1308.

Zhu, L.H., Holefors, A., Ahlman, A., Xue, Z.T. and Welander, M. (2001). Transformation of the apple rootstock M.9/29 with the *rol* B gene and its influence on rooting and growth. Plant Science 160: 433–439.

Zhu, L.H. and Welander, M. (2000) . Growth characteristics of the untransformed and transformed apple rootstock M 26 with *rol*A and *rol*B genes under steady-state nutrient supply conditions. Acta Horticulturae 521: 139–146.

Zhukovsky, P.M. (1965). Main gene centres of cultivated plants and their wild relatives within the territory of the U.S.S.R. Euphytica 14: 177–188.

Zhukovsky, P.M. (1968). New centres of origin and new gene centres of cultivated plants including specifically endemic micro centres of species closely allied to cultivated species). Bot. Zh. 53: 430-460. PI. Breed. Abstr. 38 (38).

Zhukovsky, P.M. (1970). World genofund of plants for breeding. Mega-gene centres and endemic microcentres. Leningrad, 87 p. Translated by E. E. Leppik.

Zielinski, Q.B., (1955). Modern Systematic Pomology. W. M. C. Brown Company, Dubuque, Iowa, USA

Zielinski, Q.B., Reimer, F.C. and Quackenbush, V.L. (1965). Breeding behavior of fruit characteristics in pears, *Pyrus communis* L. Proc. Amer. Soc. Hort. Sci. 86:81-87.

Zohary, D. and Spiegel-Roy, P. (1975). Beginnings of Fruit Growing in the Old World. Science 187: 319-327.